Wildlife Conservation Evaluation

Wildlife Conservation Evaluation

edited by

MICHAEL B. USHER

Department of Biology
University of York, UK

LONDON
NEW YORK

Chapman and Hall

First published in 1986 by
Chapman and Hall Ltd
11 New Fetter Lane, London EC4P 4EE
Published in the USA by
Chapman and Hall
29 West 35th Street, New York NY 10001

Printed in Great Britain at the
University Press, Cambridge

ISBN 0 412 26750 0 (Hb)
0 412 26760 8 (Pb)

British Library Cataloguing in Publication Data

Wildlife conservation evaluation.
1. Nature conservation—Evaluation
I. Usher, Michael B.
639.9'5   QH75

ISBN 0-412-26750-0
ISBN 0-412-26760-8 Pbk

Library of Congress Cataloging in Publication Data

Main entry under title:

Wildlife conservation evaluation.

Bibliography: p.
Includes index.
1. Wildlife conservation—Evaluation. I. Usher,
Michael B., 1941–
QL82.W53 1986   333.95   85–26964
ISBN 0-412-26750-0
ISBN 0-412-26760-8 (pbk)

Contents

Preface   vii
Contributors   xi

PART ONE   *Introduction*   1

1 Wildlife conservation evaluation: attributes, criteria and values   3
2 Assessing representativeness   45
3 Ecological succession and the evaluation of non-climax
   communities   69

PART TWO   *Approaches in different geographical areas*   93

4 Evaluation of tropical land for wildlife conservation potential   95
5 Evaluation methods in the United States   111
6 Selection of important areas for wildlife conservation in Great
   Britain: the Nature Conservancy Council's approach   135
7 Wildlife conservation evaluation in the Netherlands: a
   controversial issue in a small country   161
8 Evaluation at the local scale: a region in Scotland   181

PART THREE   *Specific habitats and groups of organisms*   199

 9 Forest and woodland evaluation   201
10 Evaluating the wildlife of agricultural environments: an aid to
   conservation   223
11 Ornithological evaluation for wildlife conservation   247
12 Assessments using invertebrates: posing the problem   271

PART FOUR   *General principles*   295

13 Conservation evaluation in practice   297
14 Design of nature reserves   315

References   339
Author index   371
Subject index   381

Preface

In the mid 1970s two events led me to get to know the Yorkshire Dales better than I had previously. Since 1964 I had been to the Malham Tarn Field Centre with groups of students, first from the University of Edinburgh and then from the University of York, and my family very much enjoyed the summer days we spent amid this magnificent hill scenery. In 1976, the British Ecological Society and the National Trust jointly worked on a survey of the biological interest of the National Trust properties of the Kent, East Anglian and Yorkshire Regions. Malham Tarn itself, and the surrounding farms, formed one of the twenty properties of the Yorkshire Region. I spent the bank holiday, that commemorated the Queen's Silver Jubilee, at Malham, looking fairly closely at the National Trust's landholding there. Miss Sarah Priest, who also looked at the National Trust properties, and I produced a report in late 1977, attempting both to describe and to evaluate the nature resources of the National Trust in Yorkshire. In the following year, 1978, the Nature Conservancy Council wanted to survey the whole of the upland area that was known as the Malham/Arncliffe SSSI (Site of Special Scientific Interest). A contract to look at such an exciting area, considering where boundaries should go, and looking to see if there were important areas of habitat that should be brought within the SSSI, was a superb practical antidote to an office in the University. By the late 1970s I had got to know a little about the natural history of the area around Malham, and of its value for wildlife conservation.

These various surveys have brought me into closer contact with Dr Henry Disney, who was then the warden of the Field Studies Centre at Malham. Having written an account of conservation evaluation for *Field Studies*, the journal of the Field Studies Council, Henry asked me if I would lead a ten day course on conservation evaluation at the Centre. The planning for such a course was not easy, but a group of 'students' turned up in late August 1982 for ten days of study, partly based on Henry's Diptera (flies) and partly on the survey work that I had done for the National Trust and Nature Conservancy Council a few years earlier. Mr Eddie Idle came down from Edinburgh to speak one evening during the course, and other members of the Centre staff joined in to show us some of the rarities, etc. At the end of the course the 'students', many of them working for the Government's Agricultural Development and Advisory Service (ADAS), were allowed a free run of a hill farm and had to devise a conservation plan for it. This exercise ended with a broadly ranging discussion, with the farmer himself taking a lively interest in both what had been found on his farm and what the value of the species or habitats actually was.

This book owes its origins to the day of that discussion. The course was ending, and many participants asked what books I could recommend as they were going to be involved in providing advice. There were a few obvious books that I could suggest, all with such caveats as 'but it's very expensive' or 'it only covers part of the subject', etc. Hence, the concept of editing this book arose.

One thing that I learnt when I wrote *Biological Management and Conservation* (Chapman and Hall, 1973) was that a book on conservation will never please everybody. The scientific aspects of conservation encompass so much of ecology, but then there are many aspects of conservation that go further than the science of ecology – the conservationist is dealing with aspects of human society, economics, politics and law as well. Hence, I suspect that this present book will also be criticized as being selective, since I have omitted to include a chapter on this or that (mammals, herpetofauna and lower plants are three obvious omissions). However, what I have endeavoured to do in inviting contributors is to give a feel for the whole subject. A detailed introduction is given in Chapter 1 to the contents of the various chapters, but it seems appropriate here to outline why the book is divided into four sections.

The first section is introductory, and really endeavours to cover three aspects of conservation evaluation that seem to me to be fairly fundamental. The word 'evaluation' in the title implies that one is making value judgements and hence the first chapter looks at this process. Classification of nature into recognizable units, and the dynamic processes that are at work changing one category into another over time, are both fundamental to the way that ecosystems are viewed. The second section asks questions about how evaluations are carried out in different parts of the world and at different scales. I guessed, before I invited anyone to contribute, that these chapters would be slanted towards higher plants and higher plant communities, very often thinking in terms of semi-natural or natural ecosystems. Hence, I planned a third section which would pick out some less natural communities: woodlands and agricultural land provided two examples with increasing modification by human agencies. It proved impossible to include a chapter on the most man-modified communities, those of the urban environment. Also, as I wanted to think about how groups of organisms other than higher plants could be used in evaluation studies, birds and invertebrates provide two contrasting examples. The final section of the book was planned to draw some threads together and to think about what could be done when an evaluation had been carried out. The first of the two chapters looks back over the attributes, criteria and values that had been used in the preceding twelve chapters, and addresses itself to analysing the question: 'what have people actually done?' Writing about criteria is purely an academic exercise unless they are going to be used in actual evaluation exercises. One of my aims in this book has been to include criteria which have been practically useful. The final chapter looks forward, since, at the end of the day, localities which are considered to have a high value for conservation are likely to be conserved. One of the important messages in this final chapter is that conservationists need to be aware of the ecology of the species that they wish to conserve, and that it is not sufficient only to be concerned with untested theories about nature.

Inevitably, in an edited book of this nature, there is a certain amount of duplication from chapter to chapter, but I felt that it was unwise to eliminate all duplication in a book that should have practical use. Someone interested in woodlands, for example, may not read all of the other chapters, and hence the woodlands chapter should be reasonably complete without too many references to other chapters. With authors from around the world, I also did not wish to standardize the use of words, and hence differences become apparent, such as the use of 'nature reserves' in the United Kingdom for what are called 'nature preserves' or 'nature refuges' in the United States. However, I felt that it is important to attempt to standardize the use of words which are related to criteria and I have tried to do this with the various definitions in the first chapter. Some criteria, such as 'rarity', have an intuitive meaning which is hard to misunderstand or misinterpret. Other criteria, and a prime example of this is 'representativeness', do not have this intuitive meaning, and hence it becomes very important to establish the meaning of the criterion.

In preparing this book there are inevitably many people that I should like to thank, not least my wife, Fionna, who, both at home and on holiday in the Scottish Highlands, has seen me poring over piles of manuscripts. I should particularly like to thank both Dr Henry Disney, whose invitation to teach a course at Malham led directly to the concept of this book, and Dr Chris Margules, who, as a graduate student with me, helped me immensely to clarify my own ideas about conservation evaluation. As editor, my thanks also go to those contributors who, for one reason or another, have taught me more about human nature, deadlines, etc! I am also extremely grateful to my technician, Mrs Sue Lonsdale, and my secretary, Mrs Margaret Wetherill, who have undertaken much drawing and typing during the final stages of preparation.

Rather than incorporate a series of separate acknowledgement sections at the end of each of the chapters, I felt it preferable, in an edited book, to bring all of the acknowledgements into the preface. We should, therefore, like to thank the following.

Paul Harding for preparing the distribution maps, and Dick Hunter for preparing the photographs from colour transparencies (Chapter 1).

The Natural Environment Research Council for financial support during the quarry studies (Chapter 3).

Dr Hal Salwasser and Dr Paul Hamel for their comments (Chapter 5).

Dr Art Lance and Alan Vittery for their comments (Chapter 6).

L. C. Braat and Karin George-Couvret for their help (Chapter 7).

Dr A. N. Lance and Miss N. J. Gordon for their comments and for assistance with site assessments (Chapter 8).

Dr George Peterken and others in the Nature Conservancy Council (Chapter 9).

The Countryside Commission for permission to quote from the results of the Demonstration Farm Project (the wildlife recorders were Carolyn McNab, Dr Steve Long, Dr Chris Mason, Dr John Barkham, Dr John Richards, Andrew Fraser, Dr Gordon McClone, Ian Tew, David Gregson, Andrew Gilham, and Dr Michael Usher); the Nature Conservancy Council for permission to use the material in Table 5 and the Farming and Wildlife Advisory Group for permission to reproduce Table 9; colleagues in Cobham Resource Consultants (Russel Matthews, Andrew McNab, Margery Slatter, Stephen Bass, Julian Bertlin and Mary Stephenson); and Barry Herniman (Chapter 10).

Dr R. J. O'Connor and the Nature Conservancy Council (who have funded R. J. Fuller's post under a contract with the British Trust for Ornithology) (Chapter 11).

'The Malham Methodists' (vide 'Hecamede', 1982, Antenna, 6, 170–71), who are J. Biglin, Alison Crisp (née Woods), R. H. L. Disney, Y. Z. Erzinçlioğlu, D. J. de C. Henshaw, Diane Howse, D. M. Unwin and P. Withers (Chapter 12).

Staff and students of the Biology Department, University of York 1978–1981, especially Professor Mark Williamson, Dr Michael Usher, Andrew Higgs and Dr Richard Rafe (Chapter 13).

Michael B. Usher
Dunnington, York
February, 1985

Contributors

M. P. Austin
*CSIRO Division of Water and Land
Resources
P O Box 1666
Canberra City
ACT 2601
Australia*

R. O. Cobham
*Cobham Resource Consultants
19 Paradise Street
Oxford
OX1 1LF
UK*

R. H. L. Disney
*Field Studies Council Research Fellow
c/o Department of Zoology
Downing Street
Cambridge
CB2 3EJ
UK*

D. Durham
*Ecological Services Division
Tennessee Department of
Conservation
701 Broadway
Nashville
Tennessee 37212
USA*

D. C. Eagar
*Ecological Services Division
Tennessee Department of
Conservation
701 Broadway
Nashville
Tennessee 37212
USA*

R. J. Fuller
*British Trust for Ornithology
Beech Grove
Station Road
Tring
Hertfordshire
HP23 5NR
UK*

E. T. Idle
*Nature Conservancy Council
Northminster House
Peterborough
Cambridgeshire
PE1 1UA
UK*

R. G. Jefferson
*Nature Conservancy Council
Matmer House
Hull Road
York
YO1 3JW
UK*

CONTRIBUTORS

K. J. Kirby
Nature Conservancy Council
Northminster House
Peterborough
Cambridgeshire
PE1 1UA
UK

D. R. Langslow
Nature Conservancy Council
Northminster House
Peterborough
Cambridgeshire
PE1 1UA
UK

R. J. McNeil
Department of Natural Resources
Cornell University
Fernow Hall
Ithaca
New York 14853–0188
USA

C. R. Margules
CSIRO Division of Water and Land
Resources
P O Box 1666
Canberra City
ACT 2601
Australia

S. H. Pearsall III
Environment and Policy Institute
East–West Center
1777 East–West Road
Honolulu
Hawaii 96848
USA

D. A. Ratcliffe
Nature Conservancy Council
Northminster House
Peterborough
Cambridgeshire
PE1 1UA
UK

J. Rowe
52 Raglan Road
Bristol
BS7 8EG
UK

D. Simberloff
Department of Biological Sciences
Florida State University
Tallahassee
Florida 32306
USA

M. B. Usher
Department of Biology
University of York
York
YO1 5DD
UK

S.W.F. van der Ploeg
Institute for Environmental Studies
Free University
P O Box 7161
1007 MC Amsterdam
The Netherlands

PART ONE

Introduction

CHAPTER 1

Wildlife conservation evaluation: attributes, criteria and values

MICHAEL B. USHER

1.1 Introduction
    1.1.1 The scope of the book
    1.1.2 Requisites for evaluation
1.2 The contents of the book
1.3 A review of criteria
    1.3.1 Popularity polls
    1.3.2 Diversity
    1.3.3 Area
    1.3.4 Rarity
    1.3.5 Naturalness
    1.3.6 Representativeness
    1.3.7 Other criteria
1.4 An example: limestone pavements in Yorkshire
1.5 Discussion
1.6 Summary

Wildlife Conservation Evaluation. Edited by Michael B. Usher.
Published in 1986 by Chapman and Hall Ltd, 11 New Fetter Lane,
  London EC4P 4EE

1.1 INTRODUCTION

1.1.1 The scope of the book

The title of the book consists of three words: it is, perhaps, appropriate to begin by examining what each of these words means. As some words have a wide usage, it is important to define the sector of this usage which applies in the context of this book.

The first word, *wildlife*, is the easiest of the three to define. It is a collective noun relating to non-domesticated species of plants, animals or microbes. In some scientific writing wildlife tends to be restricted to animals, especially mammals and birds, but in this book the wider definition is preferred. Within this wider definition there tends to be a restriction to plants and animals, not intentionally, but because very little attention has been paid to the conservation of microbes.

Within the intuitive meaning of wildlife is the concept of a species that grows or lives 'wild' in the area, i.e. as well as being non-domesticated the species must also be non-introduced. This poses certain difficulties, since it is not always possible to be certain whether a species is indigenous or whether it was introduced a long time previously. For plants, a key can be found in the pollen record, unless there are several closely related species. An example would be *Quercus robur* (pedunculate oak) in Great Britain. The flora of the British Isles (Clapham *et al.*, 1962) states: 'Native. Woods, hedgerows, etc: reaching nearly 1500 ft (460 m) in Derry and on Dartmoor, but rarely occurring over 1000 ft (300 m) . . . the characteristic dominant tree of heavy and especially basic soils (clays and loams) . . .'. Despite such an assertion of its native status, Anderson (1950) tended to doubt that *Q. robur* was a native species, arguing that *Q. petraea* (sessile oak) was the only native species. Clearly an argument such as this is purely academic, and hence does not form an important component of the definition of wildlife (except in so far as it relates to the criterion of naturalness: see Section 1.3.5). The criteria that can be used to assess whether or not a species is native have been reviewed by Webb (1985). Within the context of this book, domestic species are excluded from the definition of wildlife, and normally all recently introduced (within living memory) species are also excluded. The word wildlife can, therefore, be equated with the term *nature resource*.

The second word, *conservation*, is the most difficult of the three to define. Usher (1973) explored definitions which ranged from the non-interventionist, 'genuine conservation forbids any interference' (Margalef, 1968), to the interventionist, 'conservation is the wise use of the country's resources of land and water and wildlife for every purpose . . .' (Countryside in 1970, 1965). Black's (1970) statement that 'there is as yet no fully worked out and satisfying philosophy of conservation' may still be correct. However, Usher's (1973) conclusion that 'conservation is essentially concerned with the interaction between man and the environment' provides one of the generalities that Black (1970) warned could be dangerous, but it is very similar to the uses of the word in the *World Conservation Strategy* (IUCN, 1980) and in national responses to this strategy (e.g. Anon. 1983 for the United Kingdom response).

This broad definition of conservation has to be narrowed for the purposes of this text. Already wildlife has been equated to a nature resource, and the word conservation as used here is also related to this resource. The chapters in this book refer specifically to the retention over time of a viable, and hence self-perpetuating, nature resource. The concern is with what there is today and with the probability that it will be there in the future. The concentration in this book is therefore on the wildlife, and not necessarily on the whole environment of which that wildlife is a part.

The third word of the title, *evaluation*, needs a certain amount of explaining as it is often a process that is more intuitive than scientific. It is probably a truism that in most parts of the world any parcel of land could be used for a variety of purposes: wildlife conservation is one form of land use, usually a minor form, alongside agriculture, forestry, fisheries, urban and industrial usage, recreational usage, etc. Attempts to attach a monetary value to land for conservation purposes have tended to fail because there is no agreed currency for comparing the presence of wildlife, or its abundance, with the creation of jobs or the production of crops. Conservationists have, therefore, to value sites in terms which are not economic but which are largely comparative: for example, it could be said that a limestone pavement was the best in Yorkshire, or that a tropical rain forest was the best in West Africa. The use of the word 'best' implies that both value judgements and comparisons have been made: this is the process of evaluation. The relation between the science of ecology and the practices of conservation, which lead to such an evaluation, are explored in the Section 1.1.2.

However, before analysing this relation, the question 'why are we evaluating?' should be considered. An evaluation is performed in response to a demand. The demand can come from two main sources: either there is a demand to conserve, say, an oakwood (i.e. to establish an oakwood nature reserve: note that *'nature preserve'* and *'nature refuge'* are treated as synonyms of *'nature reserve'*), or there is a demand for knowledge about, say, oakwoods. It is essential that, before embarking on an evaluation exercise, the objectives or aims of the evaluation are clearly understood. The methods of carrying out the evaluation, and the results achieved, are all likely to be affected by the aims. Definition of the aims is a priority that is referred to in many of the following chapters.

1.1.2 Requisites for evaluation

A schematic representation of the requisites for an evaluation exercise
are shown in Fig. 1.1, which relates to a small field on the Hopewell
House Demonstration Farm Project (see Chapter 10 for more details of
this project). It is important to distinguish three separate, but closely
related, features shown in this illustration – *attributes, criteria*, and *values*.

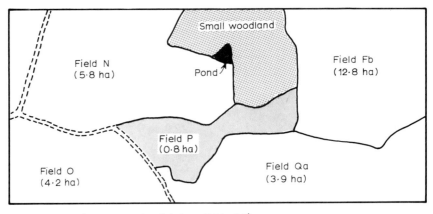

Atribute :   species list (see Table 1.1)
Criterion: species richness (65 species)
Value: large (most species rich field on farm in 1984)

Fig. 1.1 An example of an evaluation process in relation to the fields of
Hopewell House Farm, North Yorkshire, concentrating here on field P. The
attribute is the species list, shown for field P in Table 1.1. The *criterion* is the
species richness, which is 65 species for field P (and 16, 13, 26 and 29 species for
the surrounding fields Fb, Qa, O and N respectively). Because of its large species
richness, which for wildlife conservation is valued in agricultural habitats (see
Chapter 10), field P is given the greatest value of these fields. Agriculturally,
field P has very low value, and hence it has neither been ploughed nor received
fertilisers or other agrochemicals during the last decade or more.

The *attributes* are properties of a parcel of land which can be used to
reflect the conservation interest in that land. For example, an attribute of
the field shown in Fig. 1.1 would be the species present in that field: the
list of all species recorded in this particular field during 1984 is shown in
Table 1.1. Another attribute of that field would be its history: how long
is it since the soil was disturbed by ploughing or has it ever received
fertilizers or herbicides? These attributes can all be measured or
recorded by a visit (or visits) to the field, by a search of historical records,
by measurements on a map, etc.

A *criterion* is used to express an attribute in a form that can be used in
an evaluation. The attribute, the list of species, can be transformed into

Table 1.1 A list of all species occurring in the field P, which is shown in Fig. 1.1. The species names follow Clapham, Tutin and Warburg (1981), and the abbreviations for abundance represent: r (rare, not more than 10 individuals or 3 clumps in the field); o (occasional, 11–30 individuals or 4–10 clumps); f (frequent, 31–100 individuals or 11–30 clumps); a (abundant, more than 100 individuals or 30 clumps in the field).

Species	Abundance	Species	Abundance
Ophioglossum vulgatum	r	*Stachys sylvatica*	o
Anemone nemorosa	r	*Lamium album*	f
Ranunculus acris	f	*Galeopsis tetrahit*	r
R. repens	a	*Glechoma hederacea*	f
R. ficaria	f	*Plantago lanceolata*	a
Cardamine pratensis	f	*Galium cruciata*	o
Viola riviniana	r	*G. aparine*	a
Cerastium fontanum	f	*Sambucus nigra*	r
Stellaria holostea	o	*Senecio jacobaea*	o
Trifolium repens	o	*Tussilago farfara*	f
T. pratense	o	*Bellis perennis*	f
Lotus corniculatus	o	*Achillea millefolium*	a
Filipendula ulmaria	r	*Leucanthemum vulgare*	o
Rubus fruticosus agg.	o	*Arctium minus*	r
Potentilla sterilis	o	*Cirsuim vulgare*	r
P. anserina	r	*C. arvense*	f
Geum urbanum	o	*Centaurea nigra*	r
Alchemilla sp.	o	*Hypochaeris radicata*	o
Prunus spinosa	f	*Taraxacum* sp.	f
Crataegus monogyna	r	*Hyacinthoides non-scripta*	r
Anthriscus sylvestris	a	*Juncus effusus*	o
Conopodium majus	f	*Luzula campestris*	f
Heracleum sphondylium	f	*Festuca rubra*	o
Polygonum sp.	o	*Dactylis glomerata*	a
Rumex acetosa	r	*Cynosurus cristatus*	o
R. obtusifolius	a	*Elymus repens*	f
R. sanguineus	r	*Arrhenatherum elatius*	a
Urtica dioica	a	*Holcus lanatus*	a
Humulus lupulus	r	*Deschampsia cespitosa*	a
Primula veris	r	*Agrostis stolonifera*	a
Veronica chamaedrys	a	*Phleum pratense*	f
V. hederifolia	r	*Alopecurus pratensis*	a
Prunella vulgaris	o		

several different criteria. One of these criteria is species richness, simply calculated by counting the total number of species present. Another criterion is that of diversity, again easily calculated by using some form of diversity index (see, for example, Yapp, 1979 or Usher, 1983). Another possible criterion is that of rarity, estimated by counting up the number of rare species in the list. A criterion is, therefore, often a means of summarizing a lot of information. The criterion may be clearly distinguishable from the attribute (for example, the diversity index is different from the species list), or the two may be almost indistinguishable (for example, if the field had not been ploughed since 1940, that attribute of the field and the criterion, which might be expressed as the length of time since the field was last ploughed, are virtually the same). Some criteria are quantifiable, such as the species richness of the field ($S = 65$) in Fig. 1.1 and Table 1.1, whereas others cannot be quantified, such as the naturalness of that field (which would be expressed as it being 'not very natural'). Quantification of criteria is useful since it leads to repeatability between observers, and hence to a feeling that there is less bias in the evaluation process. Criteria can, thus, be viewed as gradients: in the examples species richness runs from a small number ($S = 0$ or 1) to a large number ($S = 200$ or 300), and naturalness runs from non-natural to natural. It is generally useful to divide such a continuum into an arbitrary number of discrete steps or stages.

Values can be placed upon the stages of the criterion. Thus, if one is looking at a series of farm fields, one might place more value on fields with a large species richness than on fields with a small species richness. In this example value increases more or less proportionally with increasing species richness, and the field shown in Fig. 1.1 is one of the most valuable (for wildlife conservation purposes) on the farm. It is important to realize that values are not solely determined by the scientists: they are essentially determined by the society that either owns, or some of whose members own, the nature resource. This is clearly shown by the arguments about the use of 'rarity' as a criterion in wildlife conservation evaluation. The scientific community is sometimes sceptical about its use as a criterion (e.g. Adams and Rose, 1978), although amongst the lay population rare species are probably the most interesting feature, and hence often the most highly valued feature, of any conservation area. Values, then, are determined by society, though it is often true that the scientist suggests to the society what the values should be.

It is important in this text that the three concepts of attributes, criteria and values are carefully distinguished. In general, the text will concentrate on the criteria that are used.

1.2   THE CONTENTS OF THE BOOK

In any study designating the conservationally important sites of a region, it is an essential preliminary to define the criteria which will be used to recognize those sites. The criteria should be selected a priori so that there is a minimum of bias in the actual evaluation and comparison of sites. Thus, the review of the most important sites for conservation in Great Britain (Ratcliffe, 1977; see Chapter 6) began with the definition of the ten criteria that were to be used in the site evaluation process. In a survey of nine studies of conservation assessment, Margules and Usher

(1981) listed the popularity of nineteen criteria that had been used: one criterion had been used in eight of the nine studies, and hence headed the 'popularity poll', whilst seven other criteria had only been used in a single study (see Table 1.2 for a modified form of the 'popularity poll' in Margules and Usher, 1981). Although it is important to concentrate attention on those criteria that are most frequently used, other criteria that are particularly important in special situations should not be forgotten. Section 1.3 is, therefore, a review of the criteria that have been useful in wildlife conservation evaluation.

The following two chapters in the book are concerned with relatively fundamental issues in the evaluation process. In the Developed World there is usually a good taxonomic basis for the study of organisms, and either a formal or an informal phytosociological basis for the recognition of communities. When such frameworks exist, they often provide the starting point for conservation evaluation. In Great Britain, Ratcliffe (1977) used such a framework as an initial breakdown of sites into comparable groups. However, in parts of the world where the classification of either communities or species is not well known, it is essential that a working framework be established. Within Antarctica some of the earliest conservation activities were concerned with the creation of a representative selection of communities within the developing system of

Table 1.2 A 'popularity poll' of criteria used in wildlife conservation evaluation. The table lists 19 criteria, together with the frequency with which each criterion was used in the nine studies reviewed by Margules and Usher (1981).

Criterion	Frequency of use in nine published studies
Diversity (both of species and habitats)	8
Naturalness	7
Rarity (both of species and habitat)	7
Area (extent of habitat)	6
Threat of human interference	6
Amenity value	3
Education value	3
Recorded history	3
Representativeness	2
Scientific value	2
Typicalness	2
Uniqueness	2
Availability	1
Ecological fragility	1
Management considerations	1
Position in ecological/geographical unit	1
Potential value	1
Replaceability	1
Wildlife reservoir potential	1

Specially Protected Areas (see Bonner, 1984). In Chapter 2 some consideration is given to the ways of formulating the framework so that a representative selection of sites can be made. In this sense 'representativeness' is a 'preliminary criterion', used to group together sites which can be compared, and to separate sites that are different ecologically and hence cannot be compared in the evaluation.

Most emphasis in conservation theory has been associated with the climax, or near-climax, ecosystem. However, in many situations a nature reserve or other conservation area will not be a climax ecosystem. Whittaker (1967) analysed a variety of ecological gradients, many of which demonstrate the dynamic nature of ecosystems. A whole variety of industrially made habitats can be colonized by a variety of plant and animal species, and may subsequently become suitable for nature reserve status. Although inevitably there will be a considerable emphasis in this book on the static nature of ecosystems, and particularly their description and classification, it is important to remember the ecosystem's dynamic nature both when evaluating its current importance and when predicting its future management needs. Chapter 3 has been written so as to remind the evaluator that the dynamic properties of ecosystems, especially the process of succession, should be considered during evaluation.

Different methods of evaluation have been developed in different parts of the world and when the evaluation has concentrated on areas of different spatial extent. The five chapters, 4 to 8, provide some idea of the range of approaches. In Chapter 4 the emphasis is on the Tropics, although it could be argued that the concepts apply to any territory that is not industrialized. Representativeness – the recognition of the types of nature resource in the territory – is seen to be important. However, the conservation of single species, usually those of considerable public appeal such as a panda or a parrot, is also important; in practical terms these conservation measures are likely to yield more income from the industrialized countries, both directly for conservation and indirectly via tourism.

Chapters 5 (United States of America) and 6 (Great Britain) probably demonstrate the greatest degree of refinement of the criteria that have been used for evaluating sites. Both territories, though of very different scales, have an environment that is largely modified by man. In both, a variety of criteria have been used to reflect various aspects of what is believed to have been the natural landscape. In Chapter 5 there is an emphasis on various indices which have been defined so as to encapsulate in a single number a large amount of information about a site. In contrast, in Chapter 6 the combinatorial problem has been side-stepped and greater attention is given to the various criteria, used more explicitly, though the mental integration of these criteria relies more heavily on the experience of the evaluator.

The analysis in Chapter 7 is based on a small territory that is one of the most densely populated areas of the world. The task of compiling an inventory of the nature resource has virtually been completed, and hence the sites of wildlife importance are already known (except for sites which are being created). The task of evaluation therefore becomes less important, but the concepts of environmental impact assessment take on greater importance. The sequence of chapters from 4 to 7 therefore provides a guide to conservation evaluation in a succession of countries from the underdeveloped to the extremely developed. Chapter 8, by focusing on a relatively small region within one of these four territories, serves to show how the importance of criteria depends on the geographical scale of the evaluation exercise. It is also important since it takes a multi-criterion approach, selecting sites on the basis of the best for a variety of criteria, rather than the multi-variate approach, whereby many criteria are included in a single index.

Very often in conservation planning one is concerned with the selection and scheduling of national parks, Specially Protected Areas, nature reserves, etc. In all of these, wildlife conservation is either the only form of land use or it is a dominant form of land use. However, the nature resource of a country is widely distributed and not just housed in a system of reserves, and hence it is important to contemplate evaluation in environments in which wildlife conservation will only form a secondary component of the land use. Two of these other forms of land use have been selected for discussion in Chapters 9 (forest and woodland) and 10 (agriculture). In each of these habitats the problem is essentially one of selecting criteria that will indicate the most important aspects of the nature resource. Once the most important areas have been identified, the evaluation task itself has been completed, but the integration of wildlife conservation with the dominant form of land use is likely to require a certain amount of compromise, which is clearly shown in Chapter 10. For forests, the three criteria of naturalness, diversity and rarity are likely to be the most important, whilst in agricultural environments rare species are unlikely to occur, and it is also unlikely that any habitat would be considered to be natural. Thus, in an agricultural environment, diversity and a lack of recent disturbance (or the continuance of traditional forms of management) are the most important criteria.

It is true that there have been very few evaluations of the same area, or series of areas, using different groups of organisms. Emberson (1985) reported a study of soil mites in the Yorkshire Wolds, and the ranking of several grassland sites based on their acarine species richness or diversity. This ranking was shown to be virtually independent of the ranking of the same series of sites on their botanical diversity. If the aim of a conservation evaluation is to locate sites for the optimal conservation of plant communities, of birds, or of butterflies, then it is appropriate to use just that group of species. However, the conservation ethic is increasingly directed away from the single species and more towards the totality of the nature resource. This inevitably means that knowledge about a greater proportion of that nature resource should be included in the evaluation procedure. The initial chapters, especially Chapters 2 and 3, concentrate on the botanical framework, essentially using the unproven assertion that 'if you look after the plant communities, the animals will look after themselves'. Emberson's (1985) study indicated that this assertion may not always be true. In order that as many groups of organisms as possible can be used in an evaluation, two animal groups have been selected for particular attention in this book. One, in Chapter 11, concerns the use of birds in conservation evaluation. This is a particularly appropriate group of animals because of intense public interest in birds. Mammals would also have made an appropriate group, possibly in relation to African game reserves, and fish, possibly of a coral reef ecosystem, would also have been appropriate. However, it was felt important to contrast the birds with a group of animals that do not in general have great popular support. Although Chapter 12 contains a broad review of a number of invertebrate groups, the particular emphasis in the chapter is on the flies (Diptera), a large group of insects which incorporates species with a wide variety of feeding specializations (herbivores, carrion feeders, dung feeders, leaf miners, parasitoids, etc). Whereas the emphasis with birds tends to be on population size, with the invertebrates the principal criterion seems to be diversity (species richness), though the statistical difficulties of collecting data and then making comparisons between sites are also extremely important.

The book is concluded with two unrelated chapters. In Chapter 13 the question is asked 'Can one investigate what criteria people actually used?', and it is shown that analyses of evaluators' performance do indicate some answers. There is a close agreement between evaluators on what are the best and the poorest sites, but much less agreement about more marginal sites (which are, of course, the most difficult to evaluate), and considerable differences between evaluators as to the way in which they arrive at their opinion. It would be useful to have details of some similar studies in other parts of the world so as to compare the work of evaluators with the range of criteria that are thought to be useful in making that evaluation.

The final chapter takes a different perspective. Given a free hand, how could one design a nature reserve? After all of the criteria and evaluation schemes have been discussed, is the location of reserves purely determined by non-scientific factors such as the availability of land or by political considerations? In Chapter 14 some established 'truths' are questioned and found to be lacking. The chapter serves as a reminder that the ecology of the whole biotic community needs to be understood if conservation management is to be fully effective. Criteria for evaluation and theory for design are both secondary, in the long run, when compared with a knowledge of the biology of the many species that live in the conserved area.

1.3 A REVIEW OF CRITERIA

1.3.1 Popularity polls

A preliminary 'popularity poll', which was based on a survey of nine studies, is included in Table 1.2. As more studies on conservation evaluation were published, Margules (1981) was able to revise and extend this popularity poll approach. Margules's extended poll encompassed sixteen studies, published during the decade between 1971 and 1980, in which twenty-four separate criteria had been used (see Table 1.3, in this table the criteria of 'representativeness' and 'typicalness' have been scored separately, whereas Margules grouped them

Table 1.3 A 'popularity poll' surveying the use of criteria in 17 evaluation studies published between 1971 and 1981. The maximum frequency in the table would be 17, indicating the use of that criterion in all studies. The table is modified from Margules (1981) and includes An Foras Forbartha (1981).

Criteria	Frequency of use of criterion
Diversity (of habitats and/or species)	16
Naturalness, rarity (of habitats and/or species)	13
Area	11
Threat of human interference	8
Amenity value, education value, representativeness	7
Scientific value	6
Recorded history	4
Population size, typicalness	3
Ecological fragility, position in ecological/geographical unit, potential value, uniqueness	2
Archaeological interest, availability, importance for migratory wildfowl, management factors, replaceability, silvicultural gene bank, successional stage, wildlife reservoir potential	1

together). The nine studies (1971 to 1978) in Table 1.2 are included in Table 1.3, and hence it is not surprising that there are clear similarities between the tables. However, a few differences become apparent when the two tables are compared.

Three criteria would appear to have become more popular with the inclusion of the studies that were published between 1979 and 1981. The largest increase is the criterion of 'representativeness', which has moved from a frequency of 2 to 7. It has already been argued that this is a preliminary criterion, associated with the classificational framework within which a conservation evaluation can operate. The other two criteria to increase were 'scientific value' (from 2 to 6), which is an integration of many features that contribute to the scientific interest of a site, and 'amenity value' (from 3 to 7), which represents an increasing realization that conservation will have to coexist with other forms of land use. The more practical criteria, such as 'management consider-ations' and 'availability', and the museum approach, typified by the criterion of 'uniqueness', have not increased in their frequency of use. The small increase in the criterion 'threat of human interference' (from 6 to 8) hopefully indicates that conservation is moving from the time when the conservationist only responded to threats to the nature resource, to an era when the conservationist can plan the nature resource that is to be managed. The most noticeable newcomer to Table 1.3 is the criterion of 'population size', this criterion is fully discussed in relation to birds in Chapter 12. In the popularity polls in Tables 1.2 and 1.3, four of the criteria are used in the majority of evaluation schemes, these criteria are each discussed in the following sub-sections of this chapter.

1.3.2 Diversity

Diversity can be measured either at the community (habitat) or the species level. The measurement of community diversity requires that a classification of communities exists. The number of these communities present at any site can then be counted, and indeed the communities can be mapped, when it would be possible to use the spatial extent (area) of each of the mapped categories in the diversity indices which are described below.

The diversity of species can be estimated in at least three different ways. First, diversity can refer to the species richness of the site (S), the count of the number of species known to occur, as indicated in Fig. 1.1. Secondly, it can be defined in terms of a function of the number of different species and their relative abundances (where the abundance may be measured by the numbers of individuals of each species, or by their biomass, etc.). Thirdly, diversity can relate to the number of different trophic levels and the number of interconnections between the species both within and between those trophic levels (Goodman, 1975).

Besides species richness (S), there are many indices which have been proposed for measuring diversity as a function of both the number of species and their abundance (see Pielou 1975, and Southwood 1978b, for reviews). Perhaps the most frequently used index is that which is associated with the name of Shannon, and is defined as

$$H = -\sum_{i=1}^{S} p_i \ln p_i, \tag{1.1}$$

where S is the number of species and p_i is the proportion of the ith species in a sample. H, which is usually referred to as the Shannon–Weaver index, has a minimum value of zero for a community that is a monoculture of one species, and a maximum value of $\ln S$ for a community of S species which are all equally abundant. Data for four artificial communities are shown in Table 1.4. For community U the four species are

Table 1.4 Some indices used to represent diversity. Formulae for deriving these indices are given in equations (1.1) and (1.2).

Community	Counts of the four species (in descending order)	S	H	D	d
U	25, 25, 25, 25	4	1.386	0.250	4.00
V	53, 27, 13, 7	4	1.141	0.376	2.66
W	68, 22, 8, 2	4	0.876	0.518	1.93
X	97, 1, 1, 1	4	0.168	0.941	1.06

equally abundant, and hence $H = \ln S = 1.386$. In community X, which is dominated by one of the species, H takes a small value. Community V is composed of a logarithmic series, where each species is approximately half as abundant as the preceding species: H takes a relatively large value. Community W is also a logarithmic series, but each species is approximately one third as abundant as its preceding species: here, H takes an intermediate value.

The other frequently used diversity index is Simpson's index, defined as

$$D = \sum_{i=1}^{S} p_i^2, \tag{1.2}$$

where S and p_i are as previously defined. D is approximately the probability that two individuals, drawn at random from the community, will be of the same species. Hence, in a monoculture, D would take the value 1: this is exemplified in community X in Table 1.4 where, with one very dominant species, D takes the value 0.941. In a community in which all of the species are equally abundant, D takes the value $1/S$. For community U in Table 1.4, where there are four equally abundant species, $D = 0.25$. Although D has this intuitive meaning associated with the probability of two individuals being in the same species, it is slightly confusing as the more diverse a community the smaller the value of D. For this reason, some people prefer to work with Simpson's index in an alternative form, such as $(1 - D)$, which takes the value 0 when the community is a monoculture, and approaches 1 as a community reaches its maximum diversity. Alternatively, one can define an index $d = 1/D$. This has the useful property of taking a minimum value of 1 in a monoculture, and a maximum value of S in the most diverse community of S species. Values of d are shown in Table 1.4.

As the sample size increases, there is a tendency for the number of species that are included in that sample to increase, and hence these measures of species richness and diversity can only be used to compare two sites if the sample sizes are equal. This is unfortunate since it is a truism that sample sizes are usually different. Another complicating factor is that population density, which may vary widely and independently from site to site, influences the amount of data that can be collected in any one sample unit (a core of soil, a quadrat of vegetation, a trap for insects, etc.): this is amply shown for the numbers of moths trapped in different woodlands by Taylor, Kempton and Woiwood (1976). Sanders (1968) suggested a method of standardizing data from samples of different sizes by calculating the expected number of species if all samples were reduced to the same size (obviously, the smallest size of any sample included in a series of samples). This method is known as 'rarefaction'. Williamson (1973) has criticized the method since there is a large loss of information, and Simberloff (1972) endeavoured to produce a computer simulation method, using equally sized smaller samples taken randomly from the original samples. Despite the difficulties with rarefaction, Kempton (1979) has demonstrated the value of standardizing sample sizes for various comparisons: using Taylor, Kempton and Woiwood's (1976) moth data he was able to distinguish consistent differences over time. Peet (1974) has shown that, for hypothetical communities, random sampling to form smaller samples may have different effects depending upon the dominance of one species. Thus, a community of three equally abundant species appeared to be more species rich than a community of eleven species, one of which formed 90% of the community, if samples contained less than twenty-three individuals.

The discussion of diversity, and its measurement, will clearly continue in the ecological literature for a long time. Some of the arguments both in favour and against various diversity indices used in a conservation setting have been discussed by Yapp (1979) and Usher (1983). However, in order to compare sites, Morris and Lakhani (1979) indicated that the Shannon–Weaver index was more sensitive to differences between sites than the Simpson index.

Species diversity for biological conservation is most conveniently assessed by the estimate of species richness, S. The measures of abundance which are included in indices such as H or D all require a lot of fieldwork time, and possibly also laboratory time if they are based on biomass, etc. Often there is insufficient time, or resources, to be able to measure these abundances. However, species richness should not be used uncritically, since it is known to be dependent on the size of the sample.

1.3.3 Area

It is probably a reasonable maxim that the larger the area, or extent, of a site, the more valuable it is for wildlife conservation. Although this may not be true in those few parts of the world where there are expanses of unaltered, or reasonably unaltered ecosystems (for example in some areas of tropical forest or in Antarctica), it is generally true in developed parts of the world where the natural and semi-natural habitats have been fragmented. There are three features of area that particularly concern conservationists, especially in relation to the evaluation of a site. First, there is an intuitive concept of a minimum viable size: is there some size below which the community cannot function, and hence is not worth conserving? Second, and closely related to the first, is the concept of minimum population size. Third, why does the number of species conserved increase with increasing area of habitat? These three topics are discussed below.

Relatively little work has been done on defining the minimum acceptable size for any conserved area. Implicit in a number of the studies included in the 'popularity poll' in Table 1.3 is this idea that the conservation value decreases drastically, or is totally lost, below some minimum size: the main problem is that such a size is based on a guess since there are probably no published scientific studies designed to test hypotheses of minimum area.

The most useful approach is to estimate the area requirements of the species in the reserve. Predators require larger areas than their prey; some species are better than others at dispersing and using other areas of suitable habitat for feeding, breeding, etc. Small, island-like reserves will, therefore, tend to favour species with small area requirements and species with better dispersal abilities. An estimate of area requirement is available for only a few species, and almost certainly does not exist for all the species in any one community. In a study of the ecological aspects

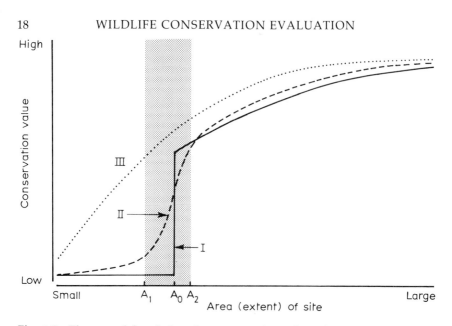

Fig. 1.2 Three models relating the conservation value of a site to its area. In model I the concept of a minimum area is valid since sites with area greater than A_o have a reasonably high value whereas sites less than A_o have a low value. Although conceptually attractive, this probably is unrealistic since A_o has not been defined for any communities. Model II is a more generalized form where the value drops greatly between area A_2, above which there is considerable value, and A_1, below which there is only a low value. A minimum area would, conceptually, lie between A_1 and A_2. Probably the most realistic model is model III, in which the value declines with decreasing area, but where there is no sharp reduction. In this model it is impossible to define a minimum area.

of a hardwood woodchip industry in Australia, Heyligers (1978) suggested 7000 ha as a minimum area for a eucalypt forest, though the only statement made with certainty was that such a forest should be as large as possible.

As predators have the largest range requirements, one approach may be to identify the top predator and to suggest a minimum area that would be appropriate for a viable population of that species. Theoretically at least, this area should be sufficient for all of the species in that food-web. There are, however, two problems. The first relates to dispersal, as it is possible that the top predator may be able to hunt in several areas, especially if it is a predatory bird. The species could, therefore, exist on smaller areas than the defined minimum. The second relates to the processes that occur in the ecosystem. Grubb's (1977) discussion of the regeneration niche for plants indicates that the species richness of communities is dependent upon the processes occurring within the community.

Minimum area, although an intuitively attractive concept, has generally been found to be deficient in practice. The fact that minimum areas have not been defined for various types of habitat probably indicates that it is an approach that has very limited use. This point is illustrated in Fig. 1.2.

The concept of a minimum population size is far more appropriate. This can be related to the concept of extinction rates or to the genetic

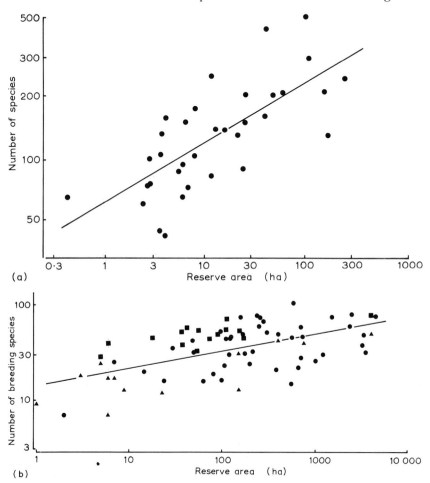

Fig. 1.3 Two examples of species–area relationships using data collected from nature reserves. (a) The data for higher plants (flowering plants and ferns) collected on reserves managed by the Yorkshire Wildlife Trust (Usher, 1979b). (b) The data for species of breeding birds on the reserves of the Royal Society for the Protection of Birds (Rafe, Usher and Jefferson, 1985). The symbols indicate woodland reserves (■), offshore islands (▲) and all other habitats (●). In both diagrams the line represents a regression equation of the form of equation (1.4).

consequences of inbreeding for the conserved population. Margules and Usher (1981) reviewed some of the literature on extinction rates, but a more analytic approach has been adopted by Wright and Hubbell (1983). They investigated the probability of extinction of a rare species that occurs on one large or on two small reserves (this point is discussed in relation to reserve design in Chapter 14). When there was no immigration (i.e. a species of low dispersal ability), models indicated that stochastic events would lead the species to extinction more quickly, and generally much more quickly, on the small areas than on the large area. When there was immigration (i.e. the species dispersed widely) the differences between large and small areas were generally negligible. Using data from the Farne Islands and Skokholm Island, Wright and Hubbell (1983) found that the probability of a bird species being present on an island approximated a model II curve in Fig. 1.2. To generalize this result from an individual species to a community, one needs to know whether there is synchrony or asynchrony in the individual species curves. Synchronous disappearance of the species from an island would result in a model II curve for conservation value of the whole community; asynchronous disappearance would lead to a curve more resembling model III, though possibly with a number of steps in it.

The relation between number of species and area is now well established: with special reference to nature conservation, Rafe (1983) has reviewed the literature and quoted a number of examples. As area increases, so the number of species recorded from that area also tends to increase. Two examples are shown in Fig. 1.3: the first relates to higher plants on the nature reserves managed by the Yorkshire Wildlife Trust and the second to breeding birds on the reserves of the Royal Society for the Protection of Birds. In both examples one can see that there is a tendency for S, the number of species, to increase as A, the area, increases. The nature of the relationship between S and A is unclear, usually because of the amount of variation about a regression line (as in Fig. 1.3). Two equations are frequently used. One, associated with Arrhenius (1921) and used by Preston (1962), takes the form of a power function

$$S = cA^z \tag{1.3}$$

where c and z are constants. By taking logarithms of each side of equation (1.3), it is transformed into

$$\ln S = \ln c + z \ln A \tag{1.4}$$

where $\ln c$ and z can be estimated by linear regression analysis. The other equation, associated with Gleason (1922), takes the form

$$S = d + b \ln A \tag{1.5}$$

where d and b are constants. Both Dony (undated) and Connor and McCoy (1979) examined large series of data sets and found that regression analysis using equation (1.4) tended to fit the data better than equation (1.5). Although the use of either model, or indeed other regression models such as S and A or S and $A^{1/2}$, is very much related to the preference of the user, the power function model of equation (1.3) has tended to be the most widely used. The two constants each have a simple biological interpretation: c provides an estimate of the number of species per unit of area, and z indicates how fast new species are added with increasing area (i.e. for the number of species to double, the area has to be increased by $2^{1/z}$; this is achieved by a 32-fold increase in area if $z = 0.2$, a 10-fold increase if $z = 0.3$, and a 5.7-fold increase if $z = 0.4$).

Habitat patches in a matrix of another habitat (usually arable land) have often been linked to islands (see, for example, the discussion of quarries in Chapter 3). Brown (1971, 1978) found that the number of species of mammals on isolated mountain ranges in the Great Basin of North America increased with area ($z = 0.33$). However, Miller and Harris (1977) found no relationship between the number of large mammal species and the area of 13 game reserves in East Africa, though the migratory habit of the animals is probably responsible for this exceptional result. Higher plants on nature reserves (Usher, 1979b) and on limestone pavements (Usher, 1980) (see Fig. 1.3 and Section 1.4), as well as in ponds (Helliwell, 1983), phytophagous insects in *Juniperus communis* (juniper) stands (Ward and Lakhani, 1977), birds in small woodlands (Woolhouse, 1981) and on nature reserves (Rafe, 1983), all show relationships between ln S and ln A that are statistically significant.

However, it should be remembered that the species–area relationship is a static description of a community. There have been very few studies where repeat surveys have been carried out so that the dynamic nature of the relationship can be investigated. Woolhouse (1981) looked at the data for 30 small woodlands, each of which had been surveyed in five consecutive years: the five regression equations were not significantly different. Usher (1973) estimated the species–area relationship for the 12 reserves in Yorkshire, for which data were available, as $S = 69.0A^{0.207}$, though as more data became available for these reserves, as well as for more recently acquired reserves, this relationship was modified (Usher, 1979b) to $S = 59.3A^{0.290}$. The change had resulted from a decade of survey work, underlining one important principle. Increased survey work in the geographical area, as well as the results from conservation management, are likely to lead to better data sets in which to undertake statistical investigations.

As well as increased survey work, two other factors contribute to the dynamic nature of species–area relationships. One is based on changes in taxonomic knowledge: in the developed world this will be relatively slow, but in under-developed areas considerable changes can take place over relatively short intervals of time. The second factor results either from natural or management-induced processes of colonization and extinction some of which are outlined in Chapter 3. It is, however, important to realize that if a site is surveyed once, the dynamic changes cannot be estimated. One must survey reserves at least twice, spaced by a sufficiently long period for immigration or extinction to occur, for the effects either of natural processes of succession or of management practices to be manifest. The 1970s saw a lot of initial surveys and publications dealing with species–area relationships: will the 1980s or 1990s see the re-surveying and assessment of the dynamic properties of those sites?

1.3.4 Rarity

Rarity is an intuitive concept that is frequently used in conservation evaluation. The definition of what actually is a rare species is difficult since there is a continuum from commonness to rareness. This is illustrated in Fig. 1.4 for breeding birds, butterflies and land snails in Britain. The species at the very left of these distributions, those that occur only in one, two or three 10 km grid squares, are clearly rare in the sense that they are very restricted. Species to the right of these

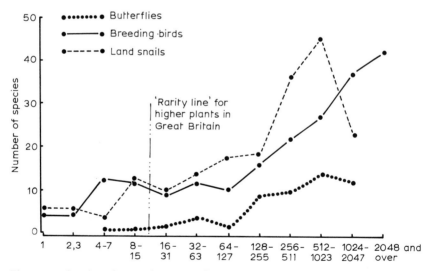

Fig. 1.4 The abundance of species of breeding birds, butterflies and land snails on the mainland of Great Britain (redrawn from Margules and Usher 1981). The horizontal axis is divided into octaves (*sensu* Preston, 1962) of numbers of 10 km National Grid squares from which the species have been recorded. A 'rarity line' is a vertical line, to the left of which all species are regarded as rare (for the vascular plants, species distributed on 15 or fewer squares are regarded as rare, and hence the line comes between the 4th and 5th abundance octaves: there is no generally agreed figure for birds, butterflies or land snails).

distributions are common, since they occur in the majority of the 10 km grid squares. However, where can one draw the 'rarity line' on these distributions such that all species to the left of the line are rare and all to the right of the line are not rare? For inclusion in the *British Red Data Book*, Perring and Farrell (1977) arbitrarily fixed the 'rarity line' between 15 and 16 10 km grid squares, i.e. the four octaves at the left of Fig. 1.4 are considered to be rare.

This definition of rarity, in terms of some feature of the geographical distribution, does not take into account either the distribution of the squares in which the species is recorded or the sizes of the populations of the species. Aspects of such features of rarity have been reviewed by Margules and Usher (1981), but many such aspects trace their origins to the dichotomy in rarity that was described by Mayr (1963): species can be rare because they are highly specialized or because they are very localized. This concept was developed by Drury (1974), who described three types of geographical distribution. One such distribution is assumed to be associated with species of stressed sites: a few individuals occur wherever a suitable habitat occurs. Species associated with serpentine rock outcrops would be an example, as would species characteristic of high nickel concentration in the soil. A second distribution is of widespread by locally infrequent species. In Britain the peregrine falcon, *Falco peregrinus*, would be an example (see also Chapter 3). Although it would fall to the right of the rarity line in Fig. 1.4, its population density is low. The third distribution is of species that usually occur in large numbers in a few localities. In Section 1.4 an example of this form of distribution is that of *Sesleria albicans* (blue moor grass). Although Mayr's and Drury's concepts provide a broad framework for viewing species distribution, population sizes and rarity, the framework omits the long term dynamic properties of many species.

Glaciations in the northern hemisphere have led to large-scale species movement, and some rare species are those that have been 'left behind'. An example of such a relict distribution would be the small areas of *Juncus alpinus* (alpine rush) or *Dryas octopetala* (mountain avens) in England: both species are more widespread in Scotland, and again far more widespread in northern Europe. These are examples of natural remnants; many species have become rare because they are man-induced remnants. The extinction of the large copper butterfly, *Lycaena dispar*, in England can largely be attributed to drainage of its fenland habitat (Duffey, 1968). The rarity of many cornfield weeds, such as the corncockle, *Agrostemma githago*, is due to the changing methods of cultivating arable land. Perring and Farrell (1977) state that *A. githago* used to be recorded in the majority of counties in the British Isles, but its decline started in the 1920s. Between 1930 and 1960 it was recorded from fewer than 200 of the 10 km grid squares, but since 1960 it has been recorded from fewer than 15 of these grid squares. In Australia, the Lesueur's rat kangaroo, *Bettongia lesueur*, was numerous before the 1850s (Finlayson, 1961), but by the 1970s it was only found on small islands off the coast of Western Australia. The Lesueur's rat kangaroo is the only macropodid that nests in burrows, and competition from the introduced rabbit for nesting sites has probably played a part in the species decline. Thus, in viewing rarity, one should think both about the distribution of the species and the effects of climatic change, land use, methods of husbandry, introduced species, etc., in reducing populations to remnants of what they were previously.

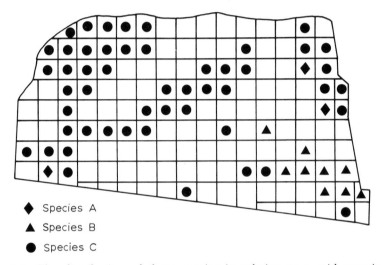

♦ Species A

▲ Species B

● Species C

Fig. 1.5 The distribution of three species in relation to a grid covering a hypothetical area (which could be a country, a region or a smaller area such as a county or district). There are a total of 150 grid squares. Notice that there are difficulties in defining 'squares' at the margins: generally a marginal area has been counted as a separate grid square if it occupies half or more of the area of a complete square. Thus the marginal squares have areas that range from 0.5 to 1.5 times a complete square. See text for discussion of species A, B and C.

Examples of the distribution of rare species are included in Section 1.4 and in Chapter 12. The concept of rarity has played a part in conservation evaluation, and it continues to play an important role. Helliwell (1982) implicitly weights his assessment for rarity, and Dony and Denholm (1985) explicitly use rarity in their ranking and evaluation of woodlands in Bedfordshire. In assessing macrophytic vegetation, de Lange and van Zon (1983) considered what they termed structural and floristic components of the vegetation, the floristic component is essentially the rarity of the species. In this discussion of rarity, it was initially suggested that rarity was a two state character of a species. Either the species is rare or it is not. The more modern use of rarity, where numbers of grid squares are counted, as in Dony and Denholm (1985), avoids the requirement of defining how infrequent a species has to be before it is sufficiently infrequent to be called rare. If there are sufficient grid squares covering the area of interest, the inverse of the proportion of the number of grid squares occupied by a species provides a reasonable quantitative index of rarity. This is demonstrated in Fig. 1.5. The area outlined could be a county, a region or a nation, and the grid lines may be at 1, 2, 10 or 100 km intervals. Species A, B and C occur in 3, 9 and 50 of the 150 grid squares, and their frequencies of occurrence are 0.02, 0.06 and 0.33 respectively. The inverses of these are 50, 16.7 and 3, which provide reasonable, geographically-based, indices of rarity.

Despite the difficulty with the definition of rarity, quantification of this criterion is simple in areas of the world where the fauna and flora are well recorded and well mapped, as in Europe (and the British Isles in particular). Rarity is conceptually easy for the layman to grasp, and hence, even though it may not be particularly important scientifically (see Section 1.3.7, however), it is of critical practical importance. Rare species cause a strongly emotive reaction in many sectors of a country's population, and hence the presence of rare species is likely to provide strong political weight if sites are to be set aside for wildlife conservation purposes.

1.3.5. **Naturalness**

The term 'naturalness' implies the recognition of some natural condition which may be difficult to define. It is often used to imply that a site has been free from human influence, or that species have not been introduced via a human agency. As most communities of plants and animals have been influenced by man, it becomes the extent of that influence that is critical. Heinselman (1971) defined a virgin forest as one that is '. . . the product of natural environmental factors and ecological processes as opposed to a forest resulting from logging, land clearing, herbiciding, planting, or similar disturbance by man'. This is a non-interventionist definition. Opposed to this are definitions which realize that man might be part of a natural ecosystem – Moir (1972) has given minimum disturbance by man as an essential feature of natural ecosystems, and Jenkins and Bedford (1973) characterized natural areas by the absence of what they called 'artificial human disruption'.

A definition of naturalness thus includes the possibility that an ecosystem may have been modified by man just as the ecosystem may have been modified by other mammals. The difficulty then becomes what is meant by 'slightly modified'. Margules and Usher (1981) considered that there were two factors of the man–environment interaction that were essential in attempting such a definition. First, the size of the human population must be limited by the size of the environment in which it lives. This implies that there is no import of food, building materials, etc. Secondly, products of the ecosystem are used locally. This implies that there is no export of biological materials. Thus, they defined a natural ecosystem as being one in which men may be present, but if they are present, then they are totally dependent on, and limited by, their environment within that ecosystem.

Another problem is knowing what natural ecosystems may be like. In areas of the world, such as Western Europe, where there is a long history of modification of ecosystems by man, it is likely that there are no fully natural ecosystems. In a country like Australia, it is relatively simple to define a natural ecosystem: Carnahan (1977), in mapping the natural vegetation of Australia, regarded the natural condition as that at the time of European settlement. There is no convenient, or generally acceptable, time in Europe before which the ecosystems could be considered natural. The analysis of pollen deposits (Godwin 1975) would indicate the advent of an iron age culture, but it is too long ago, with too many climatic changes having taken place since, for that to form a reasonable baseline for studies of the naturalness of communities. Hence, in Britain the least disturbed communities are often referred to as 'semi-natural' (Ratcliffe, 1977), or 'near-natural'.

Margules and Usher (1981) also considered a variety of ecosystems which are treated as being semi-natural, but are in fact largely the result of historical forms of land use. One example is grassland in the British Isles. Pennington (1969) has indicated that only grassland of the high mountains was not originally man-induced, and even that has been subjected to grazing by domestic stock. In the Craven Pennines of North Yorkshire, an area of carboniferous limestone, base-rich grasslands supported 13.7 species of vascular plants per 0.25 m^2 quadrat, whilst neutral and acid grasslands supported 10.8 and 6.9 species per 0.25 m^2 quadrat respectively. Pastures which have been improved by the addition of lime, generally neutral and acid grasslands, had a mean of 12.7 species per quadrat. Since the agriculturally improved grasslands had a similar diversity to the base-rich grasslands, there is a problem of the recognition of semi-natural and modified grassland habitats. The species complement was generally different: the grass *Cynosurus cristatus* (crested dog's-tail) was virtually absent from the semi-natural grassland, whereas it was abundant in agriculturally improved grassland (Usher, 1980). A second example of the effect of land use relates to the North York Moors National Park, where there is evidence for a significant contribution by prehistoric men to the creation of moorlands dominated by a virtual monoculture of heather, *Calluna vulgaris* (Atherden, 1976). In more recent times, the maintenance of the character of the moorland has been dependent upon rotational burning for grouse (*Lagopus lagopus scoticus*) management. Although the *Calluna* moors are likely to have been man-induced, they are now widely regarded as semi-natural in the sense that the vegetation has been unmodified for a long period of time.

The more natural an ecosystem, the more it tends to be valued. This is not only because of the variety of plants and animals that it contains, but also that other features of the ecosystem may have been conserved. One of the important aspects that has received rather little attention is related to the conservation of soil profiles. Ball and Stevens (1981) have discussed the importance of ancient woodlands in Britain as a feature of soil conservation. To what extent the conservation of soil profiles also relates to the conservation of the extremely diverse soil ecosystem remains unknown.

Naturalness can also be applied to species as well as to communities of species. Although it is generally felt that introduced species have rather limited conservation value, species which are long standing introductions can be valued. An example of this is the woad, *Isatis tinctoria*, in Britain. It has strong historical connections. The *British Red Data Book, I: Vascular Plants* (Perring and Farrell, 1977) contains a number of introduced species, together with a number of species of doubtful status. Although it may seem fairly simple to judge whether a species is native or introduced, there can be difficulties. An example of this is an American pondweed, *Potamogeton epihydrus*, which, in Britain, is native in the Outer Hebrides and which has been recorded from canals near Halifax in West Yorkshire. Although technically it is a native within the British Isles, because it occurs in the far north-west of the island archipelago, it is clearly an introduced species in mainland Britain.

Naturalness was included in many of the studies listed in Tables 1.2 and 1.3. There are intuitive ideas of what naturalness means, but it is one of the most difficult concepts to quantify. Usher (1980) indicated that naturalness and diversity tended to be strongly correlated, since modification of ecosystems tends to reduce the species richness. Although it is difficult to find a quantitative index of naturalness, the amount of disturbance of ecosystems may provide a reasonable basis for the quantification of this criterion.

1.3.6 Representativeness

Two criteria, both in an intermediate position in Tables 1.2 and 1.3, lead to some confusion. These are the criteria of 'representativeness' and 'typicalness'.

Chapter 2 concentrates on the criterion of representativeness. It is essentially classificatory, listing the kind of ecosystems occurring within

Table 1.5 The nine vegetation associations recognized by Margules (1984b) in the North York Moors National Park.

Association	Principal species* (with frequency, %)	Number (and %) of quadrats in survey
1. *Erica tetralix/ Calluna vulgaris*	*E. tetralix* (100), *C. vulgaris* (89), *Nardus stricta* (48)	88 (14)
2. *Pteridium aquilinum*	*P. aquilinum* (100), *Vaccinium myrtillus* (43)	80 (13)
3. *Vaccinium myrtillus/ Calluna vulgaris*	*V. myrtillus* (100), *C. vulgaris* (95)	56 (9)
4. *Festuca ovina/ Nardus stricta*	*F. ovina* (96), *N. stricta* (89), *Galium saxatile* (62), *Vaccinium myrtillus* (54), *Deschampsia flexuosa* (46), *Anthoxanthum odoratum* (42), *Potentilla erecta* (42)	26 (4)
5. *Juncus squarrosus/ Calluna vulgaris*	*J. squarrosus* (100), *C. vulgaris* (95)	41 (7)
6. *Juncus effusus*	*J. effusus* (88)	59 (9)
7. *Eriophorum* spp.	*Calluna vulgaris* (77), *E. angustifolium* (66), *E. vaginatum* (55)	47 (7)
8. *Molinia caerulea*	*M. caerulea* (95), *Erica tetralix* (45), *Anthoxanthum odoratum* (40), *Potentilla erecta* (40)	20 (3)
9. *Calluna vulgaris*	*C. vulgaris* (98)	216 (34)

*Principal species include all those species occurring with a frequency of 40% or more in the quadrats included within each association.

a geographical area. Margules and Usher (1981) included it within those criteria that could be assessed only in relation to data from surveys in the whole of the biogeographic area in which an evaluation is being carried out. In the previous section on naturalness, the *Calluna*-dominated moorland in the North York Moors National Park was referred to. Margules (1984b) took a stratified random sample of 633 quadrats over the moorland area and for each recorded the vegetation. After numerical analysis, he proposed nine vegetation associations for the North York Moors, these are listed in Table 1.5. This provides a framework within which evaluations can be carried out, and can be used in either one of two ways. It can be used in a 'stamp collecting' way, whereby a conservationist would want each of the nine associations to be represented within a collection of reserves. Alternatively, it can be used in a 'shopping around' way, whereby one knows that it is easier to find a *Calluna* association (216 quadrats) than a *Molinia* association (only 20 quadrats).

Representativeness, as outlined in Chapter 2, is often a preliminary step in any evaluation. Although Ratcliffe (1977) did not use it as one of his ten criteria, he implicitly used it by considering the national series of prime sites under headings such as woodland, lowland grasslands, heaths and scrub (with some subdivisions), etc. Representativeness is, therefore, a global concept, often constraining those sites which can be compared with each other, and hence it could be argued that it is not a criterion at all.

Typicalness, on the other hand, is not a global concept, and can be assessed for an individual site. Ratcliffe (1977) says that key sites for conservation are usually chosen as the best examples of paticular ecosystems, but their quality may be determined by features which are in some degree unusual. Although this is valid, it is also necessary to represent the typical (commonplace) in terms of habitats, communities or species. He states: 'sites sometimes have to be selected for their characteristic and common habitats, communities and species, and it is then necessary to overlook the absence of special or rare features'. He indicates that such typical habitats are particularly important for ecological research.

There is therefore a paradox between rarity and typicalness (Usher, 1980). Sites are valued if they contain assemblages of rare species, and sites are valued if they are typical. By its very definition, a site with a lot of rare species cannot be typical, and, conversely, a typical site cannot have more than an average number of rare species. To some extent then, the two criteria of rarity and typicalness are mutually exclusive. However, typicalness and representativeness as criteria, if the latter is a criterion at all, are conceptually very different, even if these distinctions have often not been made in the literature.

1.3.7 Other criteria

Reviews discussing other criteria, listed in Tables 1.2 and 1.3, can be found in Ratcliffe (1977), Usher (1980), Margules (1981) and Margules and Usher (1981). The purpose of this section is to investigate a few other criteria that have been used, usually for very specific evaluation studies.

The concept of the fragility of an ecosystem is regarded as a complex of attributes by Ratcliffe (1977), who states that it reflects the 'degree of sensitivity of habitats, communities and species to environmental change'. Since the environmental change can be due to either physical factors, such as climate, or to land use practices, Ratcliffe (1977) indicates that fragile sites are usually highly fragmented, dwindling rapidly, and difficult to re-create. However, the concept of fragility, or the inverse concept of resilience, can have different interpretations for different types of ecosystem.

As in Chapter 3, consider an ecological succession where, during a long period of time, one community follows another until some end-point, the climax, is reached. This end-point is when the community no longer changes (in general, no further change relates to a generation or two of man since records do not allow for a longer assessment, and it also represents the average condition over a reasonably large area of habitat, so as to avoid the necessity of considering cycles at a single point). Ratcliffe's (1977) definition of fragility obviously applies easily to climax communities since one is expecting them not to change unless there is a change in physical conditions or a change in management. However, during the course of a succession, the intermediate or seral communities are naturally changing, and it might be difficult to determine how much of an observed change is due to the process of succession and how much to some form of perturbation. Fragility would, therefore, appear to be more a property of climax communities than of the seral communities.

The field biologist has used the words fragility or resilience to assess something which is observable only over a long period of time. Unlike survey work, which is carried out at one time only, the assessment of fragility requires measurements at two times, at least, separated by a sufficiently long period. Case history information will, therefore, almost certainly be needed in assessing fragility or resilience as a criterion. An analogous concept is used by theoretical ecologists when they refer to 'stability'. Populations may be considered to exist at a stable size, and the theoretical ecologist might then be interested in the behaviour of a population whose size is near this stable value. If a relatively small perturbation of the population results in it returning to the stable size, then the population is said to be 'locally stable'. If the population tends to this stable size from any given size, then this stable size is considered to be 'globally stable'. This concept is developed by May (1973), and some examples of both populations and mathematical analyses are given in Usher and Williamson (1974).

Much the same property is being measured either in field assessments of fragility or in mathematical models of stability. It is unfortunate that the study of model populations has not yet advanced to a stage of being particularly useful in field assessments. Whereas the field biologist assesses the fragility of communities, the modeller has hardly started to assess the stability of simple ecosystems, and certainly not the complex ecosystems that a conservationist would be concerned with. Until such models are developed, it is unlikely that fragility can be assessed quantitatively. Its assessment will rely very heavily upon survey work and the documentation of case histories.

Another criterion that is sometimes referred to is the structure of the vegetation. The idea is essentially complex, and relates to other criteria. If a community is fragile, a small perturbation might result in its structure being affected. Usually, the less natural an ecosystem the more its structure will have been changed. These are examples of the use of vegetation structure and the links with other criteria. In general, therefore, structure is an unsatisfactory criterion: it is better to deal with structure as one facet of the more fundamental criteria. However, in assessing aquatic communities, de Lange and van Zon (1983) found community structure to be a useful concept. They defined structure as being related to the number of species (diversity correlated with structure), the percentage cover per vegetation layer, and the quantity of filamentous algae (related to the level of eutrophication). Thus, these authors could have used three separate criteria, though they preferred to add the scores for the three components together so as to derive a single value for vegetation structure.

Successional ecosystems have been discussed above in relation to fragility and are further considered in Chapter 3. However, in an assessment of urban areas, where the majority of ecosystems will be man-made or consist of natural colonists of man-made habitats, Wittig and Schreiber (1983) have used the period of development as a criterion, as well as area, rarity and diversity of habitats. Value is associated with longer periods of development since these have more species that occur naturally within the geographical area. Wittig and Schreiber used a logarithmic scale of time, with a minimum value being given to sites that are less than two years old, and then incrementing the value by one point for sites 2–5, 5–10, 10–20, 20–50 and over 50 years of age. For urban ecosystems, which are more or less totally unnatural, these time scales may be seen as the system becoming more self-regenerating and hence, in a sense, more 'natural'.

Two criteria seem to have been used very rarely, but nevertheless might be worthy of greater consideration. One relates to endemicity. Terborgh and Winter (1983) defined endemic birds in South America as having ranges of less than 50 000 km^2, and they argued that these are particularly vulnerable to changes in land use (e.g. deforestation). They located areas in Colombia and Ecuador where there was concentrated endemism, and they suggested that such areas should form the basis of reserves or national parks. Endemism is clearly related to both rarity and fragility, but it might provide a useful criterion in some parts of the world. A related concept is that of type localities. When a new species is described, the locality in which the holotype was collected is designated the 'type locality' for that species. Many species are known only from their type locality, and hence the use of type locality as a criterion is closely related to the criterion of endemicity. Bonner (1984) discussed the criteria that were used for deciding whether sites should be awarded the status of 'Specially Protected Areas' in Antarctica. Five criteria were used, and these include representativeness, endemicity (areas with unique complexes of species), type localities (or the only known breeding habitat of plants or invertebrates), and two others which are impossible to categorize (areas which contain especially interesting breeding colonies of birds or mammals, and areas which should be kept inviolate so that in the future they may be used for purposes of comparison with localities which have been disturbed by man: the former is akin to Ratcliffe's (1977) criterion of 'intrinsic appeal', and the latter reflects concepts of naturalness, typicalness and research value).

The two criteria of 'endemicity' and 'type locality' require further study. Both are related to rarity, and in fact both may be scientific, as opposed to practical or political, components of the criterion of 'rarity'.

1.4   AN EXAMPLE: LIMESTONE PAVEMENTS IN YORKSHIRE

Limestone pavements are interesting geomorphological features that have a very restricted distribution in the British Isles. The majority of them occur on the carboniferous limestone in the north of the Pennines, approximately where the counties of Yorkshire, Cumbria and Lancashire meet. A small number occur on the limestone formations in Scotland and Wales. The example used here is drawn from a collection of 49 limestone pavements that occur within an upland block of land that stretches between Malham, Arncliffe and Kilnsey in North Yorkshire. The highest point is at an altitude of 538 m, and most of the land of the Malham–Arncliffe plateau lies above 400 m.

Table 1.6 The group A species* (see text for definition) which are recorded from the Malham–Arncliffe plateau. The 49 pavements on the plateau represent 9.1% of the 537 pavements included in Ward and Evans (1976) national survey. The data are taken from Usher (1980).

| Species | Number of pavements | | Percent of national occurrencies |
	Malham–Arncliffe plateau	Nationally	
Dryopteris villarii	11	258	4.3
Gymnocarpium robertianum	17	212	8.0
Actaea spicata	8	30	26.7
Cardamine impatiens	4	13	30.8
Ribes spicatum	14	16	87.5
Polygonatum odoratum	3	57	5.3

*Group A species not recorded from the Malham–Arncliffe plateau include *Hypericum montanum*, *Potentilla crantzii*, *Dryas octopetala*, *Daphne mezereum*, *Salix myrsinites*, *Epipactis atrorubens* and *Carex ornithopoda*.

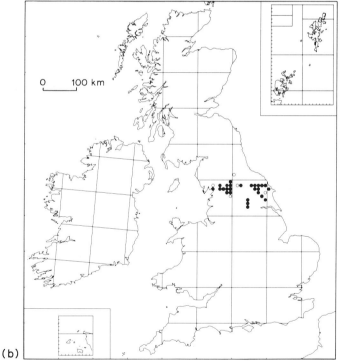

Fig. 1.6 (a) *Actaea spicata*, a group A species, growing in a deep gryke. The vertical wall of the gryke can be seen at the right. (b) The distribution map for *Actaea spicata* within the British Isles. This and subsequent maps have been produced by the Biological Records Centre, Institute of Terrestrial Ecology.

Fig.1.7 (a) *Cirsium helenioides*, a group B species, growing beside a small cliff margin to a limestone pavement. With the shelter of the cliff, the plant has grown taller (about 1.2 m). (b) The distribution map for *Cirsium helenioides* within the British Isles.

Limestone pavements themselves are flat outcrops of rock that have been weathered into series of deep, irregular crevices (known as 'grykes': the flat tops are called 'clints'). By and large the clints support no vegetation, although, exceptionally, when water and debris have collected in a concave clint, there may be communities of spring-flowering annual plants that set seed before the habitat dries in summer. In the wider grykes, that are open to the surrounding grassland, sheep and rabbits enter to graze the vegetation, which is similar to the surrounding calcareous grassland. When the grykes are narrower, and generally not open to the surrounding grassland, a woodland vegetation occurs. A few plant species, such as some of those listed in Table 1.6, are more or less restricted to the gryke habitat in Britain.

A botanical survey of the limestone pavements in Great Britain was carried out (Ward and Evans, 1976) in order to assess the conservation importance of this scarce, and threatened, habitat. The data collected during that survey were used by Usher (1980) in assessing the conservation value of parts of the Malham–Arncliffe plateau. Ward and Evans (1976) had assigned each plant species to one of four groups. Species in group A are those which appear to be dependent on the limestone pavement habitat for the maintenance of their population. All species in group A are nationally rare: the six species in this group that occur on the Malham–Arncliffe plateau are listed in Table 1.6. The distribution of one of these species, *Actaea spicata*, is shown in Fig. 1.6b. It can be seen from Table 1.6 that this plateau region is particularly important for the conservation of *Ribes spicatum* in the British Isles. The species in groups B and C are also very much dependent on the limestone habitat, but are nationally uncommon and common respectively. Examples of group B species are *Cirsium helenioides* (Fig. 1.7), which has a northern distribution, and *Thalictrum minus* (Fig. 1.8), which predominantly occurs in coastal sites. A group C species is *Phyllitis scolopendrium* (Fig. 1.9), which is widely distributed throughout most of the country. Finally, group D species are those that are essentially incidental on the limestone pavement: they occur in the surrounding grasslands. *Carex flacca* (Fig. 1.10) is an example of a group D species that is extremely common throughout the British Isles. Another example is *Sesleria albicans* (Fig. 1.11) which, although having a very restricted distribution, occurs abundantly in the grasslands surrounding the limestone pavements.

The question to answer is: 'Which limestone pavements are the most important to conserve?' Ward and Evans (1976) approached this by deriving an index based on the number, abundance and rarity of the plant species. Thus, for rarity, an A-species was given a value of 3, a B-species a value of 2 and a C-species a value of 1. D-species were considered valueless, and hence were not included in the index. Abundance was scored on a scale of 1–3. If a species was represented by one or two individuals, or was very sparsely distributed over the

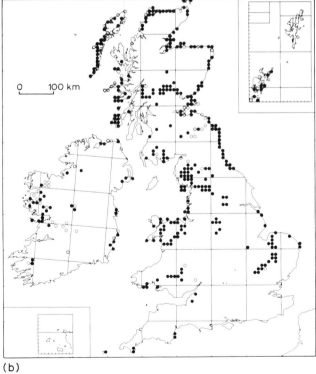

(b)

Fig. 1.8 (a) *Thalictrum minus*, a group B species, growing on shattered lime-stone rock. (b) The distribution map for *Thalictrum minus* within the British Isles.

Fig. 1.9 (a) *Phyllitis scolopendrium,* a group C species, growing from the vertical wall of a gryke. (b) The distribution map for *Phyllitis scolopendrium* within the British Isles.

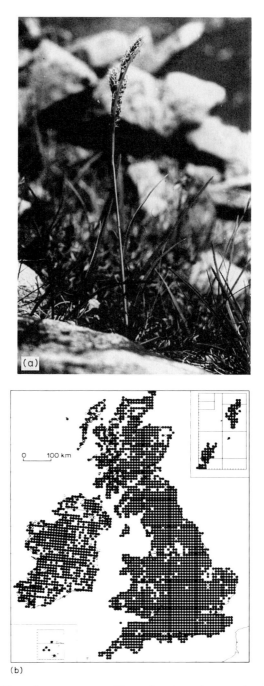

Fig. 1.10 (a) *Carex flacca*, a group D species, growing on shattered limestone rock. (b) The distribution map for *Carex flacca* within the British Isles.

Fig. 1.11 (a) *Sesleria albicans*, a group D species, growing from a shallow gryke that is surrounded by a wide expanse of clints. (b) The distribution map for *Sesleria albicans* within the British Isles.

pavement, it was given a score of 1. Species which were locally abundant on the pavement, or scattered over the whole pavement, were scored 2, whilst abundant species were scored 3. Thus, if a pavement had two A-species (both with abundance score 2), eleven B-species (three with abundance score 3, and four each with scores 2 and 1) and 20 C-species (eight with score 3, seven with score 2 and five with score 1), as well as 15 D-species, then the index is composed as follows. The A-species contribute

$$2 \times 2 \times 3 = 12 \text{ points,}$$

i.e. the product of the number of species (2), the abundance (2) and the rarity (3) scores. The B-species contribute

$$(3 \times 3 + 4 \times 2 + 4 \times 1) \times 2 = 42 \text{ points,}$$

where again the products are of numbers of species, abundance and rarity (which scores 2 for B-species). The C-species contribute

$$(8 \times 3 + 7 \times 2 + 5 \times 1) \times 1 = 43 \text{ points,}$$

and, since the D-species contribute nothing, this gives

$$FI = 12 + 42 + 43 = 97 \text{ points,}$$

where FI is the floristic index. As can be seen, large values of the index are associated with pavements that have large species richness (i.e. a lot of group A, B and C species) and with pavements that have a large number of rare species (one A-species counts for 3 times one C-species). This index was used to rank pavements from the 'best' to the 'poorest'.

It is noticeable that large pavements tend to have large values of such an index, whilst small pavements tend to have small values. Area was, therefore, clearly correlated with diversity and rarity. This prompted Usher (1980) to plot species–area relationships for the 49 pavements. Area is a difficult concept since the clints are largely devoid of life, and gryke area is impossible to measure as it has both horizontal and vertical components. Hence, in Fig. 1.12 the cartographic area of the pavement has been used. Using different combinations of species groups, one derives the following regression equations. For the uncommon and rare species (groups A and B), the relationship is

$$S = 7.85A^{0.141} \quad (r = 0.44,\ 0.001 < P < 0.01) \tag{1.6}$$

where S is the number of A- and B-species, A is the cartographic area, and significance is shown by the correlation coefficient, r. For all species dependent on the limestone habitat (groups A, B and C), the relationship is

$$S = 29.35A^{0.078} \quad (r = 0.29,\ 0.01 < P < 0.05), \tag{1.7}$$

and for all species (groups A, B, C and D) the relationship is

$$S = 37.56A^{0.086} \quad (r = 0.34,\ 0.01 < P < 0.05). \quad (1.8)$$

All three equations are statistically significant, indicating that there is a trend towards more species as area increases. There is, however, a large amount of scatter about the regression lines in Fig. 1.12, and hence the area of the pavement does not explain all of the variation in species richness. The results of using similar analyses for the 77 pavements around Ingleborough are discussed by Usher (1985b).

The important feature of these equations is that they can be used to eliminate the effects of area from a consideration of the other criteria in evaluating pavements. If *diversity* is the most important criterion, then pavement D_{max} in Fig. 1.12 is the most important for conservation purposes since it deviates the most from the regression line for all species.

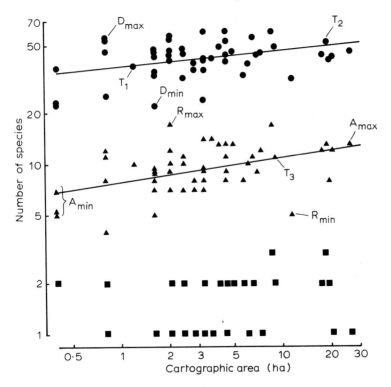

Fig. 1.12 Species–area relationships for higher plants on 49 limestone pavements on the Malham–Arncliffe plateau. Squares represent species in group A: there are too few for statistical analysis. Triangles represent species in groups A and B: the line through these triangles represents the regression equation given in the text, equation (1.6). The circles, and the line through them, represent all plant species (groups A, B, C and D) on these pavements (see equation (1.8)). Pavements indicated by letters are discussed in the text.

The pavement least valued is the one furthest below the regression line, as indicated by D_{min} in Fig. 1.12. If *rarity* (including uncommonness) is the most important criterion, then pavements R_{max} and R_{min} would have the greatest and smallest values since they deviate the most in positive and negative directions from the area–eliminated line. Obviously, if *area* is the criterion, pavements A_{max} and A_{min}, at the right and left of Fig. 1.12 respectively, are the most and least valued. If *typicalness* is the criterion of importance, then pavements such as T_1, T_2 and T_3 in Fig. 1.12 would be candidates since they are very close to the average species–area relationship.

Many of the criteria listed in Tables 1.2 and 1.3 are not appropriate when evaluating these limestone pavements. *Representativeness* cannot be used since all sites are of the same habitat-type, unless, of course, different kinds of limestone pavement could be recognized. Criteria such as *naturalness, threat of human interference, amenity value, educational and scientific value, fragility,* etc., cannot be used, since the values assigned would be virtually invariant from pavement to pavement in this restricted geographical area. It is possible that the recorded history may vary from pavement to pavement, and that some pavements may be available whereas others are not. However, in terms of basic scientific data, there are essentially only the four criteria of *area, diversity* (species richness), *rarity* and *typicalness* that can be used for comparing, or ranking, a series of pavements such as the 49 of the Malham–Arncliffe plateau.

Usher (1980) discussed these criteria in relation both to this set of pavements and the upland area surrounding them. Reed (1983) discussed the use of species–area relationships further, and he concluded that they have only limited application in conservation studies. However, as shown in the limestone pavement example, the use of a species–area relationship does highlight the fact that so many attributes of sites are correlated with area: on the pavements the larger the area the larger the species richness both of all species and of rare species.

1.5   DISCUSSION

In this chapter, the popularity polls (Tables 1.2 and 1.3) and reviews of criteria (Section 1.3) have indicated that an evaluator may be faced with a large number of criteria to choose from, although the example of the limestone pavements indicated that some criteria would be appropriate whilst others would be invariant for the series of sites to be compared and hence of little use. However, how are several criteria to be integrated so that definitive answers can be given to questions such as 'Which of the two sites is better for wildlife conservation?', or 'Can this series of sites be ranked for their wildlife conservation value?' Clearly it would be useful to be able to form an index of conservation value that had the form

$$I = f(A, D, R, T, \ldots), \tag{1.9}$$

where f represents a mathematical function and A, D, R, T, etc., are numerical values for area, diversity, rarity, typicalness, etc. (see Usher, 1985b).

Such a goal is probably impossible to achieve. For real sites, A and D are correlated, as the species–area relationships have shown. R and D are also interrelated, and, as sites with more than the average number of rare species are not typical, R and T would be inversely related. Hence, the various variables that might be included in an index such as that in equation (1.9) are not strictly independent.

Estimates of a value for each of the four variables included in equation (1.9) can be made relatively simply, but the amount of weight to give to any one of them is a matter of personal opinion. Indeed, the weights vary greatly from evaluator to evaluator, as shown in the two studies reported by Margules (1984a) and Margules and Usher (1984). There is, probably, no way of finding an objective way of weighting these components of an index of conservation value. The best that can be achieved is to make a subjective selection of weights, but to agree these a priori, i.e. before the data are collected and the sites are compared. Weighting of components of indices of conservation value is returned to in Chapters 3 and 5 of this book.

One of the major drawbacks of the example quoted in Section 1.4 is that the data were collected at one time, and hence they provide a 'snap-shot' of the series of 49 limestone pavements. Using terminology that is analogous to mathematics, the data and analysis in Section 1.4 can be likened to 'statics'. There is no information on the 'dynamics', as dis-cussed in greater detail in Chapter 3. If the series of pavements were to be re-surveyed after a sufficiently long period of time, say 10 or 20 years, would one find that relaxation (or extinction) of species had occurred on those pavements that lay far above the species–area regression line? Alternatively, would one find that the pavements far below the regres-sion line had been invaded so that their diversity had increased towards the norm? Although there are plenty of data, that can be likened to statics, available to evaluators and conservationists, the quantity of data relating to the dynamics is extremely small.

Three negative conclusions would be that (i) the interrelationships between criteria (attributes) are too complex for any meaningful index of conservation value to be formulated, (ii) there will never be agreement on the weights to give to the different attributes in formulating an index, and (iii) that the 'statics' are probably insufficient to guide conservation management.

However, to be more positive, conservation cannot wait whilst all of the dynamic processes are investigated. In the majority of chapters in this book, the evaluation is performed using data from a single survey, perhaps supported by case history studies of similar sites elsewhere. The formulation of indices is a process that most evaluators employ, occasionally explicitly, but much more commonly implicitly. One of the aims of this book is to encourage the explicit formulation of methods, a priori, so that the scientific basis to the evaluation procedure can be discussed and agreed between practitioners and users as far as is possible.

1.6 SUMMARY

As well as introducing the various chapters of the book, this chapter aims to define terms, review criteria that have been used, and give an example of conservation evaluation in practice.

Wildlife implies all species of plant, animal and microbe that are non-domesticated. *Conservation*, in the context of this book, is concerned solely with wildlife, and hence it is equated with the nature resource of a country. Before any *evaluation* proceeds, it is important to define the aims: what is to be valued, and why?

Any site can be thought of as having *attributes*, which are properties of that site that reflect its conservation interest. The *criterion* is the way of expressing an attribute in a form that can be used in an evaluation. The *values*, which are given to the states that a criterion can take, may be those of the scientist, but more importantly they should be those of the society within which the site is situated.

Popularity polls, as in Table 1.3, indicate which criteria tend to be most frequently used. These include diversity of species or habitats, area, rarity of species or habitats, naturalness and representativeness: each of these criteria is reviewed. Brief reviews are also included of typicalness, fragility, stability, vegetation structure, the successional stage of the ecosystem, endemicity and type locality.

In an example of the flora of a series of 49 limestone pavements on an upland area, only four criteria were found to be useful. These were area, diversity, rarity and typicalness. This example leads into a discussion of which sub-set of criteria are of use in any given situation, and of how several criteria can be integrated to give a final assessment.

CHAPTER 2

Assessing representativeness

MIKE P. AUSTIN

and

CHRISTOPHER R. MARGULES

2.1   Introduction
2.2   Current methods in Australia
2.3   Numerical methods for land classification
        2.3.1   Numerical classification
        2.3.2   Problems of scale, complexity and
                information
        2.3.3   Data
        2.3.4   Complexity and mapping
2.4   Methods of analysis for assessing
        representativeness
        2.4.1   The continental or between region scales
        2.4.2   Within region scales
2.5   Discussion
2.6   Summary

Wildlife Conservation Evaluation. Edited by Michael B. Usher.
Published in 1986 by Chapman and Hall Ltd, 11 New Fetter Lane,
    London EC4P 4EE
© 1986 Chapman and Hall

2.1   INTRODUCTION

The idea that a reserve or system of reserves should represent the range of biological variation in a given region has been advocated widely. In 1970 UNESCO, through its Man and the Biosphere (MAB) program, initiated a project to conserve natural areas throughout the world by establishing biosphere reserves (UNESCO, 1974). The immediate and main aim of biosphere reserves was to *represent* the range of Dasmann's (1973) global biotic provinces, as updated by Udvardy (1975), in an international system of reserves. Achievement of that aim was seen as necessary to provide sample ecosystems in a natural state, maintain ecological diversity and environmental regulation, conserve genetic resources and provide education, research and environmental monitoring (IUCN, 1978).

Representativeness does not refer simply to some notion of typicalness but rather that a reserve or system of reserves should contain biota which represent the range of variation found within some land class or region. Thus, land classification and regionalization become central problems in clarifying the idea of representativeness.

Conservation evaluation takes place at a wide variety of scales from individual sites to the global scale. Any hierarchical arrangement of land classes must include both a geographical or regional definition and a biological definition of the classes. To be consistent, the same kind of biological definition should be used at all scales. Once the biological units have been defined, there needs to be a means of both allocating land to them and of measuring how representative of those units given areas of land may be.

The assessment of representativeness requires:

(1) an hierarchical land classification of ecological units;
(2) a definition of the relevant properties of the units;
(3) a method of allocating areas to such units;
(4) a means of evaluating the representativeness of areas.

Current practice relies on subjective, often non-explicit procedures. In this chapter, current practice is examined and some alternative explicit, quantitative methods, using Australian examples at various scales, are offered. The thesis adopted is that numerical methods can offer major advantages if combined with appropriate resource information.

2.2   CURRENT METHODS IN AUSTRALIA

As elsewhere, nature conservation evaluation in Australia is dominated by pragmatism. Most reserves are created in areas where land use conflicts are likely to be minimal and on land where no tenure problems exist, e.g. vacant Crown (State owned) land. However, an attempt has been made to provide information on the representativeness of Australia's network of conservation reserves with a report entitled 'Conservation of major plant communities in Australia and Papua New Guinea' by Specht, Roe and Boughton (1974).

Specht, Roe and Boughton tried to assess representativeness using an hierarchical system of vegetation types based on structural formations. Within each structural formation three levels of vegetation differentiation (alliance, association and society) were recognized based on the dominant species in the canopy and understorey. Structural formations and their alliances (the same alliance can occur in more than one structural formation) are listed by geographical regions within each Australian State and Papua New Guinea. The conservation status of each of the 900 communities (alliance within structural formation) is then assessed by the extent to which it is conserved in existing reserves.

As a statement of the vegetation variation of a continent the report may well be unique. There are, as the authors realized, errors of both commission and omission with respect to the alliances recognized and their distributions; few national parks or nature reserves have adequate vegetation inventories. Fortunately, information has increased considerably since the report was completed with the publication of maps of the natural vegetation (Carnahan, 1977; Beadle, 1981).

Compiling the report required many pragmatic decisions that it is now timely to re-examine. One was that the biological units adopted should be plant alliances. The authors '. . . hoped that, apart from migratory animals . . ., most fauna would be conserved if a satisfactory network of reserves containing all major plant communities recorded in Australia was achieved' (Specht, Roe and Boughton, 1974). The use of vegetation communities as surrogates for ecosystems is widespread, and, given current knowledge, acceptable (see Section 2.3.4). Another decision was to combine explicitly defined structural classes (Specht, 1981) with subjectively defined plant communities, a decision not entirely satisfactory. The arbitrary class limits for continuous structural variables creates artificial pigeon-hole categories which may separate vegetation which is floristically identical. Also, the alliances were recognized on a purely subjective basis often with very limited information.

However the most significant decisions were those determining the geographic regions within which the alliances were recorded. Regional boundaries were based on different criteria in each State. One difficulty with such a regionalization can be seen in Fig. 2.1 where many regional boundaries end at State borders. The mapping also emphasizes another regionalization problem, the use of a map unit (region) within which

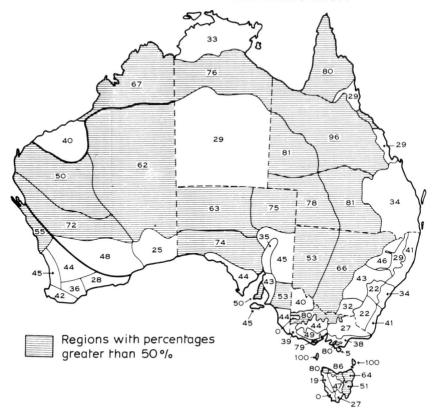

Fig. 2.1 Map showing the geographical regions of Australia used in Specht, Roe and Boughton's (1974) study of the conservation of major plant communities in Australia and Papua New Guinea. Numbers refer to the percentage of alliances not conserved within the region. Reproduced with permission from CSIRO, Canberra.

unmapped units (alliances) are described. Distributions of the unmapped units are too complex to represent or, logistically, too time-consuming to determine at the map scale used.

Another regionalization of Australia is Udvardy's (1975) map (Fig. 2.2) of biogeographical provinces. This map places Australia's ecosystems in a global context; the shape, size and location of map units are quite different from those of Specht, Roe and Boughton (1974). A comparison of published maps of Australian vegetation, e.g. Carnahan (1977) and Beadle (1981), with the two discussed above suggests that an explicit and consistent method of regionalization based on vegetation characteristics is still needed.

At the continental scale, geographical regionalization tends to reflect

Fig. 2.2 Biogeographic provinces of the Australian biogeographic realm (Udvardy, 1975).

the perceived distributions of conspicuous species rather than patterns of the co-occurrence of species in recognizable communities. At the local scale, vegetation maps now are often based on numerical classifications of floristic (and sometimes biomass) data collected using an explicit sample design (for example Austin and Basinski, 1978). It would be advantageous if a consistent hierarchy of vegetation units and map units was common to all scales. Then, inventory information at one level could be collated easily into information useful at another level, something often made impossible by confounding map unit boundaries or different definitions of the vegetation units.

Numerical classification methods offer one way of avoiding the problems created by different mapping techniques and different map descriptors, particularly if agglomerative hierarchical procedures are used.

2.3   NUMERICAL METHODS FOR LAND CLASSIFICATION

2.3.1   **Numerical classification**

Procedures for classifying objects using numerical methods suggest a means of defining appropriate hierarchies of organization of land classes and biological units, for indicating the representativeness of reserves or potential reserves. The objects (in land classification they may be grid

squares, catchments etc.; in vegetation classification they may be quadrats or sample points) are described by numerous attributes (or variables
or characters) which may be expressed in binary (presence/absence)
terms or quantitative terms.

Many numerical classification methods exist. An introduction may be
obtained from texts such as Sneath and Sokal (1973); Clifford and
Stephenson (1975); Williams (1976); and Gordon (1980).

There are four important features to be considered when employing
numerical taxonomic methods:

(1) the nature of the objects to be classified;
(2) what attributes are to be measured;
(3) what measure of similarity between objects is to be used;
(4) what strategy is to be used to sort and group the objects.

Each of these decisions is subjective and should be based on the purpose
of the classification and a knowledge of the properties of different
similarity measures and sorting strategies. Unfortunately, available
computer programmes often are used without a consideration of these
points. Numerical classifications are explicit and repeatable, not objective.

(a) *Objects*
Whether the objects are regions, plots, quadrats, or points depends
critically on the scale and precise purpose of the classification. Scale and
complexity are considered in detail in the next section.

(b) *Attributes*
The type of attributes recorded for each object also depends on scale,
purpose and sometimes on the data available. Attributes may, for example, be subjectively recognized vegetation alliances, species lists from
quadrats or rainfall and temperature records from climatic stations.
Examples of the use of these different types are presented later but it
must be recognized that broadly there are three kinds of numeric
attribute: qualitative, quantitative and multistate (Clifford and Stephenson, 1975).

(c) *Similarity measures*
A similarity measure is a quantitative expression of the degree of similarity between two objects, measured over a set of attributes.

For qualitative data, a vast array of similarity or dissimilarity (the
complement of similarity, also called 'distance') measures have been
proposed (Cheetham and Hazel, 1969; Sneath and Sokal, 1973; Hubalek,
1982). Booth (1978a) compared many measures during a study defining
floristic provinces in Eastern Australia, concluding that either Kulczyn-

ski's (1927) or the asymmetric information statistics (Dale, Lance and Albrecht, 1971) were acceptable, particularly as they were insensitive to double-zero matches.

For quantitative data there is a further wide range of coefficients or measures of similarity (Clifford and Stephenson, 1975; Wishart, 1978). The Bray–Curtis coefficient (Bray and Curtis, 1957) can be used with multistate data and is insensitive to double-zero matches.

(d) *Sorting strategies*

Williams (1971) and Clifford and Stephenson (1975) discuss different sorting strategies in detail. Three appropriate sorting strategies are UPGMA (unweighted pair group arithmetic averaging, Sneath and Sokal, 1973), 'flexible' sorting (Lance and Williams, 1966, 1967; Clifford and Stephenson, 1975) and DIVINF (Lance and Williams, 1968) which is a monothetic divisive strategy using an information statistic (Williams, Lambert and Lance, 1966).

These numerical measures and sorting strategies are robust, widely available and have been used in studying representativeness in Australia. Without the use of explicit numerical methods, the definition of vegetation (or other suggested) units will remain inconsistent, and formal means of evaluating representativeness will remain undeveloped.

2.3.2 Problems of scale, complexity and information

Complexity and scale also complicate the problem of using current methods. Plant communities form a complex pattern on the landscape, each community representing a different combination of environmental conditions. At the scale of a small area (<100 ha), a map of plant communities can be used to assess how representative parts of the reserve are of the whole. Explicit methods of vegetation analysis now exist for such studies (Whittaker, 1978a, b; Gauch, 1982). Unfortunately, in order to assess the extent to which an area represents the vegetation variation in a region, the analysis must be extended to include that whole region. A context must be provided for the potential reserve and the region must be defined.

Problems of scale and context may often be forgotten in small countries where a detailed local knowledge of their flora exists. Often, due to extensive disturbance in the country, rarity and diversity are given far more weight than representativeness in evaluating potential reserves. It might be argued that the less natural a landscape the more prominence given to diversity and rarity, while the more natural a landscape the greater the prominence given to representativeness in assessing conservation value.

Over large areas, where maps need to be at a smaller scale if they are to convey useful information, more generalized classifications of plant communities are necessary, and the region that a unit of land might be considered representative of, becomes larger. Generalized vegetation classes, if they are to be mapped, may have to be quite heterogeneous.

2.3.3 Data

When vegetation data, i.e. measurements of species abundance per defined plot, are available the numerical methods described above can be used to evaluate the representativeness of areas. When vegetation survey data do not exist representativeness can be evaluated only in relation to environment. Data are available from airphoto interpretation, Landsat images, maps or climatic records. An explicit hierarchical system of ecological units based on environmental or climatic attributes derived from a numerical classification, can provide a suitable environmental stratification for conservation evaluation.

2.3.4 Complexity and mapping

Regionalization of any kind necessarily leads to complexity of description. Regions need to be comparable in terms of their heterogeneity if areas are to be evaluated in terms of how representative their *pattern* of communities is. A variety of mapping units incorporating descriptive, unmapped sub-units been utilized widely in other disciplines but only recently for conservation (e.g. Margules, 1984b). Land systems (Christian and Stewart, 1968) i.e. subjectively defined regions having recurrent patterns of terrain, vegetation and soils, are one subjective approach to mapping landscape for land capability evaluation. Alternatively, grid cell classifications using numerical methods have been used (Cumbria County Council, 1978; Laut and Paine, 1982) while Laut, Margules and Nix (1975) classified catchments. These approaches are all viable, though examples of their use are limited.

Explicit, consistent, preferably quantitative methods should be used at all possible stages of conservation evaluation. The 'state of the art', however, is not equal to this as yet. Pragmatism is still required but this can be reduced to two stages. The first is accepting vegetation communities, biogeographic or environmental regions as surrogates for ecosystems. The second is the use of subjective judgment in evaluating the representativeness of potential reserves. The definition of an hierarchical system of units, its properties and allocation techniques can be achieved with repeatable methods. The types of approach are summarized in Table 2.1.

Table 2.1 Types of approach to classification of regions for conservation.

Attribute	Data collection	Analysis	Assessment unit	Product	Comment
Environmental					
1. Climate	Available stations (point records)	Subjective judgment and/or numerical classification	Climatic regions	Map description	Data generally available. Variety of approaches depending on climatic model assumed.
2. Combination (landform, soils, lithology, vegetation and climate)	(a) Available maps (b) Airphoto interpretation and/or satellite image interpretation plus reconnaissance field survey	(a) (as above) (b) Subjective	(a) Catchments (b) Land systems, or terrain patterns	(a) Maps of environmental regions (b) Maps of generalized regional units plus description of unmapped units	(a) Laut, Margules and Nix (1975) (b) Christian and Stewart (1968); Laut *et al.* (1977)
Biological					
1. Floristics	Species lists	Subjective judgment and/or numerical classification	Species distribution maps or occurrence in map grid cells	Maps of biogeographic regions	Practical only in special circumstances (Margules 1984b; Booth 1978b)
2. Vegetation	Quantitative measurements from defined plots	(as above)	Communities, alliances or vegetation types	Vegetation distribution maps, database and explicit analysis of conservation criteria	Practical only at local scale (Austin 1978, Austin and Miller 1978, Mitchell 1983)

2.4   METHODS OF ANALYSIS FOR ASSESSING REPRESENTATIVENESS

Because of the problem of scale and complexity it is useful to consider assessment on two scales, the continental scale (i.e. between regions) and the local scale (i.e. within a region). On the continental scale regions are often based on classifications of environmental data and assumed to be internally homogeneous. Within a region more attention can be given to internal biotic variation by, for example, establishing patterns of community distribution, utilizing major within-region environmental gradients.

2.4.1   The continental or between region scales

Land classification methods which could be used to assess represent-ativeness include those that are single factor, such as climate, or multi-factor, for example soil and landform classification. Alternatively, vegetation data can be used to produce direct biological classifications.

(a) *Climatic representativeness*
Recently, numerical taxonomic methods have been used for climatic classification (Russell and Moore, 1970; Austin and Yapp, 1978). How-ever, the multitude of climatic variables that exist means that some conceptual framework for selecting an appropriate set is required. Fitz-patrick and Nix (1970) consider that plant responses can be generalized with respect to three climatic variables, light, temperature and moisture.

Indices derived from these three variables have been used in numeri-cal classification to examine the bioclimatic distribution of mulga (*Acacia aneura*) in Australia (Nix and Austin, 1973) and to determine bioclimatic zones of arid Australia (Austin and Nix, 1978). In both cases, emphasis was on the representativeness of areas for purposes other than conser-vation. However, this approach could be modified for a variety of purposes, including conservation assessment. Walker and Gillison (1982), for example, proposed a general bioclimatic zonation of Australia which Gillison (1983) used to examine the distribution of savanna.

(b) *Multifactor environmental representativeness*
Soils, lithology and landform attributes can be combined with vege-tation data to produce a classification of biophysical regions. Two princi-pal difficulties are (i) the availability of suitable data and (ii) what should constitute a suitable unit of land for classification. One example of a biophysical regionalization was provided by Laut, Margules and Nix (1975) for the Australian continent. Their solution to the first problem was to use available maps, at scales of approximately 1:1000000, of

terrain and lithology, soils and vegetation, and to superimpose on that a climatic classification. The second difficulty was solved by using catchments which were mapped within boundaries defined by a national drainage divisions and river basins map. The catchments, it was argued, provided natural domains representing a level of ecosystem integration.

For each catchment, the presence or absence of an array of attributes, derived from the available maps, were recorded. This resulted in a data matrix of 4591 catchments and approximately 2000 attributes; much too large for economical computer analysis. The climatic classification was used therefore, to define major regions within which the catchments could be classified. The program used was DIVINFRE (Mayo, 1972), a monothetic divisive procedure using an information statistic with a reallocation phase.

The study resulted in 300 biophysical regions varying in size from a few hundred square kilometres on the east coast to several thousands of square kilometres in inland Australia. The regions were not necessarily composed of contiguous catchments. The study has been used to assess the conservation value of parts of the area involved on the south coast project described in Section 2.4.2(a) but the economic and political climate has not been favourable for its use in institutional decision making.

A second example of a multifactor regionalization is the survey *Environments of South Australia* (Laut et al., 1977). This survey was commissioned as a feasibility study for an ecological survey of Australia. It covered the State of South Australia, an area of approximately 1 000 000 km^2, and was completed in two years. No numerical classification was involved. Rather, units of land, called environmental associations, were mapped onto black and white Landsat images at scales of 1:250 000 in the agricultural districts and 1:500 000 in the pastoral districts (deserts and semi-deserts). Map boundaries generally represent changes in pattern visible on Landsat imagery. Each environmental association was composed of a number of unmapped environmental units, by which the environmental associations were described. For example, an environmental association may consist of an undulating calcrete plain with isolated, easterly trending, aeolian dunes and scattered depressions. The environmental units in this case are: an undulating plain which is dominant, dunes which are subdominant, and minor depressions. In the survey each environmental unit was described in terms of landforms, surface water (where present), soils, existing vegetative cover, native vegetation (where any remains) and land use. The environmental associations were grouped hierarchically into environmental regions, that in turn were grouped into provinces that had a distinctive climate and geomorphology. This hierarchical arrangement makes it possible to recognize regionalizations of the State of South

Australia at different scales for different purposes: aggregations of environmental associations or regions for regional problems, and environmental unit information for local problems (Laut, 1984).

The State Department for Environment and Planning has used this survey as an environmental stratification of the State for assessing the extent to which its system of conservation reserves represents the environments of the State. In some cases, the acquisition of new reserves in environmental associations not represented has been given priority over the acquisition of reserves in associations already represented.

(c) East coast biogeographic provinces

Booth (1978a, b) provides an Australian example of how biogeographic provinces may be defined numerically using lists of tree species, mainly eucalypts, for grid cells defined by 1° of latitude and 1° of longitude. Data for 111 species were obtained from the published maps of Hall, Johnston and Chippendale (1970) for 206 grid cells in Australia south of latitude 34°S and between 136°E and 152°E longitude. The numerical technique finally adopted was to combine the classification results of the two preferred methods, the Kulczynski coefficient (1927) with flexible sorting ($\beta = -0.25$) and the asymmetric information statistic (Dale, Lance and Albrecht, 1971). This was done as no single classification was judged satisfactory. Booth recognized 13 provinces (groups coincident in both classifications) and seven areas (groups not coincident) that were interpreted as representing transition zones between the provinces. The provinces are regions with characteristic combinations of species, that could be used to evaluate the biogeographic representativeness of existing reserves in eastern Australia.

(d) A worked example

Part of the data set used by Laut, Margules and Nix (1975) described in Section 2.4.1(b) is reclassified here in order to assess the extent to which biogeographic regions of the Murray–Darling Drainage Basin are represented in the existing reserve system covering that basin.

The Murray–Darling Drainage Basin covers an area of approximately 1 062 500 km^2 in the south-east of Australia. The Great Dividing Range forms the eastern boundary of the Basin in an area characterized, prior to European settlement, by tall eucalypt forests and woodlands. Moving west and north, the climate becomes drier with less reliable, more intermittent rainfall and a consequent gradation through more open, lower woodlands of eucalypts to open woodlands dominated by *Acacia* and *Casuarina* species, arid grasslands dominated by *Astrebla* spp. and shrublands dominated by members of the family Chenopodiaceae. In the south-west tall shrublands of mallee eucalypts, a growth form characterized by multiple stes arising from an underground root stock occur over extensive areas of sand plain and dunes.

The Bray–Curtis (Bray and Curtis, 1957) association measure was used for this re-analysis because it is insensitive to double-zero matches and because there were different numbers of soil, landform and vegetation attributes. The Bray–Curtis measure solves this problem by treating each attribute type, soils, landforms and vegetation, as a single attribute with different numbers of multistates. Each of the three attributes will have the same range of contribution, 0 – 1, to the similarity value.

The climatic stratification used by Laut, Margules and Nix (1975) has been retained both for convenience of data handling and to allow a comparison of the results. Thus, catchments were classified within climatic zones and not simultaneously over all catchments in the Basin.

The original classification resulted in 73 regions, or parts of regions, being recognized in the Murray–Darling Basin. At a similar level of detail, the agglomerative hierarchical methods employed here produced 64. If each of the 64 regions is accepted as a distinct environment, it is possible to assess the extent to which the existing reserve system represents those environments. Table 2.2 shows that slightly more than one third of the 64 regions are not represented in revers at all and almost a further quarter have less than 1% of their area in reserves.

Table 2.2 The number of biophysical regions and proportion of their area represented by the existing (at June 1979) reserve system in the Murray–Darling Basin, at the 64 region level and the 42 region level.

No. of regions not represented	No. of regions with <1% of their area in reserves	No. of regions with 1–5% of their area in reserves	No. of regions with >5% of their area in reserves
64 region level			
26	17	15	6
42 region level			
16	11	9	6

It might be unrealistic to expect 64 large reserves in the Murray–Darling Basin. The agglomerative hierarchical methods used in this classification means that the same classification can be used to derive a smaller number of regions. Accepting a higher level in the classification (greater heterogeneity) 42 regions can be recognized. The conservation status of those 42 regions is also summarized in Table 2.2.

2.4.2 **Within region scales**

Several existing studies illustrate methods of analysing and describing biotic variation within regions. Three which use methods suited to assessing representativeness within regions are described below.

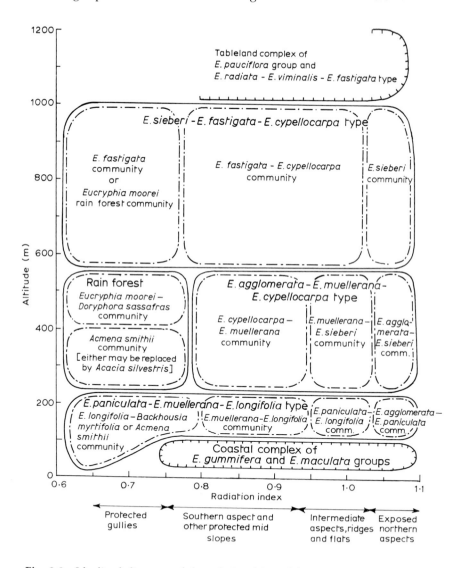

Fig. 2.3 Idealized diagram of the relationships of forest vegetation to altitude and aspect for the forests (mostly *Eucalyptus* spp.) occurring on sediments south of the Currowan Creek – Clyde River estuary line in the south coast study area, N.S.W. (Austin, 1978).

(a) *South coast project*

In a methodological study for regional land use planning (Austin and Cocks, 1978), a vegetation survey of an area of 6000 km^2 was undertaken (Austin, 1978) and methods of conservation evaluation were examined using the resulting data (Austin and Miller, 1978).

The vegetation data set consisted of 576 plots and 600 species from a survey based on a stratified sample, and was analysed using a divisive information analysis procedure with reallocation (DIVINFRE, Mayo, 1972). Using presence/absence of species, a complex array of communities was defined, many of them structurally distinct, for example coastal dune communities, rain forests, heaths and swamps. Within the eucalypt open forest structural type, however, numerous communities were indicated based on obscure ground layer herbs. The usefulness of a classification depends on its communicability and the ease with which classes can be recognized in the field. Thus, in order to make the classification a practical one for conservationists and foresters, it was modified subjectively using quantitative data (basal area) from the forest trees. The resulting communities were then described using a form of direct gradient analysis (Whittaker, 1967). This provided a model on which to base extrapolations from the sample plots to wider areas (Fig. 2.3).

The communities proved not to be useful map units on any practical scale as occurrences were often of only a few hectares. For mapping purposes descriptive units called forest types, were recognized. A forest type was defined as that group of communities which occupy a topographic sequence from warm, northerly aspects to cool, protected, southerly aspects, in a particular altitudinal zone. Other environments such as swamps and coastal dunes were treated by means of community complexes (Austin, 1978).

The estimated distributions of the forest types and other community complexes, together with their constituent communities, were entered into a computer data bank with an associated mapping system (Cook, 1978). This provided a means of estimating the area and distribution of any community, and any other recorded attribute. For conservation studies, maps of disturbance and species rarity were produced. A criterion of representativeness at the regional level was defined, using a subjective index of representativeness for each community (Fig. 2.4).

Using the data bank, the estimated area of different communities presently conserved can be compared with that achieved if an index of representativeness is used to select areas, or with that achieved by a large national park proposal (Table 2.3). The use of other criteria and of combined representativeness and rarity indices is discussed further by Austin and Miller (1978). Neither the data collected nor the criteria adopted were used to define the area or boundaries of two national

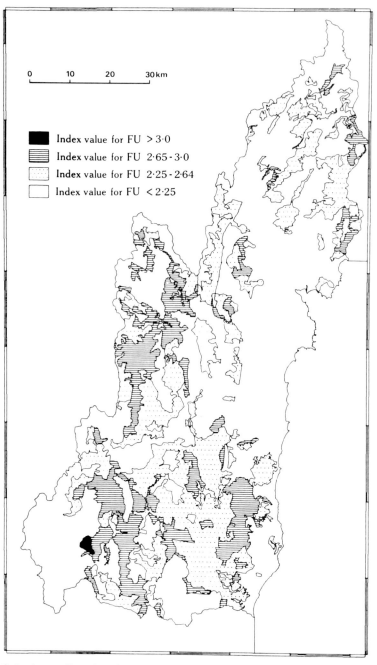

Fig. 2.4 Areas (functional units, FU) having potential for conservation based on an index of representativeness for the study area. The index used a weighting of communities based on their abundance and distribution (Austin and Miller 1978).

Table 2.3 Examples of estimated vegetation composition of the proposed Tuross–Deua National Park as in 1974 (from Austin 1984).

Forest type (Eucalyptus spp.)	Area in park (% park)	(% region)	Area currently conserved (% region)	Actual area (% region)
E. fastigata–E. cypellocarpa–E. sieberi	43.0	7.9	0.3	15.4
E. paniculata–E. muellerana–E. longifolia	3.4	0.6	0.1	10.3
E. gummifera–E pilularis–E. maculata	—	—	<0.1	2.2
Coastal forests E. botryoides	—	—	<0.1	<0.1
Estuarine complex	—	—	—	0.2

parks established since the completion of the study. Institutional politics and political decision making played the major roles. Subsequent management of the national parks and the development of environmental impact statements for the exploitation of State forests in the area have used the data.

There are two significant features to this study. First, the information context in which the study took place is totally different from the information context in Britain and Europe and probably many parts of North America. For example, during the course of the study, accounts of five new, undescribed eucalypt species were published for the area. Primary inventory data were therefore critical to any conservation evaluation exercise in the area. The survey confirmed that Specht, Roe and Boughton's (1974) estimate that 63% of existing conserved areas in New South Wales were inadequately known was correct for southern coastal New South Wales. Secondly, conservation evaluation was treated differently from the usual estimation of the 'best areas'. The approach was to exclude areas *unsuitable* for conservation. It was argued that whilst it was possible to obtain consensus on which areas were unsuitable for a particular land use, different value judgements could lead to very different concepts of 'best' and therefore it was better to provide a methodology which recognized that and allowed alternatives to be explored easily (Cocks and Austin, 1978).

An approach using a large data base and explicit algorithms is a counsel of perfection: resources are not available nor is the institutional decision-making system adapted to explicit statements of value systems in land use planning. However, the study serves as a model from which ideas can be taken and modified for practical circumstances. A new, modified approach which can be used with available data rather than requiring primary survey data, has been applied to the development of a preliminary zonation and management plan of the Great Barrier Reef National Park (Cocks, Baird and Anderson, 1982).

(b) *Gradient analysis and faunal studies*

The south coast project (described in Section 2.4.2(a)) used Whittaker's (1967) direct gradient analysis approach to establish a model of forest types. Kessell, Good and Potter (1982) have used direct gradient analysis

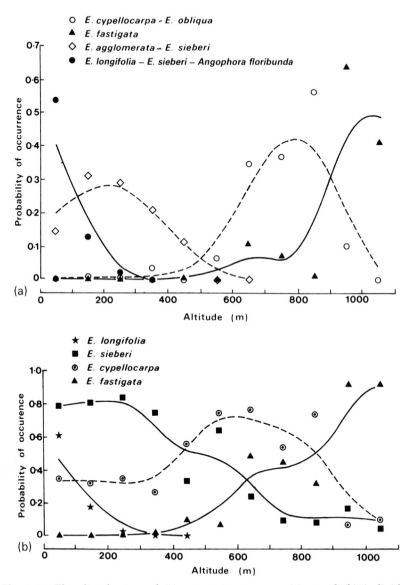

Fig. 2.5 The distribution of (a) vegetation communities and (b) individual species of *Eucalyptus* in relation to an altitudinal gradient. (From Austin and Braithwaite unpublished.)

to establish a computer data base for national park management with particular reference to modelling the impact of fire on vegetation. Direct gradient analysis could also be used in the assessment of representativeness. If a particular community occurs over a range of environmental conditions, i.e. along an environmental gradient, then in order to conserve a representative set of populations (gene pool) reserves should encompass areas with the full range of conditions along that gradient.

Figure 2.5(a) provides an unpublished example (from M. P. Austin and L. W. Braithwaite) of communities characteristic of a woodchip concession area near Eden on the far south coast of New South Wales. The communities were defined by numerical classification of presence data of the canopy species from logged plots using Czekanowski's (1932) coefficient and UPGMA sorting. The forest is a typical sclerophyll forest dominated by various species of *Eucalyptus*. The communities are named after species which occur in more than 80% of the plots allocated to that community. Each community has a particular altitudinal range. To assess how representative, of a particular community, a conservation area might be, the range of altitude within the area could be compared with the known range of the community. In this way, the occurrence of communities in separate conservation areas or possible conservation areas can be compared and the areas assessed as to whether they complement one another by encompassing different parts of the continuum of variation for that community.

Similar analyses can be performed for individual species (Fig. 2.5(b)). The analysis need not be confined to single gradients, nor need it lack statistical rigour. The application of generalized linear modelling to direct gradient analysis (Austin, Cunningham and Good, 1983; Austin, Cunningham and Fleming, 1984) demonstrated that detailed statistical models predicting species distributions can be obtained using four environmental gradients. Testing differences in community distributions in different conservation areas and areas under consideration for conservation, appears to be a distinct possibility in the near future.

Use may be made of gradients for assessing the value of areas for fauna conservation provided a correlation can be established between fauna and vegetation and, hence, between fauna and environment.

In a series of papers, Braithwaite (1983), Braithwaite, Dudzinski and Turner (1983) and Braithwaite, Turner and Kelly (1984), using the vegetation data from the survey described above together with data, obtained when trees were logged, on the occurrence of arboreal marsupials (possums and gliders) provide a striking example of the possible use of gradients. The plot data used for Fig. 2.5 were aggregated into forestry coupes (forest management zones) and a numerical classification was carried out. The dendrogram (Fig. 2.6), with the associated density data for the marsupials, shows that the animals are predomin-

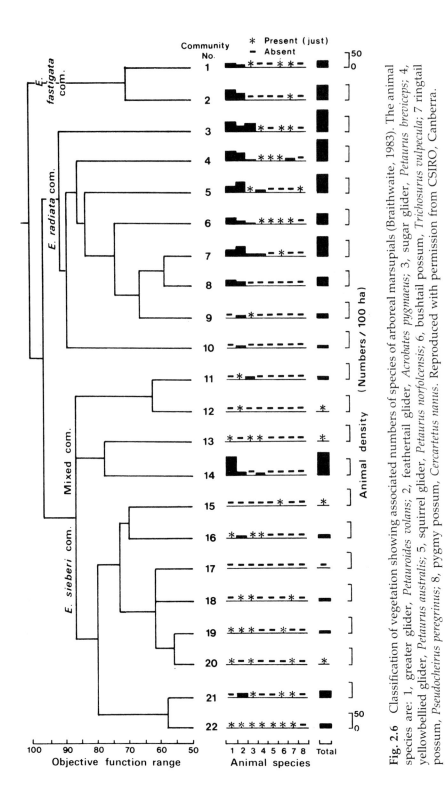

Fig. 2.6 Classification of vegetation showing associated numbers of species of arboreal marsupials (Braithwaite, 1983). The animal species are: 1, greater glider, *Petauroides volans*; 2, feathertail glider, *Acrobates pygmaeus*; 3, sugar glider, *Petaurus breviceps*; 4, yellowbellied glider, *Petaurus australis*; 5, squirrel glider, *Petaurus norfolcensis*; 6, bushtail possum, *Trichosurus vulpecula*; 7 ringtail possum, *Pseudocheirus peregrinus*; 8, pygmy possum, *Cercartetus nanus*. Reproduced with permission from CSIRO, Canberra.

antly associated with a limited number of communities (numbers 1–7 and 14). These communities are generally characterized by the presence of one or more of the three eucalypt species occurring in the area, known as peppermints (*Eucalyptus radiata*, *E. dives* and *E. elata*). Subsequent analysis indicated that the primary determinant of faunal density was a gradient in foliage nutrient concentration, which was associated with the presence of peppermints and certain other species. Foliage nutrient concentration varies with both species and lithology (rock type). By ordinating the vegetation communities on species frequency and plotting their position against occurrence on different geological types (reflecting lithology), it was shown (Fig. 2.7) that the vegetation communities, and hence densities of arboreal marsupials, are closely related to a fertility gradient in the lithology of the area.

Once a correlation has been established, gradients such as these then provide an indirect means of assessing faunal representativeness in the same way that the representativeness of vegetation communities can be assessed.

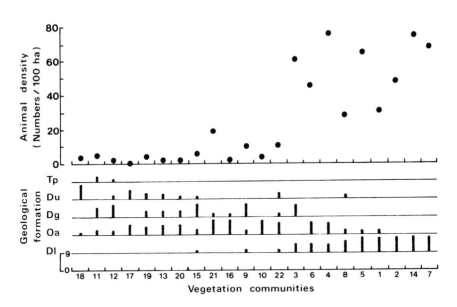

Fig. 2.7 Frequency of vegetation communities on different geological formations (Braithwaite, Turner and Kelly, 1984). The abbreviations represent: tertiary sediments (Tp); Devonian sediments (Du); middle Devonian granites (Dg); Ordovician sediments (Oa); Devonian grandiorite (Dl). The bars show the index proportion, on a scale from 0 to 9, of the total area of each community sampled occurring on each geological formation. Reproduced with permission from CSIRO, Canberra.

2.5   DISCUSSION

The four requirements for the assessment of representativeness – a hierarchical classification of land units, the definition of relevant properties, an allocation method, and an evaluation procedure – can now be considered in the light of the examples presented.

Numerical classification methods can provide consistent, explicit hierarchies regardless of whether biotic or environmental data are used. The objects to be classified, the ecological units, can take a variety of forms, for example catchments, grid cells, or landform elements, and they can be adapted to any scale from local to continental.

A method of allocating possible reserves to the ecological classes or biogeographic regions defined, and a means of evaluating the representativeness of potential reserves, are less easily determined. However, the example of regionalizations of the Murray–Darling Drainage Basin showed that it is possible to examine whether each classificatory unit on a particular level of a hierarchy is represented by a reserve, and, in this way, to assess the extent to which an existing reserve system represents the range of biotic variation over a given area. Similarly, within a region, major environmental gradients can provide a means of assessing the extent to which a reserve or system of reserves represents the ecological variation within that region.

The numerical approach is emphasized for four reasons. First, numerical methods ensure that consistent, compatible data are used and, therefore, valid comparisons between possible conservation areas can be made. Secondly, subjective value judgments will always be present, but explicit quantitative methods ensure that they are readily apparent, for example the choice of sorting strategy, and that they can be modified easily. The final judgment, on how satisfactory a system of reserves is with respect to the known multidimensional continuum of ecosystems, will remain subjective, though measurement procedures will be possible once the judgment procedure has been determined. Thirdly, without numerical methods there will be no explicit means of trading off representativeness against other major measurable criteria of conservation value, for example diversity and rarity. Fourthly, simple opinions on the importance of a particular area of land for conservation have not constituted an effective approach to conflicts between conservation and other land uses. A fully developed argument, based on a clear procedure for determining relative conservation value in terms of stated criteria, with explicit methods for assessing those criteria, may prove more effective. Margules, in Chapter 13, suggests one possible procedure.

Eventually there will be universally accepted methods for measuring clearly defined criteria of conservation value. Then, the weight given to each in any particular assessment will depend on the planning context, the policies to be considered and the value judgments of the planners, whether they be professional planners, conservationists, elected representatives or the community at large.

2.6 SUMMARY

In order to select representative samples of the environment, an appropriate environmental stratification is necessary. Such a stratification requires the definition of both suitable spatial units and a set of relevant descriptive attributes, and should be based on an explicit, numerical classification procedure, which preferably produces a hierarchical classification. It is best to base the classification on distributional data for the biota of interest. Various environmental attribute sets are available as alternatives (for example climate or lithology), but any classification should be based on biologically relevant attributes.

The use of a hierarchical classification allows the level of heterogeneity, at which representative samples are to be taken, to be specified clearly. At continental, or inter-regional, scales, mapping units which are necessarily environmentally heterogeneous provide the stratification. At regional, or local, scales, gradients of environmental variation within that unit are a more appropriate form of stratification. Direct gradient analysis is one simple graphical method for examining local variability using several attributes. It can be provided with a rigorous statistical foundation.

Whereas the criteria of diversity and rarity are seen to be important in geographical areas, such as Western Europe, where the biota are well known, representativeness is seen to be important in areas, such as Australia, where many of the species may well be undescribed.

CHAPTER 3

Ecological succession and the evaluation of non-climax communities

RICHARD G. JEFFERSON
and
MICHAEL B. USHER

3.1 Introduction
3.2 Criteria used to evaluate important sites for
     wildlife conservation
3.3 Disused quarries and pits
     3.3.1 Disused chalk quarries and chalk pits in the
            Yorkshire Wolds
     3.3.2 The ecological processes
3.4 Discussion
3.5 Summary

Wildlife Conservation Evaluation. Edited by Michael B. Usher.
Published in 1986 by Chapman and Hall Ltd, 11 New Fetter Lane,
   London EC4P 4EE

3.1   INTRODUCTION

In the British Isles, management induced, non-climax or seral plant communities are of great value for wildlife conservation. Bradshaw (1977) estimated that 90% of the key sites in Ratcliffe (1977) are, to a greater or lesser extent, influenced by man's activities, especially by agricultural and silvicultural practices. Plant communities completely unmodified by human activity are consequently of extremely rare occurrence in the British Isles. The maintenance of communities such as chalk grassland or lowland heathland, for their conservation interest, is largely dependent upon arresting successional change.

Ecological succession refers to the temporal changes that occur in the species composition of a community of plants and animals after either a radical disturbance or the opening of a new patch in the physical environment (Horn, 1974; Connell and Slatyer, 1977). Two types of succession have generally been recognized: primary and secondary. The former occurs in areas initially lacking soil and vegetation (for example on bedrock, sand dunes and recently glaciated surfaces), whereas the latter occurs in areas where a community had existed prior to the disturbance. These sequences can occur over periods of time with climate and physiography being substantially stable, or they can occur when the local environment is changing under the influence of extrinsic factors such as climate, erosion, deposition or the input of nutrients (Drury and Nisbet, 1973). The possible mechanisms driving succession have been reviewed by Connell and Slatyer (1977). Succession was traditionally thought to continue until a climax or steady state equilibrium was reached: the climax was regarded as the ecological community best suited to the regional climate and site type (based on soil and geology). However, Connell and Slatyer (1977) concluded that succession never stops and that there are no examples whereby a steady state equilibrium has been reached; small scale changes in species composition continually maintain a flux in all so-called climax communities.

Habitats or communities, which are not dependent upon management techniques for arresting successional change to maintain their wildlife conservation interest, include sand dune systems, shingle beaches, high altitude dwarf-shrub and *Rhacomitrium* moss-lichen heaths. In contrast, the majority of communities depend upon traditional management practices, designed to arrest successional change, in order to maintain their existing species composition and structure. These communities include those dependent on grazing, usually by sheep or cattle (limestone and acid grassland, lowland heath), on cutting with no addition of inorganic fertilizers (hay meadows) and on traditional forestry methods such as coppicing and wood pasture (ancient broadleaved woodlands).

However, particularly in the last 100 years, industrial activity has led to the creation of totally new habitats for natural colonization by plants and animals. Some of these man-made habitats have developed a considerable wildlife interest (Kelcey, 1975; Catchpole and Tydeman, 1975; Davis, 1976; Holliday and Johnson, 1979; Kelcey, 1984). These habitats include 'holes' resulting either from quarrying into sedimentary, metamorphic and igneous rocks, or from the excavation of unconsolidated deposits such as clay, sand and gravel. The 'holes' may be either wet or dry depending on the geology of the deposit extracted and its geographical location. Mining subsidence in river valleys has also created new permanent water bodies. Other industrial habitats can be termed 'heaps' since they consist of industrial waste tips resulting from non-ferrous metalliferous mine workings from a variety of manufacturing processes. In some cases these wastes are calcareous in nature, one example being the alkali waste in south Lancashire produced by the now obsolete Leblanc process for the manufacture of sodium carbonate, and are colonized by plant communities which are species rich with many uncommon species. Often such communities resemble those found on calcareous soils over limestone (Gemmell, 1982; Greenwood and Gemmell, 1978). Since limestone outcrops do not occur in the area, the nearest source pool for calcicolous species is presumably the calcareous dune slacks on the west coast of Lancashire, suggesting that colonization has been achieved by long distance seed dispersal. Lee and Greenwood (1976) have described similar plant communities from lime beds, in Cheshire, which have resulted from the Solvay process. Lead and zinc mining has also created habitats which locally have considerable floristic interest (Ratcliffe, 1974; Bradshaw, 1977). In the northern Pennines, for example, lead mine spoil supports plant communities containing species indicative of high levels of heavy metals: these communities contain the so-called 'metallophytes' such as *Armeria maritima, Thlaspi alpestre, Cochlearia officinalis* agg. and *Minuartia verna*, all of which are known to be metal tolerant (Antonovics, Bradshaw and Turner, 1971).

The conservation interest of derelict industrial sites is known to be dependent on the retention of early successional plant communities. Increase in both soil fertility and soil depth lead to successional changes which will ultimately result in reduced floristic richness with the development of tall grassland and scrub communities. Davis (1983), for example, predicted that within a few decades, in the absence of intervention, Brockham Chalk Quarry in Surrey would proceed to woodland dominated by *Betula pubescens*. This example emphasises the need for management to arrest successional change.

3.2 CRITERIA USED TO EVALUATE IMPORTANT SITES FOR WILDLIFE CONSERVATION

In Britain, much attention has been focused on the selection of sites of wildlife conservation importance in relation to semi-natural community

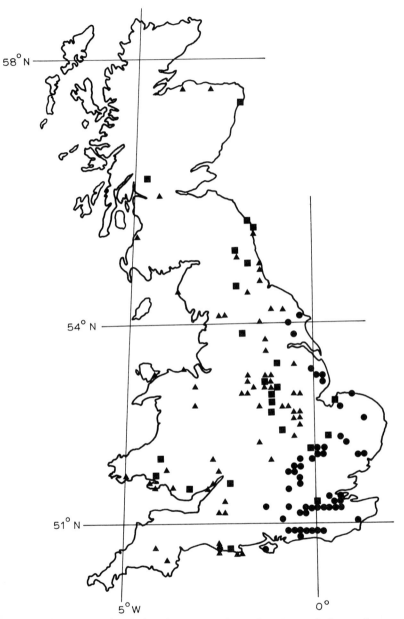

Fig. 3.1 The geographical distribution in Great Britain, and the geology, of disused quarries with some form of wildlife conservation protection. Geology is denoted by the following symbols: ●, chalk; ▲, limestone; ■, all other formations, including calcareous mudstone, sandstone, clay, dry sand and gravel, and both igneous and metamorphic rocks. The quarries are plotted on the basis of 10 km squares of the National Grid, only one symbol being shown per square irrespective of the number of quarries (modified from Jefferson 1984b).

types (Ratcliffe, 1977, and Chapter 6 of this book). However, it is pertinent to examine briefly the criteria used in identifying the industrial sites which are of conservation interest, many of which are already Sites of Special Scientific Interest (SSSIs) or County Conservation Trust nature reserves. In the case of disused calcareous quarries, the presence of plant communities which resemble the semi-natural calcareous grassland of the region is an initial criterion. Species richness, the presence of regionally or nationally rare plant species, and the size of the quarry are then usually taken into consideration. The uniqueness of the plant species composition, or the community structure associated particularly with mining and manufacturing wastes, are other criteria, particularly in areas originally lacking the substrate types created by industrial activity. The chemical composition of some wastes, for example high heavy metal concentrations in metalliferous mining wastes, also presents an opportunity to study evolutionary adaptation by species to extreme soil conditions (Bradshaw, 1977). However, in all assessments of quarries the emphasis seems to have been overwhelmingly on the plant communities rather than on the animal communities, which can show just as striking successional sequences (Parr, 1980).

3.3   DISUSED QUARRIES AND PITS

The importance to wildlife conservation of recently created man-made habitats has been emphasized by Jefferson (1984b). Figure 3.1 shows the distribution in Great Britain of disused quarries (excluding wholly flooded sites) which are either protected as County Conservation Trust nature reserves or as SSSIs. This information was derived from questionnaires circulated to the County Conservation Trusts and to the Nature Conservancy Council (NCC) regions (Jefferson, 1984b), but it represents only part of the total nature resource as there are a number of abandoned quarries, with known wildlife conservation interest, which have no protection, and further sites may be discovered as ecological survey work increases.

Jefferson (1984b) estimated that 88% of disused quarries with some form of conservation protection were on calcareous geological strata. The flora of quarries situated on acidic (base-poor) rocks is usually of little conservation interest. Such quarries are often situated in areas where there is no threat to the existing plant communities, often of grassland or heathland, which are widespread in their occurrence (Ratcliffe, 1974). Disused chalk and limestone workings, in contrast, support early successional, species rich plant communities, akin to, but not exact replicas of, existing semi-natural calcareous grassland. Regionally and nationally rare species also occur (Davis, 1979; Humphries, 1980), including many species of the Orchidaceae. Table 3.1 (modified from Jefferson, 1984b) lists the nationally rare native species found in 'dry'

Table 3.1 Rare plant species occurring in disused quarries and pits in the British Isles. The number of sites for a particular species in a county, where more than one, is given in brackets in the third column.

Species	Geology	County	Source
Adiantum capillus-veneris	Ordovician slate	Gwynedd	Day and Deadman (1981)
Ajuga chamaepitys	Cretaceous chalk	Kent	Nature Conservancy Council (personal communication)
Bunium bulbocastanum	Cretaceous chalk	Bedfordshire (2)	J. G. Dony (personal communication)
	Cretaceous chalk	Cambridgeshire	Perring et al. (1964)
Carex ericetorum	Permian limestone	South Yorkshire	Hodgson (1982)
Cerastium pumilum	Jurassic limestone	Oxfordshire	Lousley (1969)
Corallorhiza trifida	Carboniferous limestone	North Yorkshire	Davis (1979)
Crepis mollis	Permian limestone	Durham	Heslop-Harrison and Richardson (1953)
Daphne mezereum	Carboniferous limestone	North Yorkshire	Yorkshire Wildlife Trust (personal communication)
	Carboniferous limestone	Derbyshire	Johnson (1978)
Epipactis atrorubens	Permian limestone	Durham (2)	Davis (1979), Durham County Conservation Trust (personal communication)
	Carboniferous limestone	Clwyd	Day (1978)
Gentianella germanica	Cretaceous chalk	Bedfordshire	J. G. Dony (personal communication)
Herminium monorchis	Cretaceous chalk	Bedfordshire	J. G. Dony (personal communication)
	Cretaceous chalk	Hampshire	Hampshire and Isle of Wight Naturalists Trust (personal communication)
Herniaria glabra	Unconsolidated sand	Lincolnshire	Holliday and Johnson (1979)
Hornungia petraea	Carboniferous limestone	Derbyshire	Clapham (1969)
Hypochaeris maculata	Jurassic limestone	Cambridgeshire	Ratcliffe (1974)
Lotus angustissimus	Unconsolidated gravel	Kent	Holliday and Johnson (1979)
Lychnis viscaria	Andesite	Perthshire	C. Connell (personal communication)

Species	Substrate	County	Reference
Muscari neglectum	Cretaceous chalk	Cambridgeshire	Perring *et al.* (1964)
Orchis militaris	Cretaceous chalk	Suffolk	Suffolk Trust for Nature Conservation (personal communication)
Orobanche purpurea	Clay	Humberside	Holliday and Johnson (1979)
Peucedanum palustre	Cretaceous chalk	Cambridgeshire	Cambridge and Isle of Ely Naturalists Trust (personal communication)
Poa bulbosa	Jurassic limestone	Oxfordshire	Lousley (1969)
Pulsatilla vulgaris	Jurassic limestone	Cambridgeshire	Hepburn (1942)
Pyrola rotundifolia	Cretaceous chalk	Essex (2)	Jermyn (1974)
	Cretaceous chalk	Kent	Davis (1979)
Seseli libanotis	Cretaceous chalk	Cambridgeshire	Davis (1979)
Thlaspi perfoliatum	Jurassic limestone	Oxfordshire	Lousley (1969), Holliday and Johnson (1979)
Verbascum lychnitis	Sand and gravel	Clwyd (2)	Day (1978)
Vulpia unilateralis	Jurassic limestone	Leicestershire (2)	Messenger (1971), Stace (1984)
	Cretaceous chalk	Kent (2)	Nature Conservancy Council (personal communication)

disused quarries (nationally rare species were defined as those occurring in 50 or fewer 10 km squares in the British Isles, excluding Ireland). In most cases, however, the botanical interest of calcareous quarries is regional rather than national. This is reflected by the fact that there is only one disused quarry (see Fig. 3.2a) listed as a key site in Ratcliffe (1977). This is a very old Jurassic limestone quarry in Cambridgeshire,

Fig. 3.2 An ancient quarry, the Hills and Holes at Barnack in Cambridgeshire. (a) A general view of part of the calcareous grassland showing the encroaching scrub. (b) One of the species particularly associated with this quarry is *Pulsatilla vulgaris*, shown here in seed.

Fig. 3.3 Three local or uncommon species of flowering plants that occur in the ancient quarry at the Hills and Holes, Barnack, Cambridgeshire. (a) *Aceras anthropophorum*, the man orchid. (b) *Pulsatilla vulgaris*, the pasque flower. (c) *Orobanche elatior*, the knapweed broomrape, here parasitic on *Centaurea scabiosa*.

last worked in the 16th century, and supporting a species rich calcareous grassland flora (Hepburn, 1942) with many local species (see Fig. 3.2b and Fig. 3.3). The importance of calcareous quarries as refuges for local and rare plant species and grassland communities is attributable to three main factors. First, in the British Isles, calcareous soils support plant communities that are more species-rich than those of acidic soils. Secondly, in northern latitudes, calcicolous vegetation draws upon a reservoir of species which is considerably larger than the reservoir of calcifuges (Grime, 1979). Thirdly, it is the calcareous grasslands in particular which have been, or are being, agriculturally improved or developed for industrial and urban purposes. Agricultural intensification, industrial and housing development, and, paradoxically, quarrying have, for example, considerably reduced the area of Magnesian (Permian) limestone grassland in the counties of Durham and Tyne and Wear (Doody, 1977; Ranson and Doody, 1982).

The survey conducted by Jefferson (1984b) showed that 93% of the total number of quarries with conservation protection were designated wholly or partly for their botanical interest. Although open hard rock quarry floors support few species of breeding birds, quarry faces in some areas have provided substitute breeding sites for the peregrine falcon (*Falco peregrinus*). Ratcliffe (1981) estimated that at least 25 quarries are known to have been utilized by this species. Invertebrates have received little serious study in quarry habitats. Most of the available information is of an anecdotal nature (Davis, 1979), although in some instances, for example Wharram Quarry Nature Reserve in the Yorkshire Wolds, there is extensive information on file and in the reserve management plan.

3.3.1 Disused chalk quarries and chalk pits in the Yorkshire Wolds

The Yorkshire Wolds, which form the northernmost section of the English chalk of Upper Cretaceous Age, extend from the River Humber in the south to Flamborough Head in the north. The Wolds are predominantly an area of intensive arable farming, and fragments of unimproved chalk grassland are restricted to the sides of the network of dry valleys and to the north-facing scarp slopes or brows. In the last 20 years, an estimated 43% of this grassland has been changed to more intensive agricultural and forestry use, or else lack of management has resulted in scrub encroachment (Rafe and Jefferson, 1983). The extent of arable farming and the decrease in chalk grassland area highlight the contribution that disused quarries can make to the conservation of flora and fauna in the Wolds by providing habitats for species which have declined due to land use change (Jefferson, 1984a).

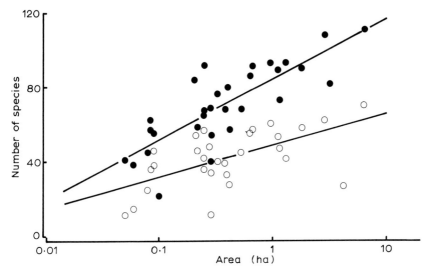

Fig. 3.4 The species–area plot for all plant species (●) and for the chalk grassland, CG, species (○). The regression lines are $S = 84.0 + 32.4 \log A$ ($r = 0.808$, $P < 0.001$) for all species, and $S = 48.6 + 16.5 \log A$ ($r = 0.589$, $P < 0.001$) for the CG species: in both cases S indicates the number of species and A the quarry's area (in hectares).

In a recent detailed study of the abandoned pits and quarries in the Wolds, Jefferson (1984a) found that larger quarries in general supported more vascular plant species, and more chalk grassland (CG) species, than smaller quarries and pits (Fig. 3.4). This relationship was to be expected since it conforms to the familiar species–area relationship described in many other studies (see Connor and McCoy, 1979). CG species include those species which are components of chalk grassland in the British Isles, and they are listed in Jefferson (1984a). Quarries closest to existing fragments of chalk grassland contained more CG species, and more restricted chalk grassland (RCG) species, than more distant sites; the species–isolation relationships are shown in Fig. 3.5. RCG species were defined as those CG species which occur in 655 or fewer of the 10 km squares in the British Isles excluding Ireland, or in 17 or fewer of the 10 km squares covering the northern chalk (Yorkshire and Lincolnshire Wolds). These figures represent one quarter and one half respectively of the number of 10 km squares in each category.

Jefferson (1984a) concluded that selection of quarry sites for wildlife conservation should not be based solely on species richness criteria, and hence neither size nor isolation are entirely suitable for conservation evaluation of ecologically similar sites. It was concluded that a better method of evaluation was to combine the number of species, local rarity,

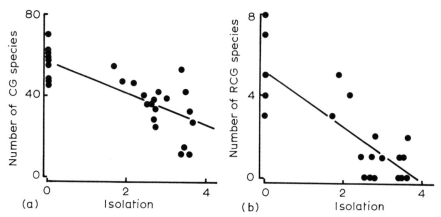

Fig. 3.5 The species–isolation plot for CG species (a) and for the RCG species (b). The categories of species are defined in the text: isolation (I) is measured on a logarithmic scale, thus $I = \log (d + 1)$, where d is the shortest distance (in metres) from the quarry to the nearest chalk grassland. Regression equations are $S = 55.7 - 7.82I$ ($r = 0.771$, $P < 0.001$) for the CG species, and $S = 5.19 - 1.36I$ ($r = 0.809$, $P < 0.001$) for the RCG species.

national rarity and the geographical distribution of species into a conservation index which could be used to rank sites according to what was perceived to be their current botanical conservation interest. The index was composed of four terms. First, there was a species richness term

$$S = (t + g)/K_1,\qquad(3.1)$$

where t is the total number of plant species, g is the total number of CG species, and K_1 is a constant that was chosen such that the mean S value over all 30 sites used in this study was 10. Second, there was a term for regionally restricted species. This took the form

$$R = K_2 \sum_{i=1}^{x} 1/c_i\qquad(3.2)$$

where c_i is the number of 10 km squares in the Yorkshire and Lincolnshire Wolds in which the ith RCG species occurs (maximum value for c_i is 17 which represents half of the total number of 10 km squares covering the Northern Wolds), x is the number of these RCG species at the site being evaluated and K_2 is again a constant such that the mean R value over all sites was 10. Third, there was a term for nationally restricted species. This took the form

$$N = K_3 \sum_{i=1}^{y} 1/d_i\qquad(3.3)$$

where d_i is the number of 10 km squares in the British Isles, excluding Ireland, in which the ith RCG species occurs, y is the number of these RCG species at a given site, and K_3 is a constant such that the mean N value over all sites was 10. Finally, there was a term for the occurrence of southern continental species. This took the form

$$C = K_4 \sum_{i=1}^{z} d_i/e_i \qquad (3.4)$$

where d_i is defined above, e_i is the number of 10 km squares north of a line from the entrance of the Mersey Estuary to the Wash (approximately the point where the River Welland enters the Wash) in which the ith species occurs (the maximum value for e_i is 164, representing a quarter of 655), z is the number of these predominantly southern species at a given site, and K_4 is a constant such that the mean C value over all sites was 10.

The four terms in equations (3.1) to (3.4) were then added together to give an index of conservation value, thus

$$I_c = S + R + N + C. \qquad (3.5)$$

Since each of the four terms has a mean value of 10, each has an equal weight in I_c, which itself has a mean value of 40. The index in (3.5) can be modified to give the two measures of rarity (R and N) the same weight as species richness and geographical distribution, such that

$$I_d = S + 0.5(R + N) + C. \qquad (3.6)$$

Few studies have attempted to make quantitative comparisons of the conservation value of sites; those that have include Ward and Evans (1976) and Day et al. (1982), who respectively developed floristic indices for assessing limestone pavements in the United Kingdom and for assessing marl pits in North Wales and north-western England. Ward and Evans' index was based on three criteria: species richness, national rarity and the abundance of the individual species, and Day et al.'s index was based on national and local rarity and species richness.

The validity of utilizing such simple indices for conservation evaluation has been questioned by Margules and Usher (1981, 1984). They discussed the difficulties in arriving at fixed weights for particular criteria that might be used in undertaking a conservation evaluation. The weights placed on particular criteria incorporated in I_c and I_d, equations (3.5) and (3.6), depend on the subjective assessment of the relative importance of the criteria. This does, however, allow some flexibility as potential users of such an index can choose the amount of emphasis that they wish to place on the four elements of the index, S, R, N and C. In I_c, the weights are equal, whereas in I_d, S and C are given double the weight of R and N. It is, therefore, possible to write these indices of conservation value generally as

$$I = w_1 S + w_2 R + w_3 N + w_4 C \tag{3.7}$$

where w_i is the weight given to the ith criterion in the index. The index is therefore weighted according to the subjective opinion of the assessor (or customer) about the importance of the various criteria. These weights should obviously be chosen a priori, and certainly not after a survey has been completed.

In practice, selection and purchase of nature reserves by conservation organizations is not always governed by ecological criteria; factors including land availability, cost, location, and the feasibility of management often have to be taken into consideration. The advantage of indices such as I_c or I, in equations (3.5) and (3.7) respectively, is that they overcome many of the problems of making comparisons between ecologically similar sites, and of communicating the results of evaluation procedures to decision makers involved in wildlife conservation (Margules and Usher, 1984). The explicit incorporation of weights goes some way in answering criticisms that evaluation is an art rather than a science!

3.3.2 The ecological processes

The indices developed in the preceding section describe a static situation: the flora (or fauna) of the quarry is surveyed, and the species list divided into those species which are locally uncommon, nationally uncommon or southern in their distribution, as well as deciding which of these species are characteristic of the semi-natural grasslands associated with the geological parent material of the quarry. However, the most important feature of quarries is that the plant and animal communities are not static.

Although it cannot be said that there is a long tradition of studying the processes occurring in quarries and pits, nevertheless the dynamic nature of their plant and animal communities has attracted ecologists (Usher, 1976, 1979a; Davis, 1982a). The dynamic nature of an isolated community can be summarized by the equation

$$S_{t+1} = S_t + I + U - E \tag{3.8}$$

where S is the number of species (at times t and $t + 1$), I is the number of immigrants, U is the number of newly evolved species, and E is the number of extinctions. U is a valid term to include in studies of islands, especially isolated (oceanic) islands, but in the quarries there is no evidence to suggest that U is other than negligible in the short time-scales being studied. Hence, the equation can be rewritten as

$$S_{t+1} = S_t + I - E. \tag{3.9}$$

Although it is likely that no studies have, in fact, estimated the immigration and extinction rates for any quarry, nevertheless it is valuable to examine these two processes in detail.

(a) Immigration

Immigration can be defined in several ways. Most generally it can be defined in terms of the presence of a species in a quarry, and hence, to take extreme examples, a seabird landing in the quarry or the arrival of a single propagule of a plant would be added to the list of immigrants. A more restricted definition would be that a species is only judged as an immigrant if it establishes a self-reproducing population within the quarry. This definition implicitly assumes some time scale, although the frequency of self-reproduction is left vague. It is perhaps best, since quarry communities are dynamic and rapidly changing, to accept the second definition with the proviso that reproduction may only occur once or twice. This then leads to a consideration of three sets of species, each in the list below being a sub-set of the set of species above it in the list. The list is:

(1) those that are potentially able to reach the disused quarry;
(2) those that are potentially able to establish themselves in the disused quarry;
(3) those that actually establish themselves in the disused quarry.

There is a fourth group of species that are particularly difficult to list. These are the species whose propagules reach the quarry and become incorporated in the soil seed bank without germination, due to unsuitable conditions. Although Jefferson (1984b) did not record *Epilobium hirsutum* growing in any quarry, he frequently found it in the viable, buried seed bank of quarry soils.

The pool of species that are potentially available will be dependent on two factors – the nature of land use surrounding the quarry and the vagility of the species which are situated further away. A quarry surrounded by intensively farmed arable land is thus surrounded by a somewhat poorer species pool than a quarry surrounded by a natural or semi-natural ecosystem (at least considering higher plant species), as is shown by the species–isolation relationships in Fig. 3.5. There is increasing circumstantial evidence that some species of plants have extremely high vagility. The example, quoted in Section 3.1, of the Leblanc waste in Lancashire being colonized by dune species, indicates that the colonists may have originated tens of kilometres from the waste heaps. Studies of colonization of Krakatau by orchids (Gandawijaja and Arditti, 1983) indicate that their light seeds may travel hundreds, and occasionally thousands, of kilometres in the air. Although such long distance dispersal is a rare event, and hence will often not be detected in seed rain in monitoring programmes, it can have considerable ecological significance.

There is clearly an 'ecological island' effect operating. Small pockets of various wild species, existing in hedgerows, field margins, roadsides, remnants of semi-natural vegetation, etc., provide a source of plants and animals that are potentially able to colonize the disused quarry. In any primary succession, the remnants of a natural or semi-natural ecosystem in surrounding areas provide the pool of potential colonizers, as shown by Taylor (1957) for the succession on recent volcanoes in Papua.

Turning to the species which are potentially able to establish themselves in disused quarries, one can ask to what extent the quarry actually resembles an 'ecological island'. Perhaps the main difference between quarries and islands is that the island lies more or less remote from the pool of potential colonizing species, whereas the quarry is situated within such a pool. The parent material of the quarry and of the surrounding communities are thus likely to be similar. The series of available species which are actually able to establish themselves will depend to some extent upon the geological nature of the quarry. Thus, if the quarry is cut into acidic rock, the calcicolous plants, which might be able to reach the quarry, are unlikely to be able to establish themselves. It is therefore appropriate to look at the list of available species in order to decide, on the basis of their known ecological characteristics, which are likely to be able to establish themselves, and which are not.

The question 'Do colonizing species share any ecological characteristics?' arises from the third group of species, those which actually establish themselves. Studies of quarries which are spatially close together, for example Wharram and Burdale Quarries in the Yorkshire Wolds, indicate that there may be a random element in the arrival of species, since these two quarries have rather different species lists (Usher, 1979a). By using sticky traps, Jefferson (1984b) was able to trap seeds blowing either within or into the quarries and again there were differences. There is thus circumstantial evidence to suggest that there is a random element in which species initially colonize a quarry, but there is no quantitative assessment of this random element. In considering the course of succession, whether one accepts a facilitation or tolerance model (Connell and Slatyer, 1977), there is a certain similarity in the life history strategies of the species characteristic of pioneer-type communities. It is generally stated that such species have light, wind-dispersed seeds (Davis (1951) has shown that the majority of cliff plants in the eastern Mediterranean have such seeds; and Rishbeth (1948) showed that the majority of the 186 species of vascular plants colonizing walls in Cambridge were wind-dispersed). Such species have a greater vagility and hence, even with a random invasion, would be expected to occur first. Also, it is certain that these initial colonizers must be able to survive the relatively inhospitable conditions offered by a quarry. Clearly, it is these species which approximate most closely to those which could be said to be 'r-selected' (MacArthur and Wilson, 1967) or to be 'ruderal' (Grime, 1979).

The nature of the immigration process is reasonably clear. There is a very large number of species surrounding a quarry which would potentially be able to reach that quarry, but only a sub-set of these, those that normally occur on the geological substrata of the quarry, are potentially able to establish themselves. The species that actually do establish themselves have some common characteristics: those early in a successional sequence are generally highly vagile, and they are able to survive in relatively inhospitable environments (climatic extremes, nutrient deficiencies, or excess of potentially toxic minerals, depending on the nature of the substrate), whilst those occurring later in the sequence are likely to exhibit different life-history strategies.

(b) Extinction

It is probably true that there is no estimate of the extinction rate of species in quarries. This is because of difficulties of estimating (or even of observing) extinctions.

The first difficulty is understanding exactly what an extinction is. Although it is simple to argue that an extinction occurs when the species no longer exists in a quarry, the example of Viola persicifolia (V. stagnina) in Wicken Fen Nature Reserve (Rowell, Walters and Harvey, 1982) demonstrates the problems. This species had not been seen in the reserve for more than 60 years, and, as the reserve is extremely well known biologically, it was deemed extinct. However, when the buried seed bank from a series of soil cores was germinated, plants of V. persicifolia developed. This species was, to all intents and purposes, extinct as it had not been seen in a vegetative condition in the reserve: however, it is now known that the species was present in a dormant condition, and that given disturbance it would have been present again in a vegetative condition after germination from the buried seed bank. Although this is a dramatic example, other examples can be found in quarries.

In 1976, a small portion of the floor of Wharram Quarry Nature Reserve was scraped so that an expanse of bare chalk was re-created. In the more open plant communities that had developed over a 40-year period,

Reseda luteola was an infrequent species. Bloomfield (1971) recorded it with a frequency of only 2% in a series of 495 1 m square quadrats. However, on the 1976 scraped area, *R. luteola* germinated rapidly and, in 1977, formed what appeared to be a virtual monoculture of tall flowering spikes (Fig. 3.6). It has become apparent that, although the scraping of the quarry floor resulted in the loss of the soil and the majority of the buried seed bank, the actual disturbance lead to the germination of the remaining *R. luteola* seeds. Being biennial these flowered in the following year, since which time *R. luteola* has returned to its former position of being a scarce species within the nature reserve. Jefferson (1984b) was able to demonstrate that *R. luteola* is one of the commonest species in the buried seed banks of the chalk quarries and pits in the Wolds.

The second difficulty with extinction is a philosophical one: a sighting proves that a species is present, but the failure to make a sighting does not prove that it is absent. A few examples from Wharram Quarry Nature Reserve can be quoted.

The grass *Desmazeria rigida* was first observed in the spring of 1977, when it was found to be abundant in the reserve. The species is small, oppressed to the chalk surface, dark in colour, and seldom forms mats more than 5 cm in diameter. Almost certainly it had been overlooked for many years prior to its discovery. On 26 July 1981 the butterfly *Melanargia galathea* was first seen in the reserve. This is a relatively large species, very obvious even with casual observation, and a rarity in the Yorkshire Wolds (Rafe and Jefferson, 1983): almost certainly it had not colonized the quarry prior to 1981. A small white woodlouse, *Platyarthrus hoffmanneseggi*, which lives in ants' nests, has been seen in the reserve for many years up to 1983. The fact that it was not found in 1984, during a search for it, is unlikely to imply its extinction: this small species, with a subterranean habitat, is likely to be seen only intermittently. However, if no bee orchids, *Ophrys apifera*, were to be seen in any one year, no doubt the extinction of that species, very much associated with the reasons for the establishment of the reserve, would be noted: *O. apifera* is an obvious species, and is the reason why many people visit the reserve.

Extinction is, then, a difficult concept. First one needs to be certain about the existence of the species, either in an active condition (for example vegetative growth) or in a resting state (for example in the

---

Fig. 3.6 The natural re-vegetation of a scrape in Wharram Quarry Nature Reserve in the Yorkshire Wolds. (a) 14 months after the scrape was created there are abundant rosettes of *Reseda luteola* (May 1977). (b) Three months later the *Reseda* is flowering and providing large quantities of seed to the buried seed bank. (c) Six years after the scrape *Reseda* is an infrequent species in the scraped community. The tall plants in the foreground are *Chamerion angustifolium*.

buried seed bank). Secondly, one needs to think about how an extinction may be recognized: this may be simple for obvious and showy species such as *M. galathea* and *O. apifera*, or it may be virtually impossible for small and insignificant species, such as *D. rigida*, or for species that generally live out of sight, such as *P. hoffmanseggi*.

3.4   DISCUSSION

In the equation for the change in the number of species, an equilibrium would have been reached in a quarry when

$$S_{t+1} = S_t$$

which could occur only when

$$I = E.$$

However, quarry ecosystems, except for very old ones such as the Hills and Holes at Barnack mentioned in Section 3.3, are in a relatively early stage of ecological succession when

$$I > E$$

will be the norm, and hence in general

$$S_{t+1} > S_t.$$

This trend for the number of species to increase with time was demonstrated by Parr (1980) for the soil fauna in Wharram Quarry Nature Reserve (Fig. 3.7). It is particularly clearly shown for the soil mites (Cryptostigmata and Mesostigmata, the latter containing many of the predatory species), but is shown to a much lesser extent by the springtails (Collembola). It can be argued that, as a quarry ecosystem starts with very few species, it is inevitable that the number of species will increase with time. This argument is based on a purely random process whereby the probability of adding a species is much greater, when there are few species present, than the probability of losing a species already present.

   However, at Wharram Quarry the number of individuals also increases. This is probably associated with the development of a litter layer and the increasing depth of 'soil', and hence this could be related to the space for the animals to live in: simply the relationship that as soil volume increases the number of animals inhabiting a unit of area of the quarry floor also increases. However, the more important aspect of Fig. 3.7 is that the diversity of the soil animal communities also increases during the course of succession. There are very obvious increases in *H* for both groups of mites, and also in the springtails there is a trend to increasing diversity. The interesting feature of Fig. 3.7 is that the

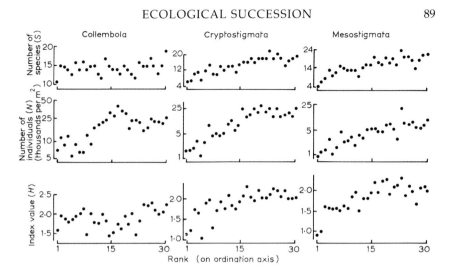

Fig. 3.7 Changes in the number of species (*S*), the number of individuals (*N*) and the Brillouin index (*H*) of three groups of soil arthropods (Collembola, Cryptostigmata and Mesostigmata) during an ecological succession in Wharram Quarry Nature Reserve. The successional sequence, indicated by a rank from 1 for the youngest to 30 for the oldest, is based on a multivariate analysis of the complete soil arthropod community. The data are from Parr (1980), and the illustration is modified from Usher (1985a).

numbers of individuals, the numbers of species, and the diversity all tend to increase during the course of succession.

Evaluation of non-climax communities for their conservation value must take into consideration two factors. First, what is present at the time that the evaluation is carried out? Secondly, what is likely to happen during the foreseeable future as a result of successional processes?

An index such as I_c or I_d in equations (3.5) and (3.6) provides a reasonable basis for deciding the value of the 'hole' or 'heap' at the present time. It is important to decide what features of the community are required – whether it is the typical species for the geological formation that has been quarried, or the presence of uncommon or rare species, or the occurrence of species on the edge of their geographical range – and these can all be simply incorporated into an index. So that an evaluation is not subsequently considered to be biased, all of these features should be decided before the evaluation exercise is carried out, and the weights given to the various components should also be agreed a priori. Given a set of attributes and given the importance (weighting) of each, an evaluation and comparison of the present nature resource in non-climax ecosystems is relatively straightforward.

However, the difficulty lies in the prediction that is associated with the ecological changes which will inevitably take place. There have been some attempts to develop models that will predict the progress of a successional ecosystem, for example Usher (1981) modelled the grassland system in the Breckland, and Stephens and Waggoner (1970) modelled the development of a mixed forest ecosystem in the United States of America. However, such models have probably only been applied twice to nature reserves (Tucker and Fitter (1981) for Askham Bog Nature Reserve, and Carter (1985) for Wharram Quarry Nature Reserve). Models are generally based on information derived from relatively short-term studies on the present nature of the community, and hence are generally of limited value in predicting major changes resulting from the colonization by new species, the change in structure from grassland to woodland, etc.

Very frequently the best that can be attempted is to predict future values from a study of case histories of similar sites elsewhere in the geographical area. However, prediction from case histories can be dangerous. A study of the papers in Davis (1982a) indicates that, on the northern English chalk and limestone, quarries are generally colonized by a variety of herbaceous plant species and often have considerable value for wildlife conservation (for example, Ranson and Doody, 1982). In contrast, on the southern English chalk and limestone, some quarries are rapidly colonized by woody vegetation (Davis, 1982b), without an intervening herbaceous vegetation phase, and they have far less value for wildlife conservation. However, in all of these studies, one generalization can be made: calcareous materials are much more likely to be colonized by communities that contribute to the nature resource of the country than acidic materials. In terms of quarries, those cut into chalk and limestone will frequently develop a wide diversity of species of plants and animals, whilst those cut into slate, granite, sandstone, etc, seldom develop the breadth of species that would lead to their conservation (the exception is in relation to certain individual species, such as protection for the falcon, *Falco peregrinus*, mentioned in Section 3.3, and choughs in some Irish quarries (An Foras Forbartha, 1981)). In terms of heaps, those with a basic substrate, such as the Leblanc waste (pH 7.5–8), develop a community that is far more diverse than the community developing on acid substrates, such as colliery spoil (pH 2.5–5). Gray (1982) quotes a species–area relationship for Leblanc waste heaps of

$$S = 47.9A^{0.16}$$

and for colliery spoil heaps of

$$S = 13.8A^{0.71}$$

where area, A, is measured in hectares. The values of the regression constant indicates that, on an area of 1 ha, the Leblanc waste is more than three times as species rich as colliery spoil. It is perhaps a truism that the more basic the substrate the greater the probability that a newly created site will have, or will develop, communities of wildlife conservation value.

In most studies of successional environments, including most of the examples quoted in this chapter, the emphasis is on the *natural* colonization of habitats by microbes, plants and animals. However, with increasing autecological and synecological knowledge, it should be possible to go some way towards re-creating the type of ecosystems that conservationists desire. A recent review (Bradshaw, 1983) discusses approaches towards the re-creation of a variety of different types of ecosystems.

However, it is important to remember that conservation requires an objective. If the objective is for a diverse community, or for a community with locally or nationally rare or uncommon species, then these successional ecosystems of base-rich substrates may yield what is required. However, if the objective is to conserve the species associated with early stages of colonization, the so-called 'r-strategists' or 'ruderals', then it may be that a greater range of substrates will be satisfactory, and that management will have continually to re-create 'new' surfaces for colonization. Definition of objectives is an essential pre-requisite for any evaluation.

3.5 SUMMARY

Although poorly represented in national series of nature reserves or key sites in Great Britain, industrial sites such as quarries and waste heaps have developed a nature resource that is valued locally. Some quarries in chalk and limestone have naturally developed communities which are diverse and which contain an assemblage of rare or uncommon species. Waste, such as that from the Leblanc and Solvay processes, has also developed diverse communities with local rarities.

Two features of these non-climax ecosystems are important in their evaluation for wildlife conservation. First, as a result of survey, the present value of the nature resource can be estimated: a series of indices which can be used for the evaluation and comparison of quarries on the northern chalk are described. Secondly, as a result of ecological succession, the present communities will change: neither mathematical modelling nor case history studies are found to be entirely successful in predicting what will happen in the future, and hence evaluation contains an element of guesswork. However, before any evaluation of such non-climax habitats begins, it is important to define the objectives of managing such habitats for wildlife conservation.

PART TWO

Approaches in different geographical areas

CHAPTER 4

Evaluation of tropical land for wildlife conservation potential

RICHARD J. McNEIL

4.1 Introduction
4.2 Are tropical areas different?
    4.2.1 Physical, biological and ecological differences
    4.2.2 Socio-political differences
    4.2.3 Differences in land use
    4.2.4 Megadevelopments are just beginning
4.3 Current approaches to evaluation of land
    4.3.1 Evaluation and single species conservation
    4.3.2 Evaluation and conservation of species richness
    4.3.3 Evaluation and conservation for representativeness
    4.3.4 Evaluation and other single-purpose objectives
    4.3.5 Evaluation and multi-purpose objectives
4.4 Improving evaluation methods for tropical land
    4.4.1 Role of foreign and international institutions
    4.4.2 Changes in values
    4.4.3 Reduction of pressures on land
    4.4.4 Rationalizing evaluation procedures
4.5 Summary

Wildlife Conservation Evaluation. Edited by Michael B. Usher.
Published in 1986 by Chapman and Hall Ltd, 11 New Fetter Lane,
  London EC4P 4EE
© 1986 Chapman and Hall

4.1 INTRODUCTION

Due to the ecological and social differences the evaluation of land in the tropics for its wildlife conservation potential is, and should be, different from that of temperate regions.

Geographically speaking the tropics include that part of the earth between the Tropic of Cancer, 23° 30' N, and the Tropic of Capricorn, 23° 30' S. Using this definition, every country in South America except Uruguay, is completely or partly tropical. Every African country, except Morocco, Tunisia and some South African enclaves, lies entirely or partly within the geographical tropics. Much of Asia and Australia, and thousands of Pacific and Caribbean islands, are also included.

Isotherms defining the tropics in terms of mean annual temperature generally follow these latitudinal lines closely. Wellman (1962) suggested defining the tropics as those areas where cold-sensitive plants such as palms could grow all year. This approach has the clear advantage of excluding regions at higher elevations which tend not to share many ecological features of lowland tropics. Since this definition excludes large regions which are politically and socially 'tropical', it is best to use the geographical definition, remembering that the discussion will tend to apply best to hot, humid, lowland areas.

A principal reason for examining the tropics closely is that by far the largest number of living species of organisms, and of both undiscovered organisms and endangered species, are to be found in the tropics. Although estimates of the total number of undiscovered species vary greatly, it is widely acknowledged that most of them live in the tropical rain forests. Myers (1979) says: 'We share the earth with at least five million other species . . . Of these, at least two-thirds occur in the tropics . . . species are becoming extinct at a rate of hundreds and perhaps thousands each year – the majority of these extinctions occurring in the tropics'.

4.2 ARE TROPICAL AREAS DIFFERENT?

4.2.1 Physical, biological and ecological differences

Tropical areas share many physical characteristics, such as high annual solar radiation, year-round warmth (in lowlands), and climatic invariability, and hence great plant productivity, rapid decomposition, and rapid nutrient cycling are typical. Tropical species have narrow niches, considerable competition, and a large degree of interdependence (Jordan, 1981). But perhaps the most notable characteristic of tropical areas, especially of tropical rain forests, is their high level of stored information. This information is found in the form of species richness, genetic variation within populations, and interrelationships or 'pattern richness'.

Tropical ecosystems are more diverse than those in other areas. Although causal explanations are inadequate it is quite clear that tropical areas contain large numbers of species in comparison with temperate areas (Pianka, 1966). The total number of species inhabiting an area (species richness) is only one measure of the total diversity but such data are widely available and easily understood. It is well known that tropical rain forests are rich in species. For example, Pires, Dobzhansky and Black (1953) found 179 species of trees on a 3.5 ha plot in Amazonian Brazil, and they estimated the total number of tree species in the community to be about 250.

High species richness necessarily implies low population density of each species which in turn is related to smaller 'effective populations' (the number of genetically different breeding adults in a population). High species richness is also related to greater specialization, to the likelihood of more complex relationships between species, and possibly to lower stability in ecosystems.

Moist tropical forests (cloud forests are included in the term 'rain forest') and coral reefs seem to be especially rich in species. Other tropical areas, such as grasslands and deserts, are biologically more like temperate regions (they are excluded from further discussion here since, as a rule, they need much less special attention in relation to any ecologically distinguishing characteristics).

Species richness alone would offer a strong basis for providing extra attention to ecosystems such as tropical rain forests when considering establishment of areas for wildlife conservation. As the understanding of ecological and evolutionary processes becomes more sophisticated, information (of which species richness is only a small part) will be used more and more as a primary measure of our conservation efforts.

Each species does represent, of course, a marvellous storehouse of information, and genetic diversity within species and populations is beginning to receive considerable attention (Soulé and Wilcox, 1980; Frankel and Soulé, 1981; Ehrlich and Ehrlich, 1981). Yet another kind of information, difficult to quantify and probably for that reason given less attention by conservationists, is the number and quality of interrelationships between and among organisms. Robinson (1978) argues that the tropics (i.e. rich areas such as rain forests and coral reefs) are qualitatively different because of 'the great and disproportionate complexity of interspecific interactions'. In regions where physical conditions are relatively 'easy', communities are primarily 'biologically accommodated' rather than 'physically controlled' (Slobodkin and Sanders, 1969). This biological accommodation has resulted in not only a species richness but also in an interspecific complexity which might be called 'interrelationship richness' (or 'pattern richness'). From the examples available (Robinson, 1978, describes several), it appears that this type of information, here called 'pattern richness', is also related to specialization and hence to a fragility inherent in rich ecosystems such as tropical rain forests.

4.2.2 Socio-political differences

When compared to temperate areas, tropical areas tend to be different demographically and socio-politically. They tend to have high human population densities, large birth and death rates (and hence young age structures), and large human population growth rates. Tropical areas tend to have a low per capita Gross National Product (GNP) and low 'quality of life' as measured by various indices. For example, Morris (1979), using his Physical Quality of Life Index, places 46 tropical countries in his list of the lowest 50 countries in overall welfare. When ranked by GNP, 45 of the 50 lowest ranked countries are tropical (1970–75 data).

Many tropical countries are not only poor, but their situation is worsening. Per capita GNP declined in 18 countries during 1970–79 (World Bank, 1982). Sixteen of these are tropical countries. Similarly, per capita food supplies have declined recently in many regions, particularly in subsahelian Africa. Pressures for development are strong and increasing; tropical forests are particularly vulnerable to urgent demands for capital and land. Tropical countries tend to be 'less developed' or 'undeveloped' and to have considerable external and internal pressures for rapid economic growth, often necessarily based on rapid extraction of natural resources.

Tropical areas tend to have low levels of literacy, a legacy of recent colonialism (and hence usually a lack of middle-level bureaucrats), unstable governments, few higher education institutions and little emphasis on long-term conservation issues. Those regions which still have large areas of tropical forest tend to hold official and popular views of the forest as an obstacle to be conquered rather than as a source of wealth and a treasure to be preserved. Tropical areas tend to have few individuals and organizations with strong interests in conservation and preservation issues, and limited histories of successful citizen intervention in natural resource issues. Popular literature and other media tend to provide extremely limited coverage of conservation-related programs and issues.

4.2.3 Differences in land use

Regions such as most of Europe, temperate North America and China have experienced centuries of intensive use of natural resources. Deforestation, farming of prairies, irrigation of deserts, high-technology, and large-scale mining, urbanization, and similar practices have resulted long ago in many adaptations by natural ecosystems, including the 'relaxation' (or reduction) in numbers of species that accompanies isolation of small units of habitat types.

In contrast, some tropical areas, including rain forests, have seen little pressure from human populations and are apparently little changed from conditions of a few thousand years ago. In contrast to deforested Scotland (see Chapters 6 and 8) or the almost completely re-shaped landscapes of Japan, Brazil's interior, for example, represents a region which has experienced modest human intervention until very recently. The low human populations of Amazonian rain forests, with access to very limited technologies, had little impact with their hunting, gathering, and shifting cultivation. Activities such as mining of gold and gemstones, tapping of rubber trees, harvest of selected tree species (such as mahogany) and of animals (such as jaguars, parrots, and otters) have left largely intact the ecosystems and most of the species of that region.

Also, in the tropics many current types of land use more closely mimic undisturbed conditions than do related land uses in temperate zones. Tropical agriculture tends to occur in smaller units, with a greater tendency toward intercropping and use of relatively 'unimproved' local species and varieties, with less application of chemical fertilizers and pesticides, less mechanization, all of which are generally less intensive and less intrusive techniques than those of temperate areas. Tree crops such as fruits, nuts, and rubber are frequently grown, often intermixed with herbaceous crops. Areas may be left fallow for many years. The net result of this style of farming seems to be that in the tropics those species which are somewhat adaptable are doing very nicely because conditions have changed only moderately.

Slash-and-burn agriculture is a farming technique, or set of techniques, still employed by hundreds of thousands of tropical farmers. Such an agricultural system, especially when practised on smaller scales, closely mimics the effects produced when senility, disease, or a windstorm brings down a giant tree in the intact rain forest, creating an opening where succession proceeds anew.

4.2.4   Megadevelopments are just beginning

Megadevelopments include various types of huge land manipulations, forest clearing, monocultures, plantation agriculture, high-technology farming, extensive highway networks, large mining operations, huge water impoundments. Temperate regions have a long history of large-scale human interventions. Huge dams for power or irrigation dot the

North American landscape. Most of the forested areas of Europe were cleared centuries ago. Highways and power lines criss-cross the majority of temperate lands. The tropics have not been immune from human influences. But cross-continental highway systems, huge water impoundments, immense mining operations, and similar vast-scale, high-technology intrusions are relatively new to most tropical regions. Physical, biological, and social adaptations to these interventions have been minimal. In comparison with more disturbed temperate regions, tropical areas are not only more vulnerable, but the values to be protected, as measured by species richness, endemism, and pattern richness, are greater. Conservation has only recently become an issue in the tropics; it has become essential to both refine criteria for recognizing what areas need absolute protection and to proceed with conservation of large areas while better methods for choosing land are being developed.

4.3   CURRENT APPROACHES TO EVALUATION OF LAND

Evaluation of land for wildlife conservation potential is conducted differently in the tropics because the tropics are different ecologically and in their social, political and economic characteristics. Evaluation depends on one or more of five principal aims of conservation: single species conservation, conservation of species richness, conservation of representativeness, other single-purpose objectives, and multipurpose objectives (including non-conservation objectives). Each of these conservation aims will be discussed separately.

4.3.1   Evaluation and single species conservation

It is clear that conservation efforts not only frequently concentrate on single species but also that those species selected represent only a very narrow sample of all species in the tropical ecosystem, or even of all species at risk in that ecosystem. Efforts have been concentrated on species which are large (or otherwise conspicuous), 'pretty' or 'beautiful' (or otherwise attractive), or those that are economically valuable (except for some butterflies, these are almost invariably vertebrates). Very few plants or invertebrate animals or even fish, amphibians and reptiles, have received attention. This phenomenon, of course, is not restricted to the tropics.

Robinson and Bari (1982) present a clear example in their discussion of the rationale for the creation of Komodo National Park in Indonesia: they state 'The area's principal conservation objective relates to preservation of the giant Komodo monitor lizard, *Varanus komodoensis*, an endangered species and the world's largest known living lizard'. They then indicate that conservation measures for the species require preservation of 'the Komodo ecosystem in its entirety'.

In Guatemala, a refuge (along with many related conservation measures) for the atitlán or giant pied-billed grebe was similarly designed for the sole purpose of attempting to save from extinction an endangered species of bird (LaBastille, 1974). Tigers, rhinoceroses, gorillas and many other species have similarly been beneficiaries of reserves set aside to protect them.

Many further examples could be cited of the decisions made to set aside lands for the conservation of single species. Quite similar is the idea of attempting to conserve a number of closely related species. Setting aside a marsh clearly affects many migrating and resident species of birds; Tortuguero National Park in Costa Rica was set aside to protect an important nesting area for several species of sea turtles. In setting up Lake Malawi National Park 'the main objective was the protection of a reasonable cross-section of the Cichlid family', a family of fish of which 30% are found only in Lake Malawi (Croft, 1981).

Typically, with migratory species such as sea turtles, many kinds of birds and insects, for example the monarch butterfly, protection is offered only in part of the organisms' range. In some cases, as with sea turtles, this is partially because life history information is incomplete. In other cases, it is only becoming clear now that migratory species such as birds are suffering pressures in one part of their range which affect their numbers elsewhere. With some species, such as sea turtles, many birds, or migratory butterflies, protection can only be offered practically during certain seasons when populations become concentrated in small areas.

Protection of taxonomic groups larger than a single species leads naturally to the consideration of still broader goals.

4.3.2 Evaluation and conservation for species richness

Until recently, attention had been concentrated on a few spectacular threatened species, but a second approach is being considered with increasing interest. Much wider recognition is now being given to the value of the information, especially the genetic information, stored in assemblages of wild plants and animals, not just in individuals.

An estimated 25 000 plant species are threatened with extinction (Lucas and Synge, 1978). Over 1000 vertebrate species and subspecies (IUCN, 1975) or 1000 bird and mammal species alone (Council on Environmental Quality, 1980) are theatened with extinction. According to *The Global 2000 Report to the President* extinction can be expected, by the end of the century, of at least 500 000–600 000 of the present 3–10 million species now living (Council on Environmental Quality, 1980).

It is impossible to identify and record the presence of all of the species in an area. Nevertheless, evaluations are beginning to consider the importance of conserving the richness (i.e. the largest possible number of species) in areas under consideration.

Perhaps the most notable example of the use of this concept is the Brazilian effort to set aside areas based on their identified role as Pleistocene refugia (Pádua and Quintão, 1982). In 1972, Brazil had sixteen national parks and four biological reserves but none were in the Amazon region. A decade later Brazil had 24 national parks and ten biological reserves, totalling 10^7 ha. Many of these new areas are in Amazonia, and their siting was in large part determined by the locations of Pleistocene refugia, 'areas where plant and animal species are believed to have been isolated for considerable periods of time during cold, dry eras when the Amazon was not completely forested . . . It is believed that large numbers of endemic species will be found where these presumed refuges (as identified for various taxonomic groups) overlap or join' (Pádua and Quintão, 1982). Since the Amazon, as with all moist tropical forests, is almost unknown botanically and zoologically, basing the choice of conserved areas on the concept of Pleistocene refugia is seen as a way of increasing the likelihood of choosing areas rich in species and thereby preserving a large proportion despite our ignorance and the pressures of rapid change.

Total size of areas to be protected has been based on Terborgh's (1975) suggested minimum area of 250 000 ha for neotropical birds in lowland rain forest, which presumably will keep extinction rates below 1% of the initial number of species per century. In an effort to gain further insights into minimum effective size of conserved areas, the World Wildlife Fund and the Brazilian National Institute for Amazon Research are conducting an experiment to determine the number of species which can be supported in isolated patches of forest of various sizes (Lovejoy, 1982).

Frankel and Soulé (1981) argue that 'it is the availability of space for adequate numbers of the larger species which is crucial for long-term survival of an ecosystem'. Large reserves automatically offer better protection for many small and mostly unknown species and for 'pattern richness' or interactions. Also, the interiors of large reserves tend to be better protected from human intrusions. Until more is known about the concepts and mechanisms involved in wildlife conservation evaluations, it will almost always be best to opt for the largest reserves possible. In the tropics, preservation of very large areas is still possible, as is the retention and management of large semi-natural buffer zones.

In a less sophisticated way, many other planners have used species richness as a basis for making decisions regarding the locations and boundaries of areas to be conserved. Similarly, marine parks, almost all of which are tropical and most of which are centred on coral reefs, use faunal richness as a primary criterion for choice of site.

4.3.3   Evaluation and conservation for representativeness

Another approach, and perhaps the most widely used today, is the evaluation of representative areas (see Chapters 1, 2 and 13 for a discussion of this concept). The biosphere reserve concept, essentially the need for a network of representative ecological areas, began under UNESCO's Man and the Biosphere Programme. The concept is also being used now by the International Union for the Conservation of Nature and Natural Resources (IUCN), United Nations Environment Programme (UNEP), the Food and Agriculture Organization (FAO), many national governments, and private organizations. Di Castri and Robertson (1982) say, 'the main objective of the biosphere reserve concept is *in situ*, long term conservation of the representative ecosystems of the world and of their component plants, animals, micro-organisms, etc . . . to safeguard the genetic material that makes up the biosphere . . . to allow evolution to continue'. There are now 266 biosphere reserves in 62 countries, many of them tropical (Laird, 1984).

The recent remarkable progress in Brazil (see Section 4.3.2) was based partly on this concept of conservation for representativeness. In their review of conservation progress in Amazonia, Wetterberg, Prance and Lovejoy (1981) discuss the creation of some 11 830 000 ha of newly protected areas in tropical forests of Bolivia, Brazil, Ecuador and Venezuela between 1977 and 1981. These protected areas represented efforts to maintain samples of lowland forest ecosystems; choices were based on Prance's (1977) identification of seven phytogeographic regions and on Udvardy's (1975) classification of the biogeographical provinces of the world. The latter classification is now being further refined for use by IUCN's Commission on National Parks and Protected Areas in their efforts to define and assist in conservation of representative areas (Milne, personal communication). Harrison, Miller and McNeely (1982) provide a useful review of current world coverage of protection for representativeness. It is clear from their discussion that most major international conservation organizations are making concerted efforts to identify representative ecosystems and habitat types, and that representativeness has become a fundamental guide to conservation evaluation.

4.3.4   Evaluation for other single-purpose objectives

In some cases, evaluation of land for wildlife conservation is based on other single-purpose objectives. For example, many East African national parks have been set up because the overall wildlife 'spectacle' is important for tourism. In Kenya, the masses of aquatic birds at Lake Nakuru National Park and the migration of hundreds of thousands of large grazing mammals (from Tanzania's Serengeti plains) into the Masai Mara typify this approach. The basis for decisions in such cases has been the potential income from tourism.

Occasionally other economic values may be foremost. In Papua New Guinea, areas have been set aside for commercial production of crocodiles (National Research Council, 1983b), for farming of birdwings and other large, attractive butterflies (National Research Council, 1983a), and of large grasshoppers and stick insects (Anonymous, 1980). Game ranching is being conducted experimentally on many lands of eastern and southern Africa.

Sometimes, individual plants, animals or stands of trees are preserved for their symbolic, religious, historical, or sentimental values. 'Sacred groves' and the homes of spirits are preserved, and individual animals such as Kenya's 'Ahmed', a huge old elephant, and New Zealand's 'Blossom', an elephant seal which frequented urban harbours, are given special protection. The government of Trinidad and Tobago has proposed a new park to protect spirits and mythological creatures (Meganck and Ramdial, 1984).

4.3.5 Evaluation for multi-purpose objectives

A national park was created in the Galapagos Islands not only to conserve the unique Darwin's finches found there, and not just to protect the many other endemic and rare species, but also because of the value of the archipelago's wildlife for tourism, for the historical values related to Charles Darwin's discovery there of the mechanism of selection and evolution, and for the scientific values presented by the unique opportunity to study evolution 'in progress' and other island biological phenomena.

Similarly, the Brazilian reserves have been chosen not only because they are particularly rich in species or representative as habitats or ecosystems but also because of their potential scientific, economic and social importance. A tiger reserve protects other species; a sacred grove may offer watershed protection; a national park set up to help preserve an endangered bird may also offer national prestige or help to define national borders. The Amboseli Reserve in Kenya is designed to conserve wildlife and simultaneously to provide both new income opportunities and a possibility of retaining a cattle herding culture for the resident Masai people.

It is becoming the rule almost everywhere in the tropics that evaluation of land use is virtually always measured against multiple objectives. The needs of wildlife must be compatible with human colonization projects, protection of indigenous people, watershed protection, agroforestry, defence of national borders, tourism, and other types of 'development'; it is only then that land will receive protection for wildlife conservation.

It is true, of course, that many kinds of land uses can be compatible, and it is sometimes very helpful to tie arguments for wildlife conservation to other issues such as the economic values of tourism. The earlier emphasis on the preservation of a few spectacular species of birds and mammals should give way to today's more sophisticated approaches: to preservation for richness, representativeness and genetic variability. Although conservation will often be opportunistic and will often involve compromise, the greatest opportunities will come from combining wildlife conservation interests with other priorities of the human society.

4.4 IMPROVING EVALUATION METHODS FOR TROPICAL LAND

The process of selecting land for wildlife conservation in the tropics will seldom rely heavily on sophisticated principles of island biogeography (see the comments in Chapter 14, however), refugium theory, or similar ecological and evolutionary concepts. These ideas have sometimes been used crudely and with limited understanding (see, for example, the critical paper by Margules, Higgs and Rafe, 1982); even as a deeper understanding is gained, the usefulness of these concepts will be limited because, on the one hand, land will be used almost always for multiple purposes, requiring an 'optimization' among criteria for choosing and managing land, and, on the other hand, social, political, legal, ethical, economic, and cultural constraints will limit the possibility of using ecologically based criteria.

However, many improvements can be made on today's methods. In particular, the following four are considered: increased effort by foreign and international institutions, changes in values, reduction of pressures on land, and the making of more explicit and rational procedures for wildlife conservation evaluation.

4.4.1 Role of foreign and international institutions

International organizations, notably IUCN's Commission on National Parks and Protected Areas, IUCN's Species Survival Commission, UNEP, and UNESCO's Man and the Biosphere Programme, have done outstanding work in the evaluation of lands for wildlife conservation and in activities leading from those evaluations. The creation and management of hundreds of parks and reserves worldwide can be directly attributed to their efforts and to those of other similar organizations.

But it is also true that in considering the dedication of resources for conservation, national governments in particular must be responsive to political and economic realities. Siebert (1984) presents an excellent example of a proposal for a national park in the Phillipines. Establishment of a park in the Leyte mountains has been recommended for conservation of certain prominent species including endangered species, as well as conservation for both richness and representativeness. In addition watershed protection, tourism and recreation development are important potential benefits. Little progress has been made, however, toward the establishment of a park. Even though the long-term values are clearly important, short-term pressures are severe and contrary. Budget restraints prevent assignment of sufficient staff to provide basic development and management services. Other government agencies see different priorities for economic development. For example, the route of a planned primary road bisects the proposed park. If completed as now planned this road will open to illegal settlement and resource extraction, areas previously protected by their inaccessibility. Finally, ever-increasing pressures by poor and landless people are causing serious and growing problems of both illegal residency and farming in the area and illegal harvest of forest products such as rattan, timber, orchids and parrots. 'Diversion of scarce cash for the establishment of a park, irrespective of its recognized long-term value, is simply an unaffordable luxury at present' (Siebert, 1984).

Similarly, creating a park from land needed by desperately poor people is unrealistic. Most tropical governments can use assistance from wealthier organizations and governments, both in providing the expertise and the necessary labour to assist in evaluations and in providing financial resources sufficient to sway decisions in favour of conservation efforts. In many cases this has provided a role for a private organization, such as The National Trust for Barbados (and its equivalent in many Commonwealth countries) which has purchased and which manages a unique natural area called Welchman's Hall Gully. In many cases these organizations act as conduits, turning over property titles to government agencies once development dangers have passed and government has found the stability and funding to provide long-term protection and management. For example, in Costa Rica, two of the largest wildland reserves are in private ownership; current proposals would combine both with adjacent public land to make large national parks (Hartshorn, 1982).

Evidence is mounting that destruction of tropical forest habitat is causing decreases in populations of many songbirds which nest in North America. A few dollars spent on preservation of tropical wintering areas would produce far greater returns than much larger amounts spent to preserve or manage temperate-region summering and breeding areas. A North American organization, Ducks Unlimited, with members mostly from the United States, has spent millions of dollars to buy and protect waterfowl habitat in Canada.

It would be wise to divert a very large proportion of the conservation funds available in temperate areas to the establishment and protection of tropical parks, reserves, and similar protected areas. This would allow tropical governments to re-evaluate those lands, and it would make possible the protection of far more and larger areas. Foreign individuals and international institutions can purchase land, to hold either permanently or temporarily, and they can provide expertise and human labour to evaluate, survey, and protect wildland resources. They can also provide funding for ongoing protection and management, and provide a good example by striving to do the best job possible of choosing and managing land for conservation in their own countries.

4.4.2   Changes in values

Evaluation of land is done not only by biologists, ecologists, and naturalists but also by economists, politicians, and others with quite different interests or value systems. The process of evaluation can be improved (and hence the potential for wildlife conservation enhanced) if certain values are explained more clearly and thoroughly, and are both heard and understood by a wider audience. Educators and others could explain the various values of nature better. In addition the cultural survival of indigenous people and the values of the various kinds of information they hold could be closely tied to wildlife conservation, to the advantage of both.

Multiple-use objectives can be described and fostered. The vital role of forests for watershed protection is virtually always compatible with wildlife protection. The harvest of resources (including tourism) from parks and reserves could make establishment of such areas economically viable. There has been a reluctance to allow imperfections in policy and in management: for example, people living in, or harvesting products from a national park have been considered as harmful, and therefore no park at all has been established when any park would be better than none. Lovejoy (1980) argued that even the act of putting a park on a map is a valuable beginning, especially if the park is in a remote area. What national parks and similar reserves should and could become, in order to fit them to both human and wildlife needs, is a philosophical consideration that needs re-examination.

4.4.3   Reduction of pressures on land

If other pressures can be reduced, the value of tropical land for wildlife conservation will be enhanced relatively. This is a job both for national governments and for those from other countries.

The pressures caused within tropical countries by ever-increasing human populations, and their increased wants and needs, must largely be dealt with internally (perhaps with outside assistance as requested). The demands for resource extraction tend to come both from external wants of consumer countries and from internal needs to develop foreign exchange funds. Political and social pressures, especially those adverse to wildlife conservation, tend to be internally generated.

4.4.4   Rationalizing evaluation procedures

Procedures for wildlife conservation evaluation can be made much more sophisticated, precise, and powerful. Many government agencies and private organizations see their task as 'preserving nature' or as 'setting aside as much wildland as possible' and have little clear rationale or specific goal orientation. The writing of clear and explicit objectives would aid many organizations in furthering their work.

Opportunity costs are not usually well examined. Is it preferable to preserve a few small (and expensive) recreation parks near urban areas, or a spectacular beach for tourism, or a remote wilderness area for wildlife? If the latter, are we interested in richness or representativeness? Can either richness or representativeness best be attained by protecting one large area or many small ones? (See the SLOSS arguments in Chapter 14.) Is wildlife conservation best enhanced by preservation of land, or by education, or some other means?

Priorities are often not carefully considered. What should be the balance between meeting urgent short-term problems and long-term health and stability of ecosystems? What are the tradeoffs between increased tourism and its effects on environment? Is richness or representativeness more important? Is a locally rare habitat type more important than a spectacular landscape? What obligations do we have either to future generations, to people in foreign countries, or to plants and animals?

Selection criteria can probably be much improved (see Margules and Usher (1981) for a review of criteria used in assessing wildlife conservation potential, as well as Chapter 1). Continuing research by biologists is giving additional and more sophisticated insights into concepts such as island biogeography theory, refugium theory, and the minimum size of refuges. Valuable guidance should come from the studies by Lovejoy and Schubert (Lovejoy, 1982) regarding refuge size and relaxation rates. The work of Soulé and Wilcox (1980), Frankel and Soulé (1981), Margules and Usher (1981), and Margules, Higgs and Rafe (1982) and many others (see references in these excellent summaries and reviews) has provided a great impetus for further work, and many biologists are finding fruitful areas for investigation in conservation biology. Increased attention could be directed toward the effects both of buffer zones and of productive activities which mimic undisturbed tropical forest. For example, what are the differences in protected tropical forests between those with buffer zones having shifting agriculture and those having buffer zones with forest plantations? What are the relative effects of buffer zones with selective logging, plantation forests, or agroforestry? Which species are benefitted, and which are harmed, by the gathering of minor forest products such as rattan or butterflies?

Similarly, the anthropologists, sociologists, psychologists, economists and other social scientists could help to refine other criteria related to evaluation of tropical lands. Is there a stage in the development of a country when wilderness is no longer an obstacle to be conquered but becomes something of value ? Does attaching an economic value to a plant or animal reduce or enhance its potential aesthetic and spiritual values? Is there a definable level of economic security at which it becomes possible to devote attention and resources to preservation of nature? How does a world-view, based on fatalism, allow for wildlife preservation efforts?

Finally, how can the art and science of decision making be improved? How can 'optimization' choices be made rationally? How can better methods and tools be created for making decisions with incomplete and faulty data? In the tropics time is short and the pace of change is quickening. Clearly the tropics represent regions where wildlife resources are both unique and valuable. Those resources are rapidly disappearing and urgent action is a necessity if their uniqueness and value are to be saved.

4.5 SUMMARY

Tropical areas are unlike other regions, both ecologically and socially. Moist, low-elevation regions tend to be relatively unstressed, they are rich in species but low in resilience, species interactions are numerous, and total information may be very high. In tropical areas where there are high human population densities, the demand for extractive uses of resources is large. Tropical areas, especially tropical forests, have evolved under limited pressures from large-scale human activities.

Conservation efforts have evolved from preservation of single species (usually conspicuous and attractive birds and mammals) towards more sophisticated evaluation of land for conservation objectives such as richness, or representativeness. Today's planners and policy makers must nearly always accommodate multiple objectives; conservation of wildlife in the tropics will depend on compatibility with other, mostly short-term, economic goals. Defining conservation objectives is clearly an important task.

Preservation of very large areas is still possible in the tropics and should receive high priority. Foreign and international organizations can do much to assist in evaluation, land purchase, and management. They can help to inform and mobilize opinion, reduce pressure on land resources, and improve selection processes by further studies of natural and social criteria.

CHAPTER 5

Evaluation methods in the United States

SAM H. PEARSALL, DARYL DURHAM
and
DAN C. EAGAR

5.1 Introduction
5.2 Evaluation of habitat for preservation
    5.2.1 Research natural areas
    5.2.2 National natural landmarks
    5.2.3 State natural areas programmes, the Nature Conservancy and the emergence of a national standard
5.3 Evaluation of habitat for management
    5.3.1 Trends in the development of evaluation methods
    5.3.2 The habitat evaluation procedure
    5.3.3 Other level two approaches
    5.3.4 Level three approaches
    5.3.5 Validating species-habitat models
5.4 Conclusions
5.5 Summary

Wildlife Conservation Evaluation. Edited by Michael B. Usher.
Published in 1986 by Chapman and Hall Ltd, 11 New Fetter Lane,
  London EC4P 4EE
© 1986 Chapman and Hall

5.1   INTRODUCTION

The aim of this chapter is to discuss several wildlife evaluation techniques which are currently used in the United States. The chapter is divided into two sections: the first deals with evaluating wildlife and habitat for the purpose of establishing areas designated principally for the protection of wildlife (nature preserves). The second section deals with wildlife habitat evaluation for management and includes environmental assessment and conflict avoidance. In each case, only a few of the more important systems are discussed.

5.2   EVALUATION OF HABITAT FOR PRESERVATION

5.2.1   **Research natural areas**

The Federal Committee on Research Natural Areas (RNAs) was formed in 1966 to promote and guide the selection of ecological preserves on federal land. Original membership consisted of agencies within the US Departments of Interior and Agriculture. By the time that it was officially chartered in 1975 as the Federal Committee on Ecological Reserves, its membership included several other federal agencies, including the Council on Environmental Quality, the National Science Foundation, and the Smithsonian Institute. The 1975 charter states

> . . . creation of a permanent Federal Committee on Ecological Reserves is considered essential. It is to provide the leadership for a coherent national program on ecological reserves which can come only at the Federal level. The responsibilities of agencies to lands and natural area programs under their jurisdiction remains unchanged; management of lands and execution of programs remain in their dominion. (Federal Committee on Ecological Reserves, 1975)

The criteria for selection of RNAs originally included opportunities to preserve 'examples of all significant natural ecosystems for comparison with those influenced by man', opportunities for education and research, and opportunities to preserve rare species of plants and animals (Federal Committee on Research Natural Areas, 1968). In 1977, the programme was revised to

> . . . preserve a representative array of all significant natural ecosystems and their inherent processes as baseline areas. This action provides a potential range of diversity, including common, rare and endangered species or disjunct populations. (Federal Committee on Ecological Reserves, 1977)

Each of the participating agencies has developed its own version of these criteria (Nature Conservancy, 1977a). Few of the federal agencies which participate in the Research Natural Areas Programme have conducted systematic searches for potential RNAs. A notable exception is the United States Department of Agriculture (USDA) Forest Service which conducts surveys through its Research Branch inventories of potential RNAs. The goal is to establish RNAs to protect an example of each of the forest types described by the Society of American Foresters (Eyre, 1980). Although the USDA Forest Service has published regulations governing the nomination of RNAs, it has not published criteria for evaluation of plant and animal populations for the purposes of the nomination process. The various agencies have designated 442 areas comprising 5.7×10^5 ha as RNAs (Thibodeau, 1983; United States Department of State (USDS), 1984).

5.2.2 National natural landmarks

The National Natural Landmark (NNL) Programme of the United States Department of the Interior (USDI) National Park Service identifies and registers land, other than the national parks in the United States, which is representative of the American landscape. Land which is registered under this programme is not legally protected, and does not change ownership (Nature Conservancy, 1977a; USDS, 1984). Criteria for inclusion in the registry were published in 1973. They included:

3. An ecological community significantly illustrating characteristics of a physiographic province or a biome.
4. A biota of relative stability maintaining itself under prevailing natural conditions such as climatic climax community.
5. An ecological community significantly illustrating the process of succession and restoration to natural conditions following disruptive change.
6. A habitat supporting a vanishing, rare, or restricted species.
7. A relict flora or fauna persisting from an earlier period.
8. A seasonal haven for concentrations of native animals . . . (Anonymous, 1973)

Selection of NNLs has been conducted somewhat more scientifically than the selection of research natural areas. Initial NNL studies were conducted by professional scientists for each of several 'themes', for example caves and springs (Nature Conservancy, 1977a). Theme studies proved to be too broad in geographic scope to be very useful, so a second set of studies was commissioned for the potential ecological and geological NNLs in each of the major physiographic provinces of Fenneman (1948). Reports from both sets of studies consisted of introductory material followed by abbreviated analyses of each site which was thought to qualify in the study region. Many NNLs were registered on the basis of these studies. More recently, NNLs have been intensively analysed by individual ecologists and geologists. Registration has depended on their recommendations or on scientific committee review. The NNL Programme has never established an explicit procedure for evaluating flora and fauna of potential NNLs, but has, instead, relied on the professional judgement of evaluating scientists. More than 500 areas have been designated by the National Park Service (USDS, 1984).

In addition to the RNA and NNL Programmes, several other programmes in the federal government attempt to identify public and private lands for protection (cf. Nature Conservancy, 1977a; USDS, 1984). None of these programmes requires a standardized approach to evaluation of either plants, animals or their habitat, so no nationally accepted method has evolved. As an ironic result, the most effective programme for the preservation of animal and plant habitats has been the National Wilderness Programme which was created for a substantially different purpose (Thibodeau, 1983). Support for wilderness on undisturbed Federal lands has been so popular with the American public that 3.4×10^7 ha have been protected (primarily in the western states), with the consequence that both species and habitats have been protected as well (Cahn, 1982; Thibodeau, 1983).

5.2.3   State natural areas programmes, the Nature Conservancy and the emergence of a national standard

Almost all of the 50 states now manage one or more programmes designed to identify and protect natural areas. The first legislatively mandated state programme was established in Illinois in 1963, quickly followed by others in the Midwest (Nature Conservancy, 1977b). By the early 1970s, a natural areas movement had begun in New England, and these two regions led the nation in state sponsored nature preserve activities for several years.

Various approaches to evaluating potential natural areas in the United States have appeared in the literature, for example, Tans (1974), Gehlbach (1975), Adamus and Clough (1978), but none of these have been adopted by any programme outside its state of origin. By the mid-1970s, it was apparent that the federal government was not able to develop a national approach, and most states were working independently to establish inventories of potential natural areas. The Nature Conservancy, a private, land-purchasing organization dedicated to the preservation of natural diversity, began a series of collaborations with state agencies, initially in south-eastern states, that has resulted in the emergence of a standard approach to evaluating potential natural areas on the basis of threatened or endangered species or critical natural community types (Halcomb, Boner and Sites, 1976; Nature Conservancy, 1977b, 1982).

The Nature Conservancy entered contracts with Mississippi (1974) and Tennessee (1975) to establish Natural Heritage Programmes. Staff

Table 5.1 The information included on the element ranking form for the Natural Heritage Programme data base.

Item Number	Information
1	The element's common and scientific names (including synonymy)
2	A brief synopsis of habitat for plants and animals
3	Descriptions of community constituents and topo-edaphic substrate for natural communities
4	Definition of an element occurrence, for example minimum viable population size, hibernacula for migratory bats
5	A statement of taxonomic distinctness, for example monospecific family
6	Special status as assigned by federal and/or state agencies e.g. threatened or endangered
7	Degree to which the element is protected under current federal and state laws (including, but not limited to, endangered species statutes)
8	Estimated total global element occurrences
9	Estimated global abundance
10	Number of element occurrences recorded in the state database
11	Estimated state abundance
12	Global and state ranges (accompanied by range maps)
13	Estimated number of element occurrences protected in preserves globally and in the state
14	Known threats to survival
15	Ecological fragility (sensitivity of the species or community to variations in its habit)

biologists of these programmes were charged with creating and maintaining computerized databases of the 'lowest common denominators' (LCDs) of nature preserves. These LCDs were initially defined as known locations of plants and animals listed as endangered, threatened, or of special concern by either state or federal agencies; locations of natural communities believed to be the best representative examples of their types; significant geological features and formations; and 'champion' trees (later omitted). Any LCD was considered an element of natural diversity, and any mapped and computerized location was considered an element occurrence. Each element occurrence was accompanied by an attribute file which listed both natural characteristics, for example the size of population, and record characteristics, for example the herbarium where the voucher specimen is kept. Furthermore, abstract or element attribute files were maintained on the known ecological requirements for each plant and animal species and on the characteristics of each natural community (Halcomb, Boner and Sites, 1976; Jenkins, 1976, 1977, 1982; Burley, 1983).

Between 1974 and the early 1980s, many nature preserves were created under the aegis of state natural areas programmes and through land acquisition by the Nature Conservancy. Priorities were established on the basis of professional judgement of staff biologists and on searches of the databases for clusters of element occurrences. By the early 1980s, this approach proved too inexact for the Nature Conservancy's purposes, especially because the process of assigning endangered and threatened status to rare species was sometimes subjective. The format of Natural Heritage Programmes was amended to include element ranking.

Element ranking is a procedure for selecting elements for protection in nature preserves. For each plant, animal, or natural community element in a Natural Heritage Programme database, an element ranking form is filled out. The element ranking form includes the information shown in Table 5.1. Separate element ranking forms deal with state rank and with global (planetary) rank, both of which are assigned (Nature Conservancy, 1983). State ranks are assigned by the state Natural Heritage Programmes based on the information in their databases. Global ranks are currently assigned by the Nature Conservancy (Table 5.2) based on accumulated state ranks (Nature Conservancy, 1983).

State ranks parallel these global ranks, so that elements are also ranked from S1 to S5. For both state and global ranks, codes are available to indicate elements which are accidental or exotic species in North America (for global ranks) or in the state, elements for which only historical records exist (global or state) but which are expected to be rediscovered, elements believed to be extirpated from the state or extinct, and elements for which ranks cannot be determined because insufficient information exists. Elements which are considered as exotic or accidental in either the state or North America, or which are only

Table 5.2   The five categories into which the Nature Conservancy (1983) groups species, ranging from the most endangered (rank G1) to the least endangered (rank G5).

Rank	Rank title	Number of estimated occurrences	Description
G1	Critically endangered throughout range	1–5	Critically imperilled globally because of extreme rarity or because of some factor in its biology making it especially vulnerable to extinction
G2	Endangered throughout range	6–20	Imperilled globally because of rarity or because of other factors demonstrably making it vulnerable to extinction
G3	Threatened throughout range	21–100	Either very rare and local throughout its range or found locally (even if abundantly in limited locations, as with certain successional endemics) in a very restricted range
G4	Apparently secure globally	>100	Apparently secure globally, though it may be quite rare in parts of its range, especially the periphery
G5	Demonstrably secure	>100	Demonstrably secure globally, though it may be quite rare in parts of its range, especially the periphery

known historically in the state, are not assigned state ranks. Elements believed to be extinct are not assigned state or global ranks. Elements with unknown status are ranked as GU and/or SU. Global and state ranks are then combined, and elements are ranked in order (Table 5.3) for use in nature preserves programmes or for other conservation purposes (Nature Conservancy, 1983).

Table 5.3   A method of combining global (G) (see Table 5.2.) and state (S) ranks into a combined ranking system. Note that the table is read from left to right and, vertically within any column. The Nature Conservancy establishes its priorities on the basis of global ranks, using state ranks only to break ties within states. State natural area programmes tend to follow suit but may place a higher emphasis on state ranks, especially in the G3–G5 range. U ranks, for species of unknown status, are placed between ranks 3 and 4: this is based on assumptions about the probable risk of extinction of the average case U species.

G1S1	G2S1	G3S1	GUS1	G4S1	G5S1
	G2S2	G3S2	GUS2	G4S2	G5S2
		G3S3	GUS3	G4S3	G5S3
			GUSU	G4SU	G5SU
				G4S4	G5S4
					G5S5

For plants and animals infraspecific taxa are ranked in a different way. The species is ranked first, and then the infraspecific taxon is ranked and assigned a 'T' rank for global and state status. For example, a variety ranked as G3T1S2T1 would be a critically endangered subspecies of a species ranked as G3S2 (Nature Conservancy, 1983).

Element ranking of plant and animal species has become, within the limits of available information, a fairly exact and consistent process, and the Nature Conservancy is now maintaining a central database of the global and state ranks of all plants and animals listed as endangered or threatened by any of the 50 states or by the federal government. The ranking process is adequately documented so that assignment of species ranks is replicable. Element ranking of natural communities varies substantially from state to state, depending on how natural communities are defined, for example by dominant species or genus in the overstorey.

In the 17 states without Natural Heritage Programmes, the Nature Conservancy is contracting with other professionals to rank endangered and threatened species. There is thus every reason to believe that within a few years all species classified as endangered or threatened in the United States will have been assigned state and global ranks. Unfortunately, neither the federal government nor any state has done an adequate job of identifying endangered or threatened invertebrate animal or non-vascular plant species. A few of these have been listed by federal and state governments, and element ranks have been assigned and preserves created for some of those listed. Many separate nature preserves have already been established on the basis of rare plant and vertebrate species and natural communities, using the element ranking process together with the Natural Heritage Programme element occurrence databases.

5.3 EVALUATION OF HABITAT FOR MANAGEMENT

5.3.1 Trends in the development of evaluation methods

Prior to 1969, American wildlife policy emphasized management for game species and migratory waterfowl. In 1969, the passage of the National Environmental Policy Act shifted the emphasis of government action to the evaluation and conservation of all species of plants and animals. As a direct result of the Act's passage, federal agencies have developed various habitat evaluation models and databases in an effort to understand the effects of management on the living environment and to minimize negative management impacts on wildlife habitat (Thomas, 1982). Since 1969, a number of federal laws have firmly established the requirement that habitat and the effects of management on habitat be understood and incorporated into all federal land management planning and practice (Hirsch *et al.*, 1979; US Army Corps of Engineers, 1980; USDI Fish and Wildlife Service, 1980a; Salwasser *et al.*, 1983).

Seven trends have dominated this shift from managing game species to managing habitat.

(1) Habitat is treated as synonymous with the environment, cf. 'the sum total of all physical and biological factors impinging upon a particular organismic unit' (Pianka, 1974, p. 2).
(2) Habitat quality for a species is equated with the carrying capacity of its environment (USDI Fish and Wildlife Service, 1980a).
(3) Numerous models have been developed that describe species–habitat relationships based on a priori judgments supplemented by empirical data (USDI Fish and Wildlife Service, 1980a; Marcot, Raphael and Berry, 1983).
(4) Modelling has generally proceeded through four phases – conceptual, diagrammatic, symbolic, and computer-based (Hall and Day, 1977) with the present emphasis on computer-based.
(5) A tendency to establish rigid forms and exact procedures for evaluating habitats has developed (USDI Fish and Wildlife Service, 1980a).
(6) The field is well integrated, with new methods based on previous experience.
(7) There is considerable interest in developing a single standardized approach for habitat evaluation based on a general agreement about the information requirements of habitat evaluation (USDI Fish and Wildlife Service, 1980a; Salwasser and Tappeiner, 1981; Cushwa, 1983). These include (i) information on animal diversity and distribution, (ii) habitat definition and life history requirements, (iii) species–habitat relationship models (working level approaches to niche definition), (iv) habitat availability and quality (inventory), (v) habitat location, (vi) habitat responses to alternative management practices, and (vii) accounting systems that facilitate selection of management practices based on resource-goals. Salwasser and Tappeiner (1981) add the stipulation that monitoring and periodic revision of management practices are essential.

Van Horne (1983) states that all approaches to non-relative habitat evaluation can be sorted into three levels. Level one is the level at which the quality of the habitat for a given species is directly evaluated on site, almost invariably based on some measure of population density. Level two includes indirect evaluation of habitat quality for a given species or group of species based on data and conclusions drawn from level one, for example Thomas *et al.* (1975), Thomas (1979), USDI Fish and Wildlife Service (1980b). Level three includes approaches which attempt to evaluate habitat for the entire wildlife community.

Level one work is the most purely scientific, and is generally not acceptable to land managers who must make many decisions in a context of limited resources. Furthermore, level one approaches do not include mechanisms for predicting and evaluating future conditions (Schamberger and Krohn, 1982).

Level two approaches focus on species selected because they:

(1) are representative of their guild (for example Short, 1982);
(2) are especially sensitive to fluctuations in environmental quality;
(3) are considered endangered or threatened in part or all of their ranges;
(4) are game or commercially valuable;
(5) are considered to be of special interest to people, for example songbirds.

Level three approaches often consist of aggregations of individual species data without regard for the effects of predation, competition, symbiosis, and other interspecific relationships (van Horne, 1983), and for this reason many feel that level three approaches are impractical (USDI Fish and Wildlife Service, 1980a). The Habitat Evaluation System (HES), a non-aggregated level three wildlife habitat model (see Section 5.3.4), was developed by the US Army Corps of Engineers (1980). This model shows considerable promise for incorporating level two approaches with community analyses in a level three system.

Thomas (1979) stated:

> The knowledge necessary to make a perfect analysis of the impacts of potential courses of forest management action on wildlife habitat does not exist. It probably never will. But more knowledge is available than has been brought to bear on the subject. To be useful, that knowledge must be organized so it makes sense both biologically and silviculturally.
>
> Perhaps the greatest challenge that faces professionals in forest research and management is the organization of knowledge and insights into forms that can readily be applied. To say we don't know enough is to take refuge behind a half-truth and ignore the fact that decisions will be made regardless of the amount of information available. In my opinion it is far better to examine available knowledge, combine it with expert opinion on how the system operates, and make predictions about the consequences of alternative management actions. (Thomas, 1979 p. 6.)

5.3.2   The habitat evaluation procedure

The most popular selected species approach is the Habitat Evaluation Procedure. HEP is used to predict the carrying capacity of an ecosystem for a particular animal species by using an understanding and inventory of the species' habitat requirements. The original purpose of HEP was to provide a simple, general approach for environmental impact assessment (USDI Fish and Wildlife Service, 1980a). HEP has not been used for preserve selection or design, although its potential for the latter is substantial. HEP was originally based on the work begun by Hamor (1970) and Daniel and Lamaire (1974), and is currently restricted to terrestrial and inland aquatic animals (USDI Fish and Wildlife Service, 1981). The most useful HEP reference (USDI Fish and Wildlife Service, 1980b) details the procedure for habitat evaluations, and is a revision of the first version (USDI Fish and Wildlife Service, 1976) and expanded by Flood *et al.* (1977). The major change between 1977 and 1980 was the development of a requirement for documented habitat models as a component of HEP.

The six steps in applying HEP are shown in Table 5.4. In step 2 one or more species (which may include individual life stages, for example tadpoles) are selected, often on the basis of guild representation (see especially Short, 1982; Short and Burnham, 1982; Short and Williamson, in press). The most ecologically sensitive species in each guild should be selected. In step 4, a Habitat Suitability Index (HSI) is derived from development of certain key habitat components from the literature and available expertise on each evaluation species. The presence and quality of these components in the evaluation area are then compared with optimum habitat, and the result is expressed as a number between 0 and

Table 5.4 The six steps in applying the Habitat Evaluation Procedure (HEP).

Step	Description
1	Divide the evaluation area into stands containing relatively homogeneous cover types
2	Select species on which to base the evaluation
3	Estimate the total habitat area in the evaluation area for each selected species. The area of available habitat is the total area of all cover types used or potentially used by the evaluation species
4	Calculate the Habitat Suitability Index (HSI) for each evaluation species for each stand in the evaluation area. Examples are given in the text
5	Conduct a baseline assessment. The baseline assessment developed using HEP is the total area of the evaluation area multiplied by the HSI score calculated for a given evaluation species. This yields Habitat Units (HUs) which are always expressed in terms of a given species, group of species, or guild
6	Use HEP accounting procedures. These are explained in the text

1 as an indicator of carrying capacity (after Inhaber, 1976) (USDI Fish and Wildlife Service, 1980a). This process is typically accomplished using transformation curves (see Fig. 5.1). A number of HSI models are presently available (for example see Sousa, 1982), as are detailed instructions on the preparation of HSI models (USDI Fish and Wildlife Service, 1981).

An HSI model consists of a set of key habitat component descriptions with instructions for assessing the quantity and quality of key habitat components for each habitat present in the evaluation area. Each key habitat component is also assigned a weight based on its relative importance to the species. HSI scores for each stand are based on the key habitat component values multiplied by their relative weights. Every model provides a procedure for normalizing the resulting HSI score to the standard 0–1 scale.

Schroeder (1982) has prepared one of the simpler HSI models for use with HEP. This model, for the Yellow Warbler (*Dendroica petechia*), is based on the reproductive requirements of the species, assuming that an environment providing satisfactory breeding habitat will also provide adequate food, water, and cover for other, less sensitive, less energy dependent activities. This assumption is drawn from a review of 17 references yielding a brief description of the species' habitat requirements. Schroeder's model is limited to application in deciduous shrubland and deciduous shrub/shrub wetlands (USDI Fish and Wildlife Service, 1981), and only three transformation curves are employed. The first of these (Fig. 5.1) plots the percent deciduous shrub crown cover on the x-axis against the suitability index for that value on the y-axis. The percent deciduous crown cover for a stand of common cover type in the evaluation area is measured using the line intercept technique, the value is found on the x-axis, a vertical line is drawn to the transformation curve, and a horizontal line from that point to the y-axis yields the suitability index (V_1) for percent deciduous crown cover. Average height of the deciduous shrub canopy is the second variable. The initial data are collected using a graduated forester's rod. The data are plotted on the x-axis, which is scaled from 0 to 2.0 metres. The y-axis is identical to that in Fig. 5.1 and yields a suitability index between 0 and 1 (V_2) for this variable. The third variable is percent of deciduous shrub canopy comprised of hydrophytic shrubs. Again the line intercept method is used, and the transformation curve, although differently shaped, is plotted on an identical pair of axes as those in Fig. 5.1, and yields index V_3. In this model, the three indices have equal weight, and so no adjustments using key variable weights are required. The final HSI score for the stand is

$$HSI = (V_1 \times V_2 \times V_3)^{1/3}. \tag{5.1}$$

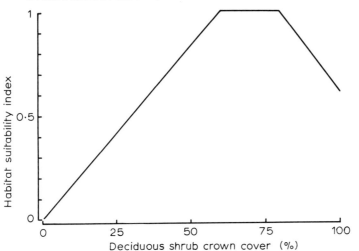

Fig. 5.1 The Habitat Suitability Index curve, used in step 4 of the Habitat Evaluation Procedure (the use of this curve is explained in detail in the text). The curve for the yellow warbler is drawn (after Schroeder, 1982). This model applies to the following cover types; deciduous shrub, and deciduous shrub/shrub wetland. This curve yields the Suitability Index for percent for deciduous shrub crown cover (v_1 in the text).

Thus, a deciduous shrubland with 90% deciduous shrub crown cover (giving $V_1 = 0.8$), an average shrub canopy height of 1 m ($V_2 = .5$), and a 30% hydrophytic shrub component ($V_3 = 0.35$) yields

$$HSI = (0.8 \times 0.5 \times 0.35)^{1/3} = 0.52. \qquad (5.2)$$

In this case, the relatively low percentage of hydrophytic shrubs, indicating more mesic than hydric conditions, overcame a relatively high crown cover value and an average canopy height value to yield a below average habitat value for the stand. Optimum habitat, according to the model, requires from 60% to 80% crown cover with an average canopy height of 2 m and 100% hydrophytic shrub species (Schroeder, 1982). The steps in reaching an HSI score are summarized in Table 5.5.

When available habitat for a selected species in the evaluation area includes more than one cover type, i.e. more than one stand, the final HSI for the evaluation area is calculated:

$$HSI = \{\Sigma(HSI_i \times a_i)\}/A \qquad (5.3)$$

where HSI_i is the index for the ith stand, which has area a_i, and $A = \sum a_i$.

An HSI model must have an assigned geographic range of application which may include some or all of the evaluation species' range. The USDI Fish and Wildlife Service prefers Bailey's (1976, 1978) ecoregions

Table 5.5 The six steps in reaching the HSI score for a stand

Step	Description
1	*Key habitat components:* key habitat components are determined for the stand type and the species in question.
2	*Key habitat variables:* key habitat variables are selected which will assess the relative habitat value of the components in the stand for the species.
3	*Key variable weights:* key variable weights are assigned to distribute importance among the key habitat variables.
4	*Field methods:* key habitat variables are measured in the field yielding certain values.
5	*Transformation curves:* these values are plotted on transformation curves which then yield suitability indices for the key habitat variables.
6	*Mathematical relationships:* the suitability indices and their appropriate weights are then related mathematically to yield a final HSI score.

for terrestrial models and the hydrologic units of Seaber *et al.* (1974) for aquatic models. Relationships between key habitat component variables may be limiting (a limiting threshold in one variable may eliminate any value for other variables), cumulative (additive), compensatory (a low value for one variable can be compensated by a high value for another variable), and spatial (in three dimensions including interspersion indices). Key habitat component values are usually related through transformation curves to the range of data collected. Many of the more recent models are computerized. All HSI models should be thoroughly documented so that every key habitat component and its relative weight can be directly related to literature references and expert opinions.

An environmental impact assessment for each species or guild can be accomplished by comparing current Habitat Units (HU) and projected future HUs using benefit–cost analysis techniques, for example Howe (1971), substituting HUs for dollars. The USDI Fish and Wildlife Service prefers calculation of Average Annual Habitat Units of the area both for the life of the project and for the area without the project so as to provide an annual assessment (Schamberger and Krohn, 1982). One flaw with this approach is the lack of any procedure for discounting future HU values (HU values should probably be discounted negatively, i.e. increase in value by current standards, according to Krutilla and Fisher, 1975). This approach limits evaluation to the projected useful life of a project.

As with all level two approaches, HEP is extremely sensitive to the selection of species for evaluation, and selection of generalist species may result in very large ranges of HSI values in suitable habitat. On the other hand, species with highly specialized habitat requirements should be selected for model development only if their habitat requirements are known. HEP models are also sensitive to the selection of variables. For example, the USDI Fish and Wildlife Service model for the veery (a common North American forest songbird, *Hylocichla fuscescens*) (Sousa, 1982) is based on 23 references on the veery's habitat requirements. It was reviewed by two experts whose comments were incorporated into the final version. The published model states that shrub crown cover is the most critical key habitat component and establishes a lower threshold of 20%; any habitat below 20% shrub crown cover is considered to have an HSI of zero for veeries. This is probably an accurate synthesis of the literature reviewed. However, one of the densest veery populations ever reported occurred in a virgin spruce–fir forest in western North Carolina (Alsop, 1970): such old growth spruce–fir forest has virtually no (certainly less than 20% cover) shrub understorey. Generalized models can obscure remarkable exceptions which should, as in the case of the veery, provide new insights about habitat requirements. Sousa (1982) states that HSI models are not 'statement[s] of proven cause and effect relationships' but 'hypotheses of species–habitat relationships'. HEP's utility depends entirely on the skill and knowledge of the model user and is limited to the most characteristic portion of the selected species' range.

5.3.3   Other level two approaches

(a) *Life-form models*
Thomas' approach to protecting and enhancing wildlife habitats in managed forests (Thomas *et al.*, 1975; Thomas, 1979) includes models developed before the HEP models were published. Thomas collected information on the relationships of all resident vertebrates in the forest communities in the four national forests of the Blue Mountains in Oregon, and then used this information to sort species into 16 life-forms based on feeding and breeding requirements and behaviour. These life-forms are very similar to the guilds of Short and Burnham (1982) except that species responses to habitat characteristics are classified as weak to strong, and thus membership in a given life-form has a qualitative aspect. Feeding responses are also given twice the weight of breeding responses. Hirsch *et al.* (1979) stated that using Thomas' method it is 'possible to predict likely impacts of changes in plant communities on life-forms and hence on individual wildlife species'. Thomas' approach is management-orientated rather than assessment-orientated, so models are used to guide management practices. For example, a woodpecker life-form model is capable of projecting the number of snags required

per unit area for a given site so as to support its optimum number of woodpeckers. Thomas somewhat arbitrarily proposes that no less than 40% of the optimum population should be the management goal, and so is able to develop a minimum snag/area figure for woodpeckers in the managed forest (Thomas *et al.*, 1975).

(b) *Pattern recognition models*

Pattern recognition models simply plot the density of species in a given habitat against a large array of habitat variables, and then identify patterns of coincidence (USDI Fish and Wildlife Service, 1981; Seitz, Kling and Farmer, 1982). Williams, Russell and Seitz (1977) developed PATREC, a pattern recognition model which incorporated Bayesian analysis so that prior probabilities (i.e. assumptions) that certain habitat factors produce high population densities can be compared with a sample set of field data and a set of posterior probabilities developed. PATREC models can be incorporated as HSI models.

(c) *Wildlife and fish habitat relationships system*

Following the development of Thomas' life-forms approach and the development of HEP, and as a direct response to the National Forest Management Act of 1976, the US Forest Service began development of the Wildlife and Fish Habitat Relationships System (WFHRS). WFHRS is a national system based on Thomas' work and incorporating HEP and other approaches into the management of 7.7×10^7 ha of national forests and grasslands (Nelson and Salwasser, 1982; Sheppard, Wills and Simonson, 1982). In the WFHRS approach to management, a data base is built on the general habitat requirements of all resident vertebrate species in a forest or forest region. Species are selected for evaluation (generally as life-form representatives) and patterns of habitat component/population density are identified. Land is managed to enhance habitat components where these are identified with high densities of selected species. WFHRS now incorporates HSI models (Nelson and Salwasser, 1982).

Nine WFHRS programmes have been established in the United States, one in each of the US Forest Service regions. Each WFHRS will eventually include: (i) an inventory of fish and wildlife habitat data in a mapped and automated data management system to include taxonomy, habitat classification, species ecology and life history, species distribution maps, species habitat relationships coefficients, special habitat quality criteria and management principles, and systematic data management procedures; (ii) habitat evaluation procedures; and (iii) management decision making procedures (Nelson and Salwasser, 1982). Every effort is made to be inclusive, so that no useful species or habitat data is omitted from the process (Salwasser and Laudenslayer, 1981). For example, the California WFHRS now includes the California Natural Heritage Programme Database and the forest production optimization model DYNAST (Boyce, 1977; McClure, Cost and Knight, 1979) as well as the California version of Cushwa's Procedure (Cushwa *et al.*, 1980; Salwasser and Laudenslayer, 1981; Dedon, Smith and Laudenslayer, 1983).

WFHRS employs three levels of habitat classification. These are: dominant attribute of the site, for example palustrine; community physiognomy; and special habitat features, for example snags. Three levels of species–habitat relationships models are also incorporated. Level one is a high–medium–low ranking of species to habitat relationships for cover, breeding, and feeding for a single species in a given stand. Level two includes information on between stand diversity. Level three includes information relating to large area diversity (Nelson and Salwasser, 1982). WFHRS output is expressed as a Habitat Capability Coefficient from 0 to 1.

(d) *A database for management*
An effort was begun in 1978 to develop a national standard for wildlife habitat databases which would serve the HEP and WFHRS models and other wildlife habitat evaluation needs (Schweitzer and Cushwa, 1978; Schweitzer, Cushwa and Hoekstra, 1978; Hirsch *et al.*, 1979; Cushwa *et al.*, 1980). A standard data dictionary was first proposed in 1980 (Cushwa *et al.*, 1980). The first State level implementation of Cushwa's 'Procedures' was in Pennsylvania where species–habitat relationships for 844 species are now on file (DuBrock, 1983). Additional databases established on this model are implemented in Missouri, California, Colorado, and Virginia. 'Procedures' contains information on the life history requirements and distribution of selected species. So far, State applications have been restricted to terrestrial and freshwater aquatic vertebrates, but the USDI Fish and Wildlife Service's National Wetlands Inventory has experimented with managing plant data, and the Office of Endangered Species has adapted 'Procedures' to create a national endangered species database to be called the Endangered Species Information System (ESIS) (Brown *et al.*, 1983). At the 1983 National Workshop on Computer Uses in Fish and Wildlife Programmes, Cushwa argued for (i) a common set of data elements to describe individual species attributes (clearly the goal of 'Procedures'), (ii) the use of a common set of land use/land cover classifications (for example Anderson *et al.*, 1976), (iii) the establishment of a central, Federal wildlife database, and (iv) the promotion of these standards to States, local governments, and national and international organizations.

5.3.4   **Level three approaches**

(a) *Habitat evaluation system*
The Habitat Evaluation System (HES) developed by the US Army Corps of Engineers' Lower Mississippi Valley Division (US Army Corps of Engineers, 1980) is a good example of a level three, or generalized, habitat evaluation approach. HES models have been developed for several wildlife habitat classes in the Lower Mississippi region, for example swamps, bottom land hardwoods, lakes and upland forests. HES models are used to generate Habitat Quality Indices (HQIs) similar to the HSIs of HEP for the general habitat of these wildlife habitat classes on sites proposed for management and development. HES follows the HEP example very closely except that habitat quality models are generated for habitat classes (natural communities) instead of for a particular species. Key habitat variables are derived from the literature for general habitat quality, presumably for all species, although much of the think-

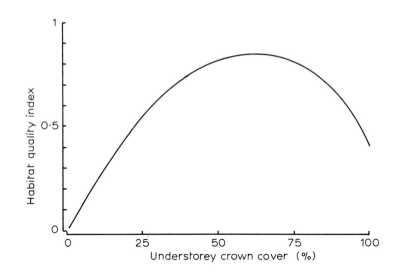

Fig. 5.2   The Habitat Quality Index curve, used in a Habitat Evaluation System (see text for details). The original version of this curve (United States Army Corps of Engineers, 1980) actually consisted of four curves, one each for desirable (preferred forage) species within reach, palatable species within reach, emergency species within reach (digestible but not palatable), and species not within reach. Since the terms 'desirable, palatable and emergency' and the definition of 'within reach' are animal species dependent, the four curves represent a serious departure from the level three intent of HES. The curve reproduced here is the original 'palatable within reach' curve. This particular model applies only to upland forest.

ing that went into the original HES document seems to have been orientated towards game species. Transformation curves are then developed by aggregating this information for each of the wildlife habitat classes (see Fig. 5.2). Key habitat variables are then assigned weights. Weighted variables are summed to produce final HQI scores which can then be used in an accounting system for environmental assessment and management (Fig. 5.2).

HES is a controversial system in that a good deal of scepticism persists as to whether a generalized habitat quality model can be formulated without modelling predation, competition, and other interspecific relationships (van Horne, 1983). HES has been applied with some success to the evaluation of potential natural areas (Smith *et al.*, 1983) and to the evaluation of general habitat quality for management. A consensus indicates that HES must be re-formulated using carefully selected guild representatives, accounting for interspecific relationships, and thoroughly documenting the formulation of models. A project is currently underway in Tennessee to develop HES models and to apply them in a community classification system and in the selection of potential nature preserves in the coastal plain physiographic province of West Tennessee (Fenneman, 1948). A variety of statistical methods will be used to correlate vegetational and terrestrial vertebrate data so as to determine the usefulness of this approach.

(b) *Evaluating and managing for diversity*

A second level three approach consists of evaluation and management for diversity (Siderits and Radtke, 1977; Salwasser, Thomas and Samson, 1982). The National Forest Management Act of 1976 specifically requires that national forests be managed to protect and enhance diversity. This has been interpreted by the Forest Service to include primarily tree and terrestrial vertebrate species (Salwasser, Thomas and Samson, 1982). The diversity indices of theoretical ecologists such as the Shannon–Weaver Index (Shannon and Weaver, 1963) are not particularly useful for land use management decisions (Southwood, 1978a; Boyce, 1981; Salwasser, Thomas and Samson, 1982). Diversity-based evaluation and management systems therefore rely on a combination of diversity concepts including variety, relative abundance, and distribution of species, habitat, and structure (Siderits and Radtke, 1977; Salwasser, Thomas and Samson, 1982).

Two approaches to evaluating habitat diversity have been proposed recently. The USDI Western Energy and Land Use Team's Rapid Assessment Methodology (Asherin, Short and Roelle, 1979) uses multiple regression models and remotely sensed land cover data to develop predictive models for bird species diversity. Short's (1982) habitat gradient model correlates plant communities (defined at the level of dominant species in the cover) in any one potential natural vegetation unit (Küchler, 1964) with guilds defined in terms of plant community structure and based on expected wildlife use for breeding and feeding. In both cases, the researchers demonstrated a high correlation between their method and actual wildlife diversity.

Whittaker's (1972) three levels of diversity (alpha – within stand, beta – between stands, and gamma – total area diversity) have become well established as management goals in national forests, although enhancement and management for all three levels is not always consistent. For example, focusing on alpha diversity discriminates against species which are specialized for relatively lower diversity habitats (Faaborg, 1980), and increasing beta diversity by increasing ecotone area will eventually lead to lower gamma diversity as generalist species become dominant (Salwasser, Thomas and Samson, 1982).

Samson and Knopf (1982) and Salwasser, Thomas and Samson (1982) recommend a 'top–down' (or gamma–beta–alpha) approach, and suggest that the 'baseline diversity goal' should be to maintain viability of all species populations, i.e. 'the adaptive fitness of plant and animal populations and evolutionary fitness of species' (Salwasser, Thomas and Samson, 1982, cf. Frankel and Soulé, 1981). The Wildlife and Fish Habitat Relationships System (WFHRS) of the Forest Service is a combination of diversity-based and featured species-based approaches designed to implement this goal (Salwasser and Tappeiner, 1981; Nelson and Salwasser, 1982).

5.3.5 Validating species–habitat models

All of the evaluation methods currently used for management of wildlife are based on models, and success is dependent on model validity. Validation of species–habitat models can include many criteria, but precision (number of significant figures), accuracy (the degree to which the model output reflects the real world), and robustness (the degree to which model outputs are sensitive to minor variations in model inputs) are the most important (Marcot, Raphael and Berry, 1983). Validation includes both hypothesis validity (the extent to which the model's structure represents real world structure) and operational validity (the extent to which the model produces results which are correct). Models should always be tested with empirical data other than that used to develop them (Lancia et al., 1982; Salwasser et al., 1983). Models developed at level two must be tested at level one, and models developed at level three must be tested at level two at least, and preferably at level one, although there is some doubt whether models developed at level three can really be tested (van Horne, 1983).

5.4   CONCLUSIONS

In the United States, two national standards have emerged for wildlife data management: the Nature Conservancy's Natural Heritage Programme approach, and a procedure for standardizing species and habitat element definitions (Cushwa et al., 1980). The Nature Conservancy's element ranking procedure is the only accepted standard method for evaluating wildlife for preserve selection. There is a definite and demonstrable trend toward the use of species–habitat evaluation models based on hypotheses about species relationships to habitat parameters. There is also a growing interest in generalized habitat models such as HES.

Validation of models remains a serious problem and a massive challenge. Prospects for success in validation are limited by the problems of collecting very large volumes of data using standard approaches to ensure replicability and compatibility.

Defining habitat quality in terms of carrying capacity and validating species–habitat models by measuring population densities is problematic. Causal relationships between selected habitat characteristics and carrying capacity are extremely difficult to demonstrate (Marcot, Raphael and Berry, 1983). Stochastic and temporal effects are nearly impossible to account for adequately in experimental design. Van Horne (1983) postulates three kinds of habitat where density cannot be reliably correlated with habitat quality: (1) highly seasonal habitat, (2) habitat with a high degree of temporal unpredictability, and (3) habitat supporting a high degree of patchiness and, therefore, large numbers of generalist species. Furthermore, site tenacity and social interactions which prevent subdominant animals from using suitable habitat may result in habitat sinks, or areas of poor habitat and high density. Some populations may include immigrating individuals which do not subsequently reproduce. These variable situations may render measured densities useless for validating density-based habitat quality models (Hirsch et al., 1979; Lancia et al., 1982; Rice, Ohmart and Anderson, 1983; Salwasser et al., 1983; van Horne, 1983).

Information about the more specialized species (especially the rare ones) is often inadequate to develop models, and yet these are the very species for which habitat evaluation and management may be the most significant (Salwasser et al., 1983; van Horne, 1983). Synergistic effects between habitat components are especially difficult to detect and measure, and lower thresholds are generally unknown (allowing totally unsuitable habitat to score higher than zero).

Sampling errors can be caused by accidental species, under-sampling, area effects, artifacts of species behaviour (as with trap-shy, silent, or motionless species), and stochastic and temporal variation (Marcot,

Raphael and Berry, 1983; Rice, Ohmart and Anderson, 1983). Sampling must be 'intense enough to discern statistical differences in populations between areas within sampling periods and between sampling periods within areas' (Thomas, 1982). Sampling at this level is costly, and when Government agencies are making decisions, it is often impossible.

None of the species–habitat models deals with the minimum population sizes required to ensure viability of species (cf. Frankel and Soulé, 1981, and see discussion in Chapter 14). They ignore internal, genetic limitations associated with population viability (van Horne, 1983). As a result, an island of acceptable habitat (according to the model) may be incapable of sustaining the species for an extended period because it cannot support the minimum genetically viable population. Salwasser, Mealey and Johnson (1984) proposed a planning process for national forests which would include minimum population requirements.

Van Horne (1983) proposed that habitat quality at level one should be defined as the product of density, mean individual survival probability, and mean expectation of future offspring for residents in one area, as compared to other areas, for a given period of time. All of the areas used in the comparison must be within the distributional range of the species and sampled at the same time (cf. Salwasser and Laudenslayer, 1981).

In all of the evaluation methodologies, the tendency is to use the most advanced hypotheses in conjunction with incomplete inventories. The absence of good, replicable data in sufficient quantities and adequate formats is impeding implementation of many of these methods. Salwasser et al. (1983) suggest that the cost of incorrectly measuring the future populations of significant species, threatened or endangered species in particular, can be very high.

> If inadequate technology, natural random events, or lack of personnel or dollars prevent the acquisition of accurate and precise data on highly valued resources, then decisions should be conservative and include monitoring activities appropriate to the decision risks. (Salwasser et al., 1983)

There are significant uncertainties associated with the outcome of all evaluation and management decisions. Only an adequate programme of monitoring can provide the necessary feedback to correct inadequate evaluation and poor decisions. Monitoring must include both populations and habitats: monitoring only habitats ignores the effects of cryptic habitat factors on populations and the effects that wildlife have on habitats, and it does not contribute data to the validation of models.

In the United States, two distinct approaches to wildlife evaluation have emerged. One, developed to assist with the selection and design of nature preserves, begins with evaluations of species significance *per se* and couples these with databases of known locations of highly significant species. The other, developed to evaluate for management, begins with models designed to predict good habitat for particular species, for groups of species, or for all species (narrowly defined to include terrestrial vertebrates and, perhaps, vascular plants) and couples these with databases of attributes associated with optimum habitat *per se*. Neither approach adequately addresses the problem of minimum viable population size nor the relationship between this variable and the concepts of island biogeography (see Chapter 14). The potential for integrating the two approaches and incorporating genetic and biogeographic considerations is very large. This integration is currently hindered in the United States by a general lack of dialogue between the wildlife management and biological conservation disciplines.

5.5 SUMMARY

The chapter begins with a discussion of evaluating both wildlife and habitats for the purposes of establishing areas designated principally for the protection of that wildlife (i.e. nature preserves). Two national programmes – Research Natural Areas and National Natural Landmarks – provided some of the stimulus for many systems in individual states of the USA. The development towards a national standard incorporates five classes of species ranking, from 'critically endangered' to 'demonstrably secure', based on both global (worldwide) and state information.

The main part of the chapter explores the adaptation of methods, used in environmental impact assessment, to the evaluation of land for wildlife conservation. The analysis progresses from level one, which is species-orientated, to level three, which is habitat-orientated. Emphasis is given to a level two approach, the Habitat Evaluation Procedure (HEP), with an example of the yellow warbler in scrub woodlands.

The review is concluded by highlighting two fundamentally different approaches in the United States. One approach, developed to assist with the selection of design of nature preserves, begins with evaluations of species significance *per se*. The other, developed to evaluate for management potential, begins with models designed to predict suitable (or good) habitats for individual species, for groups of species, or for guilds of species.

CHAPTER 6

Selection of important areas for wildlife conservation in Great Britain: the Nature Conservancy Council's approach

DEREK A. RATCLIFFE

6.1 The practical background
6.2 Basic concepts
6.3 Classification of the field of variation in wildlife and survey of candidate sites
6.4 The basis for site selection
6.5 Assessment of individual sites
6.6 Selection of the national series of key sites
6.7 The changing requirement for site safeguard
6.8 International importance
6.9 Recent developments in site evaluation
6.10 Summary
Appendix

Wildlife Conservation Evaluation. Edited by Michael B. Usher.
Published in 1986 by Chapman and Hall Ltd, 11 New Fetter Lane,
London EC4P 4EE
© 1986 Chapman and Hall

6.1   THE PRACTICAL BACKGROUND

This chapter deals with evaluation which is integral to the selection, by the Nature Conservancy Council (NCC), of a countrywide (national) series of sites, each deserving protected status by some statutory designation. The National Nature Reserve (NNR) was formerly the main instrument of site safeguard available to the NCC but, since 1981, the Site of Special Scientific Interest (SSSI) has largely assumed this function. Although 'site' connotes a specific location, it has become equated with 'area' in meaning, and this usage is adopted officially in Great Britain (England, Scotland and Wales – Northern Ireland is excluded). There are four background considerations integral to this site safeguard strategy.

(1) Need for nature protection stems largely from the damaging effects of various human uses of land and water: agriculture, forestry, urban–industrial–transport development, water use, energy generation, mineral extraction, military training (sometimes) and certain forms of recreation.

(2) Protected areas, with nature conservation as the primary land use, can, conversely, cover only a small proportion of any country with a large human population.

(3) In a heavily populated and developed country such as Britain, only certain parts are now worth considering as protected areas, and even they are often highly fragmented and isolated. In districts where semi-natural habitat with its wildlife is still extensive and continuous, mainly in the uplands, the principle of (2) still limits the number and size of areas which can be protected.

(4) Most other modern uses of land and water will continue to have adverse effects on the remaining wildlife and physical features from a nature conservation viewpoint. Non-protected areas are nearly all at risk and further deterioration and loss from the remainder of the 'nature resource' can be assumed, as future development proceeds.

It follows that the areas chosen for protection should be the best and most important fraction of the remaining nature resource, and that they should represent as great a variety and quantity of this resource as society is able to allow, within a balance of competing land uses. Their purpose is to guarantee, so far as that is ever possible, the survival of the most highly valued element of our remaining wildlife and its habitat. The rest of the wildlife resource outside the protected areas is much larger in total, but its conservation has to be achieved by other means, notably advice to, and persuasion of, other land users.

6.2   BASIC CONCEPTS

Although the process of selecting nature reserves has evolved over several decades (Sheail, 1976), the classic major exposition of rationale and actual site proposals was by the Huxley Committee (1947). Their report enunciated the principle of choosing a national reserve series to represent reference points within the countrywide field of variation in ecosystem types, each with its distinctive assemblage of plant and animal communities and controlling environmental conditions. The actual number of sites obviously depended on the number of types distinguished within the associated classification of ecosystems. There was an implication that at least one example of every unit of the lowest rank in the classification should be represented and, preferably, by the best examples. The required number of sites also depended on how often combination of types occurred within a single site.

Later, there was an attempt to set minimum standards so that *all* sites above a prescribed level qualified for selection. This began with agreement between European wildfowl experts that areas supporting more than certain arbitrary percentages of global, continental or subcontinental wildfowl populations should be regarded as internationally important, requiring specific conservation measures. This approach gained wide acceptance and similar procedures identifying nationally important wildfowl sites followed. They were later applied to selection of areas important for other bird groups (see Chapter 11). Animal species or groups which are both conspicuous and highly aggregated lend themselves to this approach. Colonial breeding coastal birds are another good example. The application of this principle only to some but not all biological features is highly undesirable since it introduces imbalance in the selection process. It should be applied 'across the board', to other sites valued for their plant communities and/or flora, or for more dispersed animal populations, but the definition and general acceptance of critical minimum standards for these other biological features are more difficult to achieve, and so far only limited progress has been made (see Sections 6.6, 6.7 and 6.9).

Within these two selection principles, a more precise definition of criteria is needed. This was felt to be important in *A Nature Conservation Review* (NCR) (Ratcliffe, 1977) which was launched in 1966 to reappraise candidate sites for an expansion of the existing national reserve series. Further survey had shown that previous recommendations had missed numerous high quality sites, and also that the rate of attrition of wildlife and habitat through human impact had accelerated markedly since 1947. There was thus an urgent sense of need to justify the protection of a much larger number and extent of key sites so as to ensure the survival of an adequate proportion of this fast-dwindling heritage of nature. The previous standards of judgement in reserve selection were not in question, but a more closely argued rationale for a substantial expansion seemed desirable. The urge to try to identify some natural or absolute criteria for prescribing the choice of sites was strong, much as the original systematists such as Linnaeus sought 'natural' principles of classification. Yet such an approach is elusive and perhaps impossible, because of the large element of value judgement integral to the whole process. The rationale of selecting protected areas has to rest on understanding of their purpose, which is essentially to do with human concerns and hence values in this sense.

These judgements of value rest on two kinds of concern. First, there are the different kinds of interest which people find in the phenomena of nature. The Huxley Committee (1947) expressed the matter lucidly in stating the functions of a National Nature Reserve Series (para 50, quoted in part by Ratcliffe, 1977). The emphasis in that statement was on scientific study, and especially on the research potential of the reserves, to the benefit of both advancement of knowledge and practical ends. There was, however, also recognition of the importance of these areas for education, and for simple enjoyment and inspiration in 'the peaceful contemplation of nature'. The enormous growth in public concern for nature conservation during the intervening period has especially involved an increase in this last, more 'popular' end of the spectrum. The most fundamental values in nature conservation are expressed in this range of purpose – scientific, economic, educational, recreational and aesthetic.

The second set of values derives from the conservation principle that priority for protection needs to be given for essentially pragmatic reasons, namely, to those features and sites which:

(1) are intrinsically most fragile and sensitive to human impact;
(2) have already lost most ground through human impact;
(3) are predictably most vulnerable to further damage and loss through a combination of (1) above and probable expansion of human impact;
(4) would represent the greatest loss to nature conservation if they were damaged or destroyed;
(5) would be the most difficult to restore or re-create if they were damaged or destroyed.

In illustration, ecosystems totally unmodified by man (i.e. natural) satisfy all these conditions subsumed under the guiding principle of conservation pragmatism. Their important characteristics are: lack of anthropogenic features; usually but not always a considerable fragility in the face of human disturbance; actual depletion and scarcity in this country today; extreme vulnerability to further human impact; unusualness in species composition and community structure; and virtual irreplaceability in original form once lost or seriously damaged. This simple analysis points to six interrelated characteristics relevant to site assessment. Vulnerability is, arguably, a truly basic factor. Given that nature conservation is about the control and mitigation of human impact on nature, if there is no threat, there is no need for conservation action.

There is, in fact, a convergence between the two sets of values – intrinsic and practical conservation need – which is obviously apparent in (4) above. This interrelationship will become clearer in the discussion of criteria for evaluation whereby the best sites are identified (Section 6.5). The rationale and procedure of the NCR will be summarized, and subsequent developments in thinking on site selection in NCC and elsewhere will then be examined briefly.

6.3   CLASSIFICATION OF THE FIELD OF VARIATION IN WILDLIFE AND SURVEY OF CANDIDATE SITES

The reference framework for National Nature Reserve selection in 1947 was essentially Tansley's (1939) vegetation classification based on ecological formations embodying environmental and life-form characteristics as well as community floristics and species dominance. Formation sub-divisions (for example woodland into oakwood, etc.) equate with major habitat types for constituent species of plants and animals. Vegetation is a major determinant of animal distribution and gives a convenient reference system for most zoological features, though some of these have to be identified largely with physical habitats. Reserves were also sometimes selected for the importance of their plant and animal species, either individually or as groups, such as rarities and biogeographic assemblages.

Countrywide formation-based surveys were launched in 1966 to fill gaps in information about the occurrence of important sites and about the national range of variation in biota. Tansley's classification was revised to include new information, especially on northern and montane vegetation. A modified treatment using environmental and floristic features was adopted for each main formation: coastlands, woodlands, lowland grasslands, heaths and scrub, open waters, peatlands, and upland grasslands and heaths. Only that for the last group approached a formal phytosociological system, since relevant information for the others was too patchy. The framework dealt mainly with natural and semi-natural ecosystems, but artificial types were described briefly. It was supplemented by lists of species characteristic of each formation type for better known groups of flora and fauna.

Field recording of sites included main environmental and land use features, plant communities, species lists of vascular plants, bryophytes, lichens, vertebrates and a few invertebrate groups. While cover was patchy for some taxonomic groups and non-existent for others (algae, fungi and many lower animals), the aim was to locate and describe consistently as many as possible of the sites of known or potential importance during a two year period. All existing reserves and Sites of Special Scientific Interest were included in the survey.

6.4   THE BASIS FOR SITE SELECTION

Selection has involved the careful sifting out, through a comparison of similar types, of those sites which represent the best examples of reference points within the Great Britain field of ecological variation. Comparative evaluation can only be applied to sites of essentially similar character: an oakwood cannot be compared with a saltmarsh nor even with an ashwood. The vegetation classification has revealed that each of these main ecosystem types shows important variations, especially related to geographical climatic gradients. The oakwoods become increasingly rich in ferns, mosses, liverworts and lichens in a westerly direction, as rainfall and humidity rise. And the zone of western oakwoods also shows a south to north gradient of variation within this oceanic flora as mean temperature decreases. Selection should aim to represent adequately these two directions of variation by a countrywide network of oakwood sites.

The number of sites needed to represent such gradients depends on the actual amount of variation and on the classification of this variation (cf. the taxonomic problem of 'lumping' versus 'splitting'). As a working principle, latitudinal or longitudinal distance of 100 km is usually sufficient to produce perceptible differences in ecological/biological phenomena attributable to regional climatic differences. Spacing between contiguous sites representing a climatic gradient should thus not exceed this distance. Any further local variations attributable to topography, geology, soils and management also need to be included, so that clusters of sites, or large and varied sites, are often needed to represent an ecosystem type within each climatic region, for example the Snowdonia oakwoods, Lake District oakwoods, and so on. Although formal biogeographic regions were not identified, the Nature Conservancy's administrative regions were used as a broad geographical basis for selecting sites. The emphasis on representation against regional climatic gradients necessarily gives an overall biogeographic perspective.

Formation survey teams had discretion to decide what degree of variation around each ecosystem type should be represented. When more than one possible candidate site for each regional variant was found, comparative evaluation was needed to identify the best. It was also necessary to ensure that only sites above a certain minimum quality were selected, so that evaluation against a standard was involved in each case.

6.5   ASSESSMENT OF INDIVIDUAL SITES

This is the stage in site selection to which the term 'evaluation' most particularly applies. Sites are compared with other similar sites and/or against accepted standards for the particular type of site. Criteria for evaluation used in the NCR were: size (extent), diversity, naturalness, rarity, fragility, typicalness, recorded history, position in an ecological/geographical unit, potential value and intrinsic appeal. They have been discussed at length by Ratcliffe (1971, 1977), and present reference will be confined to the most relevant points. Each criterion represents some biological attribute or aspect to which one or other of the values discussed in Section 6.2 is attached, to give a 'biggest equals best' yardstick of quality, for example diversity as a scientific attribute is value-free, but in site evaluation the greater the diversity the better.

Size (measured as extent) tends to be most important for the highly fragmented, 'island' types of ecosystem. In the extensive and continuous expanses of upland it does not apply to definition of the whole site, but may apply to scarce upland habitats, such as calcareous outcrops. Size (measured as number or density) is also the main criterion for assessing highly aggregated species' populations. *Diversity*, in the sense of variety or richness, is an important but often misunderstood criterion. It is an abuse of the concept to suggest that almost any kind of human disturbance (such as the planting of alien conifers on open moorland) enhances site value by increasing diversity. The control or mitigation of human disturbance is what nature conservation is all about, and the more that such intrusion is recent (i.e. post-1900) and introduces alien features and species, the more likely it is to be damaging rather than enhancing. While the number of species is one useful measure of diversity, it is thus the number and abundance of 'indicator' species characteristic of the particular ecosystem type, which is especially relevant to this criterion. Diversity has a practical advantage in that it also tends to express the concentration of interest within a site: the greater the concentration, the greater the return per unit area protected.

Naturalness is a relative term in Britain, where over 70% of the land is totally transformed from its original state. Wildlife interest is mainly in the remaining 30% which has extensive semi-natural vegetation in the sense of Tansley (1939). Wildlife value rises with naturalness but, since many important ecosystems are partly man-made, antiquity and lack of recent disturbance are often the relevant factors. The Norfolk Broads were created through mediaeval peat digging but developed an almost completely natural character. Recreatability tends to be inversely related to naturalness. *Rarity* nearly always has an historical context. Rare species tend to be either recent arrivals or dwindling relics of once larger populations. During the last two millennia, human impact has replaced climatic change as the main factor causing fragmentation and restriction of range for both communities and species. Rarity attracts interest in itself and is also a measure of extinction proneness, and hence need for conservation (see also Section 6.9). *Fragility* also relates to practical conservation: the most fragile features are those most likely to be lost and thus to need protection. Fragility equates with intrinsic sensitivity to damage: combined with threat of actual damage it becomes an expression of vulnerability.

These are the most important criteria, though fragility tends to have a 'generic' rather than a 'specific' quality, i.e. nearly all ecosystems rated highly on other grounds are especially fragile, but those of comparable type have similar levels of fragility. While fragility may in practice be largely subsumed under other criteria (naturalness and rarity), it is valuable in emphasizing that sensitivity to human disturbance is a fundamental measure of nature conservation concern, and closely related to priority for action. The other criteria are less important and usually give secondary enhancement of value to ratings based on the primary criteria. Typicalness is only occasionally used to choose a site which fills a gap in the representation of ordinary features. Position in an ecological/geographical unit refers mainly to situations where good examples of more than one formation occur within a single area, i.e. it is the enhanced value of juxtaposition, for example, when a wood is connected to a mountain or lake. In practice it is a reminder to avoid the strictly compartmental approach to which selection for particular formations can lead. Recorded history is a 'plus' in some sites which happen already to have been intensively studied. Potential value belongs mainly to the field of habitat re-creation, but may apply when no good examples of an ecosystem have survived (see Chapter 3).

Some treatments recognize 'representativeness' as a criterion, but the possibilities of semantic/conceptual confusion over the use of this word require that its meaning be defined by the user. Some writers clearly equate it with 'typicalness' in the sense of the NCR and denoting the idea of average or ordinary. In countries where little conservation survey or selection has previously been done (see Chapter 2), and where there are vast areas of species-rich, natural biotope, such as forest, the concept of selecting a series of examples representing (in the above sense) the whole field of variation has obvious relevance. To go beyond this and seek perfection by trying to find all the outstanding examples would be a daunting prospect. It might also be misconceived, because notions about the existence of areas of outstanding quality tend to apply to developed countries where natural ecosystems are now at a premium. In Britain, where so much is known about the field of ecosystem variation, and where human impact has produced such a wide range in nature conservation value within this field, it is both feasible and desirable to focus more particularly on selecting the best sites still extant. Experience in the NCR showed that the best examples of ecosystems usually contained typical or average features anyway, so that they can, in many instances, be regarded as representative as well. Once any site is identified as belonging to a particular ecosystem type (or types) it can be said to represent that type. But when the selector is trying to find the closest match between actual sites and the abstract model of a particular ecosystem type, in possession of characteristic communities and species, he or she is, in a sense, seeking the *best* representative of this type.

Representation of the field of variation is one of the two basic principles of site selection (p. 137) and it is thus debatable whether 'representativeness' can be used as a criterion in its own right or whether it necessarily subsumes other criteria, such as aspects of diversity. In the process of matching actual sites against abstract models of ecosystem reference points, it is difficult to avoid comparisons based on attributes such as number and extent/abundance of communities and species, so that aspects of the criteria of diversity and size are being applied. On the other hand, because the average or typical often runs counter to diversity and even more so to the criterion of rarity, *typicalness* is regarded as a valid criterion in its own right.

This discussion of representativeness may seem academic, but it is intended to show the need for people to clarify their use of the concept.

6.6   SELECTION OF THE NATIONAL SERIES OF KEY SITES

In selection of key sites for the NCR, coastal sites other than those of mainly ornithological interest were chosen according to the scoring system developed by D. S. Ranwell (personal communication) (see Appendix to this chapter). For other formations, choices involving multiple criteria were made by a simple process of comparison without recourse to a formal scoring procedure. Each candidate site was compared with those of similar character and quality already known to represent good examples of a particular ecosystem type. Such comparisons were based on those criteria regarded as especially relevant. Scores for different criteria, or combinations of them, were calculated only in so far as this was necessary to establish differences that were not abundantly clear from field inspection of the sites. When all the sites representing the particular reference point in the ecosystem classification had been considered, a short-list was produced which aimed to identify the best and the next-best site of the type. During this sifting process, it was also judged whether a particular site was sufficiently different from its nearest equivalents to justify being regarded as an additional reference point in the classification framework, i.e. a new type requiring separate representation. Sites which had no 'competitors' were assessed for

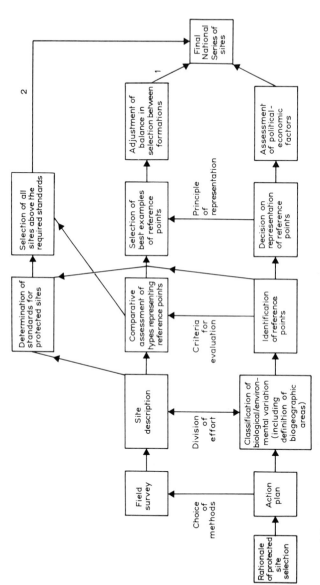

Fig. 6.1 Diagram of the sequential stages in the process of site evaluation and selection. This is a conceptual model: some stages are not necessarily separated in a time sequence, but tend to proceed *pari passu*, especially when the programme does not start *de novo* but builds on earlier approaches. On the right two strategies are recognized. Strategy 1 is analogous to the flower show procedure: prior awareness of gardeners' interest in flowers leads to definition of a number of classes, within which competitive entries are invited; within each class, entries are judged according to agreed criteria of quality, and the top few arranged in order of merit; the number of prizes actually awarded would depend on the scale of support for the show, including not only the number of entries and classes, but also the funds available. Strategy 2 is analogous to an examination to select from a field of candidates; standards are set by the examining body and all candidates exceeding the arbitrary pass mark are accepted; but the pass mark itself can be adjusted upwards or downwards as circumstances, especially financial, require or allow. In practice, strategy 2 has to be related to strategy 1 so as to allow for a sliding scale of national value to be used: otherwise, for example, nearly all the upland sites chosen would be in the Scottish Highlands.

quality according to their possession of a certain minimum of the distinctive features of their ecosystem type.

In the developed lowlands, sites tend to be readily delimited islands representing highly fragmented formations, so that choice is narrowed at the outset. Initial site identification is much more problematical in the mountains and moorlands of the north and west, with their large, continuous expanses of semi-natural vegetation. Here, the difficulty is in choosing areas of appropriate size – neither too large nor too small – when the drawing of boundaries is so often an extremely arbitrary business. Upland sites should follow the general principle of encompassing a whole topographic unit from the highest summits to the lowest surrounding valleys, with all its aspects, and containing a complete example of the particular ecosystem complex, including its largest predators. Boundary definition is quite basic to site comparison and evaluation (see Ratcliffe, 1977, vol. 1, p. 16).

Selection of the series in Great Britain is a cumulative process, building up from the regional choice to give a countrywide cover within each ecological formation, and then amalgamating totals for all the formations. It is schematized in Fig. 6.1 and the results, updated to 1985, are summarized in Table 6.1. A potential defect is that imbalance in selection 'intensity' may result from the operation of separate survey/ evaluation teams for the various formations. This could happen if the reference classification for one formation were more finely (or more coarsely) subdivided in relation to others, or if survey teams differed markedly as between a conservative or a liberal interpretation of the need for representing reference points. To reduce these risks, the NCR had a Scientific Assessor whose function was to ensure that common standards were applied as far as possible.

The Great Britain reference classification of biological features is all-important in setting a basic minimum to the number of sites needed to represent the field of variation. The operation of the principle of type representation in the Huxley Report set *de facto* national standards through the particular qualities of the sites actually chosen. The subsequent elaboration of the reference classification to a more comprehensive system then led to a search for other sites of comparable quality to fill the gaps. For instance, in East Anglia, parts of the Norfolk Broads were already reserves, as classic areas of eutrophic fen. Wicken and Woodwalton Fens, though having common features with the Broads, were also regarded as nationally important remnants of the original fenland. The NCR upheld the original evaluation of these sites, but recognized the additional national importance of certain other extensive reedswamps with outstanding ornithological interest (Minsmere and Walberswick Marshes) and of some of the hitherto neglected valley fens of Norfolk. These valley fen sites represent important variants within

Table 6.1 Distribution and extent of key (NCR) sites in Great Britain in 1985*

Country	Coastlands	Woodlands	Lowland grasslands and heaths	Open waters	Peatlands	Upland grasslands and heaths	Total† numbers of sites	Total area (thousands of hectares)
			Number of sites					
England	59	148	153	50	62	21	432	382.3
Wales	15	29	12	12	14	15	83	101.9
Scotland	64	65	14	47	50	67	269	476.6
Great Britain Total	138	242	179	109	126	103	784	959.8
Total area (thousands of hectares)	289.2	67.6	67.7	29.0	68.4	437.9	—	959.8

* The column for total number of sites allows for overlap/duplication in those cases where a single site contains different formations which are listed separately in the appropriate columns. The figures for numbers and areas of sites in the last two columns are thus less than the totals of those given under the separate formations.

† The selection of key sites follows geographical bias in the representation of formations, for example lowland grasslands and heaths belong mainly to southern and eastern England, whilst uplands are most extensive in Scotland. Woodlands and lowland grasslands and heaths are the most fragmented of the formations and are represented by a larger number of sites than coastlands or uplands, though their total area is much smaller. The area of river sites has not been measured and is excluded.

this class of peatland, but the reedswamps were added mainly because of the national rarity and the richness of this wildlife habitat, illustrating application of the second selection principle – of choosing all sites above a certain level of quality.

For some reference points (for example coastal bird habitats) it has been possible to define certain lower limits of quality for national importance. For the rest, such decision has rested on the judgement of assessors whose knowledge of the particular biological features is countrywide, and whose experience stems from close involvement in the relevant part of the community of scientists and natural historians. This might seem an especially subjective procedure but it is validated by the record of remarkably little disagreement on the importance of particular sites within this concerned community. The original choice of sites by the Huxley Committee has stood up extremely well to reappraisal, and is generally regarded as having set a high standard in selection. In the NCR standards were not lowered to deal with regional gaps, for example the Southern Uplands and especially their eastern half are now so deficient in ancient semi-natural woodlands that this formation was under-represented by reference to the national series for this formation.

A problem stemming from the formation-based approach to evaluation is that it easily leads to the overlooking of sites which, although not outstanding in terms of any one formation, are important for their combination of different ecosystem types, for example combination of woodland, scrub, heath, grassland and wetland. The operational procedures must allow for this, and in the NCR it was the job of the Scientific Assessor to ensure that such composite sites did not slip through the selection net. Many sites with nationally important examples of one formation also contain or adjoin less important examples of other types, and it was often felt desirable to include these within the area chosen, for their contribution to the ecological diversity of the whole (the criterion of position within an ecological/geographical unit, see p. 142).

The other aspect of selection which requires careful thought in terms of the national series is the question of the individual extent of sites and, hence, their total area as a proportion of our land surface. In the highly developed lowlands many nationally important sites are critically small, in terms of vulnerability to edge effects and the population viability of rarer species (see Chapter 14). The reappraisal thus considered opportunities for enlarging the area of some such sites and, for example, recommended the addition of other parts of the Norfolk Broads to the existing reserves. In the uplands and, to some extent, the coastlands, there is both need and opportunity to choose much larger sites (see earlier in this section). This is where it becomes all too clear that the selection process is not simply a technical, quasi-scientific matter, but one with a background of resource and political factors. However rational and rigorous the technical criteria and standards for drawing up the national list, the number and extent of sites which need to be safeguarded is affected by the pressure of other land and resource use developments, while the availability of powers and funds for achieving their conservation is determined by the balance of public interest. The various elements in this complex equation also change in time. While it is desirable to keep the technical and non-scientific aspects of selection as separate as possible, practitioner organizations such as the NCC cannot allow academic considerations to override practical realities.

6.7   THE CHANGING REQUIREMENT FOR SITE SAFEGUARD

When the NCR was launched, in 1966, the predominant selection principle was still the conservative one of choosing the best examples to represent reference points in the ecosystem classification. Allied to this was the experience that the best sites were not always available when negotiations for reserve acquisition were opened, or that they sometimes suffered damage or even destruction before safeguard action could be taken. It was therefore felt prudent to identify, whenever possible, an alternative example to each of the best sites. Sometimes, no comparable sites existed but, conversely, it was quite often found that there were equivalent examples of virtually identical quality, so that it was difficult to choose the best. And since no two sites are identical in character, many of these 'next best' examples could also be regarded as representing an intermediate type between two reference points in the classification. The best sites were designated grade 1 and the alternatives or additions as grade 2, though all grade 2 sites also satisfied the standards of national importance. The NCC itself later adopted the policy of treating both grade 1 and 2 sites as equally deserving of protection, when threat or opportunity for acquisition arose. The distinction has been abandoned, and all the NCR sites are now regarded as equivalent to existing NNRs in quality, and as requiring the same level of safeguard: they are termed key sites.

The interface between the selection and actual protection of important sites was further changed by the Wildlife and Countryside Act, 1981. By this time, concerned public opinion was thoroughly aroused by the alarming rate and scale of loss of semi-natural habitats and their wildlife, especially through agricultural and forestry developments. It was strongly felt that the number and extent of protected sites had to be substantially enlarged, to compensate for these losses and to provide a bulwark against further inroads. Government accordingly accepted that, in return for the principle of financial compensation to owners and occupiers for development profit foregone, there should be a statutory mechanism to allow the defence of any SSSI against damage by any

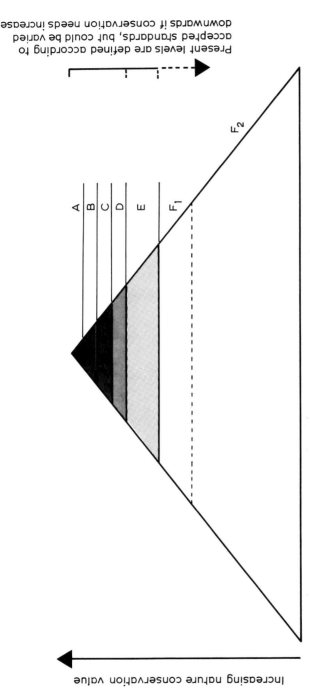

Fig. 6.2 The designation of categories of protected area for wildlife. The pyramid represents the 'nature resource' of Great Britain: categories are not intended to be to a precise scale. The following slices of the pyramid are recognized: World Heritage Sites (A); internationally important sites (A and B); grade 1 NCR sites (A, B and C); grade 2 NCR sites (D); all NCR sites, that are equivalent to actual or potential National Nature Reserves (A–D); Sites of Special Scientific Interest, covering about 10% of the total land surface (A–E); other reserves and protected areas not scheduled as SSSIs (F_1); and the wider environment of Great Britain not subject to special designation for wildlife (F_2). The base of the pyramid represents the most developed urban/industrial land with minimal wildlife value. This illustration is modified from Fig. 12 in Moore (1977).

harmful activity. The SSSIs form a Great Britain network of over 3700 biological sites presently covering about 6% of the country's land area; they include all the NCR sites but also a larger number originally chosen for their regional importance. NCC's selection process was reviewed in 1979 when it was recognized that all SSSIs had a national function in forming an interrelated countrywide network of protected areas.

The 1981 Act gives all SSSIs equal status, reinforcing the national significance of each. This strengthening of site safeguard measures shows clearly that the conservation requirement is not limited by any absolute measure of intrinsic quality, but is modified by the prevailing balance of public opinion on these matters. This public opinion is itself influenced by such factors as the continuing scale of both habitat and species loss. The decisions on number and extent of sites needing safeguard thus depend on political judgements as well as on definition of a scientific datum of quality; but the latter should be a stable bench-mark until or unless scientific opinion itself undergoes substantial shift. The process of site/reserve evaluation and selection has come to embody certain long-accepted and irreducible standards of technical procedure, and is, to this extent, 'scientific'. But the concept and standards of national importance in nature conservation have evolved steadily through a shifting balance between this scientific component and value judgements of a different kind (the second class of values discussed on p. 138). The increased scarcity value of the resource of nature has made it appropriate to relax selection limits so as to extend the label of national conservation quality to a larger number of sites as a compensatory measure.

The strategy for relating scientific evaluation to conservation practice has thus been to place sites in layers of importance, so that the priorities for safeguard accorded to the top layer can be extended downwards to other layers as changing circumstances require (see Fig. 6.2). The procedure for selecting the other SSSIs is thus no different in principle from that applied to choosing NCR sites, though it moves farther in the direction of defining minimum standards, above which all sites automatically qualify for selection. The practical outcome of the 1981 Act for the NCC is that the NCR sites are still regarded as those for which the highest safeguards should be sought, through NNR designation, whilst they and the rest of the SSSI series will be defended *ad hoc* against damaging activities through the 1981 legal provisions.

6.8 INTERNATIONAL IMPORTANCE

The biological distinctiveness of Britain results from its insular, highly oceanic position at the north-west fringe of the continental European land mass. Island isolation and lack of summer warmth account largely for a flora and fauna representing a rather impoverished version of that in mainland Europe, but isolation is too recent for any significant degree of endemism to have developed. Britain's position in relation to Quaternary climatic shifts has, however, given a unique combination of biogeographical elements, often within close proximity, such as the blend of southern Atlantic, Lusitanian, Mediterranean, Arctic, Alpine, and Arctic–alpine (Matthews, 1955) species. The juxtaposed survival of such contrasting groups has also depended on the thermo-stabilizing effect of the oceanic position, but perhaps the most conspicuous vegetational features are those associated with the high humidity of this Atlantic climate. In the extreme west, the hyper-oceanic conditions support a rich and varied flora of ferns, bryophytes and lichens, and these plants become an important component of many plant communities. Some vascular plants with markedly Atlantic distribution also form community dominants to a degree unusual in continental Europe, for example ash (*Fraxinus excelsior*), ling (*Calluna vulgaris*), bluebell (*Endymion non-scriptus*), the three gorse species (*Ulex europaeus, U. gallii, U. minor*) and heath rush (*Juncus squarrosus*). Moss heath, especially that dominated by *Rhacomitrium lanuginosum*, is an extensive and distinctive feature in parts of the north and west.

The cool, wet climate favours soil leaching and acidic peat accumulation, so that there is an unusual extent of acidic grassland and heath, and even more of blanket bog with extensive cover of *Sphagnum*. Deforestation within the submontane zone of the uplands has created a large extent of anthropogenic grasslands and dwarf-shrub heaths, some of them with a floristic composition virtually peculiar to Britain.

The birds of Britain are of outstanding importance. Mildness of winter climate provides relative refuge, so that there are important wintering areas for many migratory species. The long coast-line with large estuaries features important holding grounds for wintering and passage wildfowl and wader populations, while the seacliffs, sand dunes, shingle beaches and salt marshes provide numerous breeding places for a diverse coastal bird breeding population, including many species with an extremely restricted world distribution. Although the variety of breeding raptors is rather limited, some species occur at unusually high density. Many birds of submontane moorland also have large populations, and include various Boreal/Subarctic species otherwise scarce in, or absent from, Europe south of the Baltic.

Extra weight has been given to the representation of these features in the series of key sites. Where international criteria of importance exist, as they do for migratory waterfowl populations, they have been applied. For other features an attempt has been made to identify international importance according to views expressed by European and other overseas conservationists on the importance of British vegetation, species' aggregation and populations. The category of international importance was treated in the NCR as a bonus quality, grade 1*. Subsequently, international conventions and programmes requiring the protection of sites for specific interests have come into force and provided their own criteria for selection (Ramsar Convention on conservation of wetlands, Berne Convention on threatened species and their habitats, UNESCO Man and the Biosphere reserves and EEC Birds Directive and its provisions for special protection areas). At present several of the most outstanding of the British key sites are being proposed for inclusion within the list of World Heritage Convention sites, though this designation also takes account of exceptional scenic, landscape and historical interest, and has required the extension of boundaries beyond those drawn for nature conservation interest.

6.9   RECENT DEVELOPMENTS IN SITE EVALUATION

These recent developments include especially the quantification of attributes and criteria in scoring systems, to make the measurement of site quality more objective and scientifically respectable. Spellerberg (1981) and Margules and Usher (1981) have reviewed the literature, and numerical scoring systems are dealt with elsewhere in this book (for example see Chapters 3, 5 and 7). Compared with a purely subjective approach they provide a more consistent procedure, repeatable by different people to give similar results, and are therefore more robust to criticism. Their five main shortcomings are:

(1) when survey data for sites being compared are uneven in cover of attributes, biassed assessment results unless allowance is made;
(2) several main criteria lack independence of each other (see Chapter 1);
(3) scoring procedure and weighting between criteria themselves have an element of arbitrariness and subjectivity;
(4) they do not answer the frequent need for a measure of *difference*, as well as relative merit, between compared sites, in selection along a gradient of variation (with the principle of selection to represent reference points, it is important to know when two related sites are sufficiently different to justify choosing both);
(5) some of the values (as distinct from attributes) built into criteria are exceedingly difficult to quantify, and scoring systems in themselves do not contribute to the judgement of standards of nature conservation value.

Indeed, scoring systems tend at times to invest the whole process of selection with the appearance of an objectivity that it can never possess. Their chief merit is in arranging similar sites in order of quality by a systematic and standard procedure. This obviously assists the selection process, but it does little to help decisions needed to determine the size and adequacy of a whole series of protected areas.

One of the most useful quantitative treatments is that of rarity in vascular plants, made possible by comprehensive data on distribution in the *Atlas of the British Flora* (Perring and Walters, 1962). This mapping scheme recorded species presence in 10 × 10 km grid squares, and those occuring in 15 or fewer squares were regarded as 'rare' species: 317 in all representing around 18% of the presumed native flora. In presenting an account of these 317 species in the first *British Red Data Book* (Perring and Farrell, 1983) other information was added on magnitude of decline, number of extant localities, attractiveness to collectors, representation on nature reserves, remoteness of populations, and accessibility. By scoring these factors an aggregate threat number was calculated for each species, to show its degree of endangerment and hence need for conservation measures. This evaluation system allows rare species to be treated on their own (for example in legislation) but also contributes towards numerical site assessment.

Helliwell (1973) has attempted to evaluate wildlife in monetary terms, so that its conservation can be placed within a conventional economic framework. Quite apart from methodological considerations, there must be strong reservations about such an approach. The great strength of nature conservation is that it deals with things which are beyond pricing as market place commodities. Their ultimate values are transcendental and cannot be compared with anything beyond themselves. To attempt otherwise is, literally, to de-nature them – to the level of material values and the world of artefacts.

Moore (1982) has refined the NCC procedure for selecting SSSIs, especially by giving guidance on minimum acceptable standards. The first step was the subdivision of Britain into biogeographical districts of approximately 2500 km² each. Within each such district, sites classified into main 'habitats' (ecosystem types), according to the NCR system, were then assessed by NCR criteria; the aim being to select the best example of each habitat per district. For fragmented habitats, selection of up to five of the largest remaining examples was suggested, larger numbers being considered according to a scale of rarity and threat. Minimum extents of habitat qualifying for selection were identified. Definitive guidelines were also given on the need to select sites additionally for species conservation: rare and endangered species, largest populations of local species, outstanding species assemblages or large aggregations of colonial species, and those with miscellaneous scientific importance, for example inland colonies of coastal breeding birds, good examples of certain land use effects. It was stressed that inadequate knowledge of the status of the less familiar groups of lower plants and animals limits assessment at present, and indicates the need for subsequent additions or reassessments.

This approach rationalizes the selection of a countrywide geographical/ecological network of sites, in which each site is perceived to have both regional and national significance and can be defended on both counts. It makes an explicit statement of relationship between SSSIs and the wider countryside, making clear that selection is largely restricted to the 30% of Britain still consisting of semi-natural habitat, and that representation is related to threat to each habitat. By setting standards through cut-off levels, site selection is made more consistent nationally. The exercise of discretion and judgement still remains, but NCC staff have welcomed N. W. Moore's internal guidelines as a valuable step in standardizing SSSI selection. His approach is an important recent advance in the subject, steering a middle course between the pursuit of objectivity and quantification as an intellectual goal and the adoption of a pragmatic and flexible system closer to the hard realities of conservation experience.

The applications of island biogeography to site selection seem to have produced conflicting and confusing results, with little if anything of practical relevance in real situations (see the discussions in Chapter 14). Related concepts also remain too indeterminate to be of practical utility, for example that protected areas act as wildlife reservoirs, helping to maintain species populations in other reserves or in the wider countryside. Species are so variable in their dispersal powers and needs for population viability that general rules are elusive. Subsequent management seems often to be a more crucial factor for species survival than the exercise of initial options over factors such as size and spacing of sites. Similarly, the admonitions of Frankel and Soulé (1981) on the need to select large reserves with populations of large animals and plants of at least 50 individuals, to avoid problems of 'inbreeding depression', are sometimes an unrealistic counsel of perfection. For species such as our larger raptors, a reserve holding 25 breeding pairs would have to equal a national park in extent. This nevertheless emphasizes the need for conservation measures for such species outside the reserves. Concern to ensure population viability can also overlook the point that this is often only a bare minimum requirement for survival, which many conservationists will regard as less than adequate in terms of desirable wildlife abundance.

The formalization and quantification of an evaluation and selection system are hindered by the present lack of any generally agreed classification of vegetation in Britain. The NCC has supported the production of a National Vegetation Classification (NVC), which should be available in 1986. The NVC should provide a basis for evaluation, especially by defining more comprehensively the field of variation requiring representation in the protected sites, and also in indicating the minimum standards of sites by providing more detailed measures of diversity in both communities and species. The search for greater rigour in evaluation should continue, but with a certain humility and caution, for this is a field in which objectivity is necessarily an elusive goal, if not a Holy Grail. The greatest of the shortcomings in site selection is not the limitation of procedure but of resources: the simple point that we have still not surveyed the whole of Britain in sufficient detail to have been able to identify all the sites of NNR and SSSI quality, even by the subjective approach described. The number of NCR sites now stands at 784 and will be likely to reach 800 when survey and evaluation are completed. Several hundred other prospective biological SSSIs are known, and the eventual total will probably approach 5000 sites covering possibly up to 10% of Britain's surface.

6.10   SUMMARY

This chapter describes the Nature Conservancy Council's rationale of evaluation for selection of a countrywide series of protected areas. The two main principles of selection are (i) representation of reference points within a classification of wildlife features, and (ii) definition of minimum standards for protected area quality. Both approaches require the application of criteria incorporating value judgements about the intrinsic interest of wildlife features to people and also their vulnerability to damage and loss.

Survey has provided a classification of the national range of variation in semi-natural/natural vegetation and fauna. This was based primarily on vegetational/physiographic formations with regional variations according to main geographical climatic gradients, and more local variations depending on smaller scale differences in climate, topography, edaphic and anthropogenic factors. Within the reference system, similar sites were compared using appropriate groupings of criteria, which were applied to quantitative measures of attributes when these were available. The best examples of reference points were chosen for the Great Britain key site series: in 1985 there were 784 sites (approximately 959 800 ha or 4.2% of Britain) of National Nature Reserve quality, including 200 existing NNRs (Table 6.1). Emphasis was given in selection to features of international importance. The Wildlife and Countryside Act 1981 effectively extended the label of national (Great Britain) importance to include all Sites of Special Scientific Interest (SSSI), at present numbering over 3700 (biological sites) and covering about 6% of Britain's land area.

Recent developments include elaboration of scoring systems for evaluation. The NCC has developed a semi-quantitative system for selection of SSSIs, and a numerical evaluation of conservation needs of rare vascular plant species has been made. Greater standardization of evaluation procedure is a desirable goal, but is limited by an inherent element of subjectivity in the values themselves. Selection decisions over numbers of important sites are related as much to political and resource factors as to underlying scientific principles.

APPENDIX

Details of Ranwell's semi-quantitative index for comparative biological value of sites.

This index was developed in 1969 for the assessment of coastal habitats, and hence it relates to sand and mudflats, salt marsh, sand dune, shingle beach, and cliffs. The Comparative Biological Value (CBV) Index incorporates nine aspects of biological 'value' as follows:

S	– Size
Ph	– Physico-chemical features
O	– Optimum populations
D	– Diversity
G	– Geographical limits
P	– Purity
E	– Education and Research use
C	– Combinatory value
X	– Unknown factors

Size

Sand/mudflats

	Area(ha)		Rating(S)
	≥	4000	5
1600	–	3999	4
800	–	1599	3
400	–	799	2
	<	400	1

Salt marsh and sand dunes

	(Area(ha)		Rating(S)
	≥	800	5
400	–	799	4
200	–	399	3
80	–	199	2
	<	80	1

Vegetated shingle beach

	Area(ha)		Rating(S)
	≥	200	5
80	–	199	4
40	–	79	3
20	–	39	2
	<	20	1

Cliffs

Undisturbed run of cliff length in which site is situation (km)			Rating(S)
	≥	80	5
40	–	79	4
24	–	39	3
8	–	23	2
	<	8	1

Physico-chemical features

Type	Rating (Ph)
High speciality	3
Some special features	2
Type example	1

This is a qualitative rating based on the range of knowledge available about each habitat group. It includes both physiological features and physico-chemical soil features.

Optimum populations

Population type	Rating(O)
Best populations of one or more local species	4
Large populations of local species	3
Large population of common species and/or small population of local species	2
Representative populations	1

In general only higher plants and animals are considered and 'local' species are those restricted geographically to regions (as opposed to places) in Britain, and restricted numerically, but not to the point of imminent extinction.

Diversity

Type	Rating(D)
Outstanding diversity	3
High diversity	2
Species range small	1

This qualitative estimate is specific to each habitat type and allows for the fact that sand dunes will inevitably have a much more diverse fauna and flora than salt marshes. It could be quantified as information on sites increases.

Geographical limits

Sand/mudflats, and cliff
vegetated shingle

Estimate	Rating (G)
Many species at limits	3
Some species at limits	2
Few or no species at limits	1

Salt marsh

Number of flowering plant species at limits	Rating (G)
5	3
3 or 4	2
1 or 2	1

Sand dunes

Number of flowering plant species at limits	Rating (G)
> 15	3
5 – 14	2
< 5	1

In the case of mud and sand flats and shingle beach, qualitative estimates are made in the light of existing knowledge of sites known to be meeting places of regionally distributed faunas and floras. An analysis of the limits of species of the coast derived from Perring and Walters (1962) allows quantitative assessments to be made for salt marshes and sand dunes.

Purity

Type	Rating (P)
Little disturbance	3
Moderate disturbance	2
Much ground disturbed or polluted	1

This is purely qualitative and based on evidence of human usage and disturbance of the site.

Education and Research use

Type	Rating (E)
Much used	3
Some use	2
Potential use	1

Combinatory value

Category	Rating (C)
Adjacent to another habitat site or sites of likely national value	3
Adjacent to another habitat site or sites of likely regional value	2
Adjacent to another coast habitat type not spoilt by development	1
Surrounded by developed coastline	0

Unknown factors

The 'X' factor is only applied where map and documentary evidence are so completely inadequate that a reasonable estimate of rating for the various factors cannot be reached.

CBV Index rating

This is obtained by adding up the total obtained for each of the 9 factors considered (maximum 27), thus

$$CBV = S + Ph + O + D + G + P + E + C + X.$$

It is important to keep in mind that the CBV Index is no more than a useful sorting mechanism and that final selection of a series of sites in any particlar habitat must take into account their proximity to each other and their representativeness in relation to the coastline as a whole.

CHAPTER 7

Wildlife conservation evaluation in the Netherlands: a controversial issue in a small country

S. W. FLORIS VAN DER PLOEG

7.1 Introduction
7.2 Concise history of conservation evaluation in the Netherlands
    7.2.1 The pioneers, 1965–1970
    7.2.2 Zenith of evaluation studies, 1970–1976
    7.2.3 Criticism and doubt, 1975–1981
    7.2.4 The present, from 1980 onwards
7.3 Some criteria used in the Netherlands
    7.3.1 Rarity
    7.3.2 Representativeness and Authenticity
    7.3.3 Replaceability
    7.3.4 Naturalness
    7.3.5 Miscellaneous
7.4 Evaluation procedures
    7.4.1 Botanical and zoological evaluations
    7.4.2 Integrated evaluations
7.5 General discussion of criteria
7.6 Summary

Wildlife Conservation Evaluation. Edited by Michael B. Usher.
Published in 1986 by Chapman and Hall Ltd., 11 New Fetter Lane,
  London EC4P 4EE
© 1986 Chapman and Hall

7.1   INTRODUCTION

The surface area of the Netherlands is approximately 37 000 km^2, the country being among the smallest in Europe. With more than fourteen million inhabitants, the average population density of more than 4000 km^{-2} is among the highest of the Continent. 4000 km^2 of the Netherlands are used for urban, industrial and transport purposes and 25 000 km^2 for agriculture. Only 2000 km^2 are clearly designated as wildlife areas: these are distributed over more than 3000 sites, of which only very few are larger than 10 km^2 (see Duffey (1982) for a description

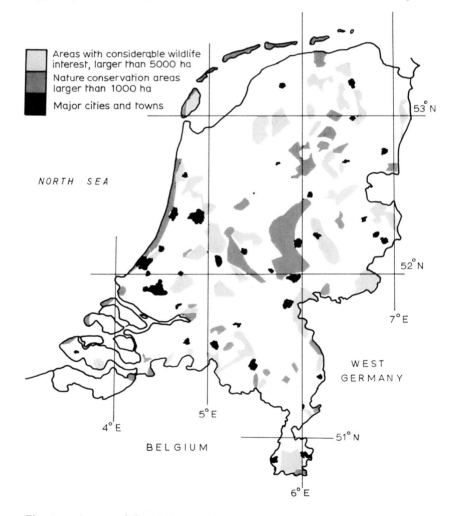

Fig. 7.1   A map of the Netherlands showing the larger conservation areas, as well as the location of major cities and towns.

of some of the larger sites). The locations of large wildlife conservation areas and other areas with important wildlife interest, are indicated on the map of the Netherlands shown in Fig. 7.1.

From the above figures it is easily understood that in this small mosaic-like country land use planning has become a controversial issue, particularly as regards environmental health and wildlife conservation. In contrast with large European countries like Great Britain, France or Germany, most problems concentrate on *changes* in land use rather than on designating the 'best sites for conservation', as these have already been selected in the past. Consequently, attention has shifted from attempts at ecological value judgements to studies concerning the effects of various forms of land use on ecosystems and landscapes.

The history of wildlife conservation evaluation in the Netherlands will be described first. Some evaluation criteria will be dealt with in detail, and trends and prospects in conservation evaluation in the Netherlands will be discussed.

Most publications on this subject have appeared in the Dutch language and are therefore not easily accessible. In the English language, an extended review, by van der Ploeg and Vlijm (1978), paid attention to evaluation aims and methods, their use and misuse, to criticizing evaluation studies and to the role of evaluation in land use planning. Van der Weijden and van der Zande (1980) have elucidated the controversy about evaluation aims and their use among Dutch scientists. Van der Zande *et al.* (1981) broadened this discussion to the situation in both the Netherlands and Belgium. Other relevant contributions in English have been made by van der Maarel (1978) and by Braat, van der Ploeg and Bouma (1979). Some details about conservation evaluation in relation to landscape ecology can be found in Tjallingii and de Veer (1981).

7.2   CONCISE HISTORY OF CONSERVATION EVALUATION IN THE NETHERLANDS

7.2.1   The pioneers, 1965–1970

After the Second World War, the Netherlands rapidly changed from an agricultural to an industrialized country. This process implied an increasing threat to the semi-natural and near-natural landscapes which originated from centuries-long stable management of the agricultural land. Westhoff (1968) stressed the importance of the richness of habitats in the country; van Leeuwen (1966) developed his 'relation theory' based on small-scale gradients of environmental factors and human influence; Mörzer Bruyns (1967) listed the benefits of nature conservation for society.

As the perceived expansion of the harbour of Rotterdam ('Europort') would seriously threaten the invaluable coastal dunes of Voorne, Adriani and van der Maarel (1968) wrote their classic 'Voorne in de Branding'. They used the species–area equation, $S = cA^z$ (Preston, 1962), for the first time in the Netherlands, comparing thirty dune areas in Western Europe and using a Dutch standard regression line for the species–area relationships.

The question of which criteria should be used for judging important sites for wildlife conservation became inevitable. Westhoff (1970) presented his ideas to an international forum.

7.2.2   Zenith of evaluation studies, 1970–1976

The year 1970 ('European Nature Conservation Year') induced much scientific and political effort in formulating the importance and benefits of wildlife conservation and environmental health. Many studies were devoted to ecological evaluation of mostly semi-natural areas; the most important ones, namely the *Environmental Survey of the Netherlands* (Kalkhoven, Stumpel and Stumpel-Rienks, 1976), *The Landscape of the Kromme Rijn* (Kromme Rijn Projekt, 1974), *The Values of the Forelands* (de Soet, 1976) and *Green Space Arnhem-Nijmegen* (Werkgroep GRAN, 1973) have been described in van der Ploeg and Vlijm (1978). A huge variety of criteria came into use; Burggraaff *et al.* (1979) list 171 criteria, concerning 17 aspects, in their review of only 13 studies. However, two dominant questions remained unanswered. (i) Is it sufficient to evaluate an area using only botanical (flora, vegetation) data? (ii) How should partial evaluations (for example soil, vegetation, avifauna) be integrated so as to make the final value judgement?

During this period the concept of 'Functions of the Natural Environment for Human Society' was also developed (explained in the English language by van der Maarel and Vellema (1975), and Braat, van der Ploeg and Bouma (1979)). This concept was meant to provide a tool for evaluating areas or ecosystems based on their contribution to human welfare. It was also a basis for the 'General Ecological Model of the Netherlands' (GEM), developed for land use planning purposes (van der Maarel and Dauvellier, 1978).

7.2.3   Criticism and doubt, 1975–1981

In 1975, the Dutch ecological scene was shocked by an article in a leading newspaper under the heading 'Environmental evaluation: biologists precisely select which landscapes may be ruined'. This criticism focused on studies '. . . often indicating on a beautiful colour map the relative "ecological values" of the various parts of a given region or even the whole country' (van der Zande *et al.*, 1981). A summary of the early criticisms has been given by van der Ploeg and Vlijm (1978). Van der Weijden and van der Zande (1980) discern three categories of criticism.

(1) *Scientific objections*. Evaluation is essentially a subjective activity; even selection of areas or species is subjective. Thus weighting of areas or species and (weighted) integration of their scores cannot be argued scientifically. Maps of values have little predictive significance as they ignore effects of processes in areas, thus emphasizing ecological patterns only.
(2) *Political objections*. Scientists should confine themselves to description, analysis and prediction; they should leave evaluation and selection to the public, to politicians and to conservation organizations.
(3) *Strategic objections*. Evaluation studies actually accept certain socio-economic developments ('defensive strategy') without cross-examination of the reasons for such developments ('offensive strategy'). Thus spatial concessions are made in advance and the case is reduced to an allocation problem.

In their review of environmental surveys in the Netherlands, Burggraaff *et al.* (1979) amply discussed these objections. Apparently there was agreement on the subjectivity of evaluations and on the scientific meaning of weights and integration formulae. Differences in opinions persisted concerning the political and strategic aspects. There was also consensus on the usefulness of environmental surveys (including maps) as such, without evaluation attempts. The number of evaluation studies declined sharply; and only a few large studies still used evaluation criteria (for example the 'Midden Brabant' study, Harms and Kalkhoven, 1979). Attention shifted towards a new tool, environmental impact assessment (EIA), to be used in most land use planning situations. However, doubts arose as to whether EIA would be a better strategic instrument than the original ecological evaluation maps, considering the proposed legislation on EIA in the Netherlands.

7.2.4 The present, from 1980 onwards

Environmental impact assessments have gained more and more importance and the EIA bill has recently passed through Parliament (though major agricultural activities were excluded from it). Impact studies (for example dose-effect relationships) are being advocated rather than environmental surveys; these, however, are still being done mainly at the regional (provincial) and at the local scale, although generally without any evaluation procedure. At present, a comparative study of relevant survey methods is being carried out under the aegis of the

Netherlands Society for Landscape Ecological Research. Another recent issue is 'ecological monitoring' which can help in detecting and explaining changes in the environment (Meijers, ter Keurs and Meelis, 1982). Nevertheless, the need for criteria for assessing qualities or properties of sites remains. Rarity is still considered important for conservation. Some other criteria used are representativeness, diversity, replaceability and naturalness (Anonymous, 1981a; Everts, de Vries and Udo de Haes, 1982; Werkgroep Methodologie, 1983).

7.3   SOME CRITERIA USED IN THE NETHERLANDS

7.3.1   Rarity

Rarity is the best known of all evaluation criteria used in wildlife conservation. It has been used, in different ways, in almost all evaluation studies in the Netherlands. The following list, based on van der Ploeg and Vlijm (1978), and Burggraaf et al. (1979), shows a diversity of uses:

(1) regional, national or international;
(2) rarity of plant species, plant alliances or vegetation complexes;
(3) rarity of species of various animal groups (birds, butterflies, mammals, herpetofauna, aquatic organisms);
(4) rarity of geomorphologically important elements, of virtually irreplaceable physiognomic (visual) elements of the landscape and of cultural monuments and other historical objects;
(5) rarity of 'natural elements' (for example certain types of woodland), of ecosystems and of ecotopes (i.e. spatially limited, homogeneous units consisting of biotic and abiotic components (Zonneveld, Tjallingii and Meester-Broertjes, 1975).

Unfortunately most evaluation procedures do not rely on hard, measurable (and therefore reproducible) facts, but on expert judgements only. This is very clearly demonstrated in the report *Natural and Cultural Values in the Countryside* (Bolwerkgroep, 1979); rarity is explicitly mentioned but it is only evaluated in a quantitative procedure in relation to nine species of meadow birds (lapwing, redshank, godwit, etc.). The overall value of the area under consideration, W, is given by

$$W = 100 \sum_{i=1}^{9} w_i n_i / q \qquad (7.1)$$

where w_i is the value of species i ($i = 1, 2, \ldots, 9$), n_i is the number of breeding pairs of species i in the area, and q is the surface area in hectares. Multiplication by a factor of 100 converts all areas to a standard of 100 ha. Areas with a final score of 75 to 150 are considered to be rich.

In many studies the rarity of plant species is evaluated on the basis of recorded distribution in the Netherlands. Distribution has been recorded since about 1900 in so-called 'hour-squares', measuring 5 × 5 km each. After 1950, all units of this grid have been revisited. Van der Maarel (1971) has produced a 'sociological-ecological' classification in which 19 groups of species are recognized and each species is classified in one of nine 'hour-square frequency categories', ranging from very rare to very common (i.e. occurring in more than 1210 of the 1677 hour-squares).

Mennema (1973) has used these 'flora statistics' in his evaluation study of the flora on the banks of Merkske stream. Field recordings were aggregated to 24 grids of 1 km^2. The following equation was used for the evaluation

$$W_f = \sum_{g=1}^{19} W_g = \sum_{g=1}^{19} \frac{\sum_{i=1}^{9} 100\, a_i\,(10 - f_i)}{\sum_{i=1}^{9} n_i\,(10 - f_i)}, \qquad (7.2)$$

where W_f is the floristic value of a particular square, W_g is the floristic value of sociological-ecological group g ($g = 1, 2, \ldots, 19$), a_i is the number of species belonging to the ith hour-square frequency category, f_i is the hour-square frequency category ($i = 1, 2, \ldots, 9$), and n_i is the total number of species in the Netherlands in the ith hour-square frequency category. As only the presence of plant species is used, only floristic (rarity) value is assessed. Moreover, this value applies to the national scale only; there is no comparable detailed information at the international level, for example for the north-west of the European continent. At the regional scale within the Netherlands this information can be used by considering exclusively the hour-squares belonging to that region.

Mennema (1973) also uses multiplication factors for each sociological-ecological group of species. These are based on vulnerability to disturbing influences and on the rarity, at the European scale, of these groups. These multiplication factors are, of course, subjectively chosen using best professional judgement. An example of the whole evaluation procedure is given in Table 7.1.

The multiplication factor implies a generalization as the sociological-ecological groups are not completely homogeneous. Arnolds (1975) therefore proposed multiplication factors for each plant species but warned against the subjectivity of these methods in general. He also stated that rarity assessment at the national scale should be done within a square kilometre grid rather than within hour-squares (25 km^2).

Table 7.1 An example of floristic-ecological evaluation (FEV) of an area of ground 1 km², using the method of Mennema (1973). The following abbreviations are used: sociological-ecological groups of plant species (SEG); total number of species recorded in each SEG (TNS); rarity value in the square, $\sum a_i (10 - f_i)$ (RV); rarity value in the Netherlands, $\sum n_i (10 - f_i)$ (RVN); the floristic value, 100 RV/RVN, rounded to the nearest integer (FV); and the multiplication factor as described in the text (MF). The two summations are from $i = 1$ to $i = 9$.

SEG	Number of species in hour-square frequency classes									TNS	RV	RVN	FV	MF	FEV
	1	2	3	4	5	6	7	8	9						
1	.	.	.	.	.	.	.	.	.	0	0	93	0	1	0
2	.	.	.	.	1	1	3	1	2	8	22	667	3	1	3
3	.	.	.	.	.	1	3	6	13	23	38	844	5	2	5
.															
.															
.															
17	.	.	.	.	.	.	2	2	7	11	17	249	7	3	21
18	.	.	.	1	2	4	3	3	.	13	47	613	8	6	48
19	.	.	.	.	.	1	5	.	.	6	19	273	7	4	28
Total*	.	.	4	4	13	32	51	49	51	204	—	—	151	—	718

* The total is based on all 19 groups of species

Recently the species information system on the basis of the 5 × 5 km grid squares has been updated (van der Meijden *et al.*, 1983). This version includes the frequencies of each species in both 1930 and 1980. This enables quick detection of endangered species (i.e. species which have been moved to a lower frequency class) in any area under consideration.

The same method has been used for breeding birds in the Netherlands (Teixeira, 1979). The *Atlas of the Dutch Breeding Birds* displays the distribution of breeding populations by presence in hour-squares, and gives estimated numbers of breeding pairs in 1979. A more detailed investigation has been reported by VAWN (1981) for the Randstad, i.e. the urbanized western part of the Netherlands.

A comparable method has been used for other animals, for example lizards (3 species), snakes (3 species), amphibians (9 species) and butterflies (24 species) in the National Environmental Survey (Kalkhoven, Stumpel and Stumpel-Rienks, 1976).

For the General Ecological Model (Van der Maarel and Dauvellier, 1978) the following index of international rarity was proposed

$$r(w) = r(g) \times r(g/n) \times r(n/w) \tag{7.3}$$

where $r(w)$ is the worldwide rarity contribution, $r(g)$ is the regional rarity (quantified as frequency), $r(g/n)$ is the regional proportion of the national rarity, and $r(n/w)$ is the national proportion of the global rarity. This method of assessing frequencies from a 5 km grid square has also been used at abstraction levels higher than the species. R. J. de Boer (unpublished) has estimated hour-square frequencies of the phytosociological associations in the Netherlands described by Westhoff and den Held (1969). These data have been used by Everts, de Vries and Udo de Haes (1982) for rarity assessment of ecotopes. Such ecotopes are roughly comparable to parts of the habitat types used by Ratcliffe (1977) for assessment in Great Britain. Everts, de Vries and Udo de Haes (1982) classified 91 ecotopes in the Netherlands, 14 being aquatic ecotopes: 34 of the 91 ecotopes are characterized as 'rare' to 'exceptionally rare'.

Until 1975, most studies used rarity of plant alliances or vegetation complexes as the basis for ecological evaluation (see van der Ploeg and Vlijm, 1978; Table 2). Since then, attention has shifted to the rarity of species or species associations. Except floristic studies such as that of Mennema (1973) (see equation 7.2), no attempts have been made to develop an evaluation procedure for an area in which all of the rarity values of plants are combined into a single index. Usually, for example in the EIA, rare species are listed without any attempt to get to a 'final value' for the collection of all those present.

7.3.2 Representativeness and authenticity

These evaluation criteria refer to the status quo of habitat types before the era of massive urban/industrial/agricultural change. They are roughly comparable to the criteria 'representativeness' in Anglo-Saxon papers (this concept is discussed in Chapters 1, 2 and 6), assuming that the reference point is (subjectively!) chosen before, for example, 1900. These criteria have been analysed in detail recently in the *Environmental Impact Statement on water extraction in South Kennemerland* (Anonymous, 1981a). In this work plant species are divided into three classes: (i) very representative species, specifically belonging to the habitat (*characteristic* species), (ii) non-representative species, not habitat-specific (*authentic* species), and (iii) 'noise' or *disturbing* species, belonging to other habitats. The impact of different activities on the plant composition of a habitat can then be calculated as

$$EF = \sum_{i=1}^{4} \sum_{j=1}^{3} P_{ij} \mid \Delta S_{ij} \mid , \qquad (7.4)$$

where EF is the resulting impact in numbers of species, P_{ij} is the weight factor for representativeness (see Table 7.2), ΔS_{ij} is the change in species composition, i are the species change classes (see Table 7.2), and j are the classes of representativeness (see Table 7.2).

The same procedure has been used in assessing impacts on vegetation

Table 7.2 Matrix of subjective weight factors (P_{ij}) as used in the *Environmental Impact Statement on Water Extraction in South Kennemerland* (Anonymous, 1981a). These weights are used in equation (7.4).

Species change classes (i)	Classes of representativeness (j)		
	1: Characteristic	2: Authentic	3: 'Noise'
1 New species	+2	+1	0
2 Species showing an increase	+1	0	0
3 Species showing a decrease	−1	0	0
4 Species that have disappeared	−2	−1	0

types, in which case ΔS is substituted by ΔA – the change of surface area (in ha). It has been used in assessing changes in 'rarity composition' of areas, using three rarity classes: rare, rather common, and common. It has also been applied to breeding birds (Anonymous, 1981a).

The classes of representativeness can also be used for ecotopes. Everts, de Vries and Udo de Haes (1982) used four classes, adding the class 'indifferent' for species which are neither authentic nor disturbing. Dutch vegetation formations have been classified into eight types, each consisting of a number of subformations, giving a total of 28 subformation types. In a 28 (subformation types) by 91 (ecotope types) matrix all ecotopes were classified for representativeness.

This criterion, when used in this rough way, can only be used in combination with other evaluation criteria. It is, nevertheless, useful because it refers to a very general conservation issue: richness of relatively intact habitats on the national scale.

In an evaluation study of the forelands of four main rivers, Rhine, IJssel, Waal and Meuse (de Soet, 1976), geomorphological aspects of the forelands were evaluated by the use of three criteria: rarity, surface area and 'perfectness'. The last criterion refers to representativeness, in the sense that an undisturbed habitat is assumed to be the most representative for its particular formation type or subtype(s).

7.3.3 Replaceability

This criterion can be defined as the ability of an ecosystem or a population to regenerate to its original state after a specific disturbance. At the end of the scale is the irreplaceability of an ecosystem or the extinction of a species. The criterion is inversely proportional to the recovery time, i.e. the period of time that an ecosystem requires to regenerate to the original state of that ecosystem prior to disturbance.

Replaceability is merely a conceptual criterion. Recovery, however, is a measurable process, as it can be derived from information about succession. The underlying value judgement is that longer recovery

Table 7.3 Classes of recovery time as used in the Netherlands (after Harms and Kalkhoven, 1979).

Class number	Recovery time (years)	Ecotope examples
1	<1	Beach pioneer vegetation
2	1–3	Pioneer vegetation on derelict arable land
3	4–10	Most brushwood vegetation
4	11–30	Chalk Grassland
5	31–100	Calcareous dune woodland
6	>100	Moorland, oligotrophic marsh woodland

times indicate more valuable systems. The criterion has the advantage of its potential worldwide use. Recovery time has been used as a judgement criterion in environmental impact studies, for example in relation to outdoor recreation (for example Willard and Marr, 1970; Bouma and van der Ploeg, 1975), or to agricultural land use (for example Odum, 1971).

As regards wildlife conservation evaluation in the Netherlands it has been used by Harms and Kalkhoven (1979) and by Everts, de Vries and Udo de Haes (1982). Both studies use the 6 class scale shown in Table 7.3. This classification into six classes relates to the habitat types occurring in the Netherlands. In alpine or tundra regions recovery times may be 500–1000 years, and hence class width may be different or further classes may be necessary. Harms and Kalkhoven (1979), in a regional study of land use in the southeast of the Netherlands, recorded 99 vegetation types of which seven fell into class 6 (these included some of the heathland, woodland and fenland vegetation types). Everts, de Vries and Udo de Haes (1982) put only three ecotopes out of 91 into class 6, considering all heathland ecotopes to be class 5. However, if the maximum recovery time, rather than the mean recovery time, is used, 12 ecotopes fall into the highest class 6, and these particularly include wet, oligotrophic habitat types.

7.3.4 Naturalness

Changes in structure and composition of habitats can be expressed in terms of naturalness, defined as being in a state not influenced by human activities. It might even be better to use terms like the degree of culturalness which can be measured by the relative proportion of new species arrived during the last 100 or 200 years (see van der Maarel, 1978).

Stumpel-Rienks (1974), in preparing the National Environmental Survey (Kalkhoven, Stumpel and Stumpel-Rienks, 1976), classified Dutch habitats on their degree of naturalness. A 10 point scale, running from 9 (natural) to 0 (completely unnatural), was used for all ecotope types (74 in both studies). This value was used in assessing values of larger land units, including more than one ecotope. For each ecotope present the surface area was estimated from air photographs and maps, discerning between surface, linear elements and point elements. For each of these, presence was scored on a 1 (= very rare) to 5 (= very abundant) scale. For evaluation, the following equation was used (as developed by van der Maarel, see Werkgroep GRAN, 1973; cf. van der Ploeg and Vlijm, 1978)

$$W = \sqrt[3]{\sum_{i=1}^{n} p_i w_i^3 \Big/ \sum_{i=1}^{n} p_i} \qquad (7.5)$$

where W is the value of land unit, w_i is the degree of naturalness (0–9) of ith ecotope, p_i is the presence score (1–5) of ith ecotope, and n is the number of ecotopes present. By varying the exponent (3 in equation (7.5)), more or less value is attributed to the most natural ecotope types. In the National Environmental Survey (Kalkhoven, Stumpel and Stumpel-Rienks, 1976) 3 was considered to be a reasonable compromise value. Of course this method as such does not account for actual quality of ecotopes, and hence additional field work is necessary in order to identify the quality.

7.3.5 Miscellaneous

Species richness seems to be a very suitable criterion when used in combination with rarity. However, most evaluation studies do not combine these two criteria relating to plant species. More often (Adriani and van der Maarel, 1968; van der Maarel, 1970; Anonymous, 1971, 1972), species richness is combined with surface area ($S = cA^z$), followed by comparison with the Dutch standard regression line for species/area relationships (van der Maarel, 1971). In contrast, most evaluations of breeding bird species do combine species richness and rarity. However, generally no comparison is made with average species richness at a national or regional scale.

Numbers of plant alliances or associations, and numbers of vegetation layers per unit of surface area, are also used. Although information on rarity of these vegetational aspects exists, it is rarely used for an attempt at integrated evaluation.

For breeding and overwintering birds, the criterion of the species abundance (in the appropriate season) is frequently used. Particularly in wetland areas (for example the Delta in the southwestern part of the

Netherlands) these abundances may significantly contribute to the wild-life value of the area (Anonymous, 1972) (see also Chapter 12). In the study on the river forelands (de Soet, 1976) special attention was paid to areas that functioned as 'sleeping places' for breeding and migrating birds.

Finally, wildlife conservation evaluation should include more than just the botanical and zoological aspects of the sites considered. Soil properties, geomorphology of the landscape and surface water quality have received much attention as they constitute basic conditions for habitat conservation, development or creation. Only in few cases, how-ever, have these aspects actually played a role in the evaluation process (see van der Ploeg and Vlijm (1978) for a brief summary).

7.4   EVALUATION PROCEDURES

7.4.1   Botanical and zoological evaluations

An extended review of botanical and zoological evaluation procedures has been given by van der Ploeg and Vlijm (1978). Some of the botanical evaluation procedures have already been mentioned: the method of

Table 7.4   Definition of quality classes for field records (relevés) as presented by Everts, de Vries and Udo de Haes (1982). HF refers to the hour-square frequency category, ranging from 1 (extremely rare) to 9 (common). The following ab-breviations are used for representativeness: noise (n), indifferent (i), characteris-tic (c), and authentic (a) species. (c/a) implies that at least some of the (c) or (a) species are present. Classes 0 and 1 differ in the percentage cover of noise species and in the presence or absence of characteristic or authentic species; most species are very common (hour-square frequency category 9). Relevés in classes 2, 3 and 4, which show a vegetation cover of less than 50% noise species, differ in respect of the number of less common or rare species, which are actually present.

Representativeness	Percentage cover of noise species	Presence/absence of rare species	Class
n/i	≥80%	≤3 species in HF 8, no species in HF 1–7	0
c/a	50–80%	≤3 species in HF 8, no species in HF 1–7	1
c/a	≤50%	2 species in HF 6 or 7 or ≥3 species in HF 8	2
c/a	≤50%	no species in HF 1–5; ≥3 species in HF 6–7	3
c/a	≤50%	≥1 species with HF 1–5	4

Mennema (1973), including rarity and vulnerability; the method of Kalkhoven, Stumpel and Stumpel-Rienks (1976), using naturalness and presence; and Adriani and Van der Maarel (1968) and others, using the Preston's (1962) equation (area related to the number of species) followed by comparison with the Dutch standard regression line for species/area relationships.

In the Kromme Rijn evaluation study (Kromme Rijn Projekt, 1974), associations of the vegetation (in 13 classes) and numbers of vegetation layers (in 4 classes) were recorded in the field. Then maps of ecotopes and ecochores (i.e. spatially limited, heterogeneous units; Zonneveld, Tjallingii and Meester-Broertjes, 1975) of the area were designed, one for each of these variables. Finally, these two maps were combined into one map by summation of class values followed by reduction to three value classes. As regards breeding birds, important areas (because of number of species and rarity of species) were indicated on the final map.

Recently, in their study of ecotope types, Everts, de Vries and Udo de Haes (1982) have tried to develop a system of quality classes for ecotopes. The system should be based on three criteria: species richness, rarity of species and representativeness of species for a particular ecotope type. Various combinations of these criteria were tried for their power of discrimination between five quality classes. The specifications of the best-fitting combination are shown in Table 7.4.

On the basis of these characteristics, field records (relevés) within areas (ecotopes) can be assigned to one of the five classes. This enables a conclusion to be reached about the relative quality of the areas (ecotopes) under consideration.

Quite recently, Clausman, van Wijngaarden and den Held (1984) published an evaluation study on the flora and vegetation of the province of Zuid-holland. From a comprehensive computer-compatible database, rarity values for plant species are established. The integral value (IV) of a species is defined as

$$IV = (R_p + R_n + R_g)/3 - TD, \qquad (7.6)$$

where R_p is the provincial rarity, R_n is the national rarity, R_g is the global rarity, and TD is the tendency (increase or decrease) of the species' occurrence. Variables and criteria used for this evaluation procedure are shown in Table 7.5. This table also gives examples for common (daisy, *Bellis perennis*) and rare (broad-leaved marsh orchid, *Orchis majalis*) plant species.

With these data at the species level, field records of sites have been evaluated by multiplying the species' integral value with the observed cover of the species. The resulting 'vegetation value' appears to range from 4 to 80. Intensively used pasture was valued 26 on average, an average road verge was valued 39, an average dune area 49, and an average oligotrophic dune slack 62.

Table 7.5   Variables, criteria and equations as used in the evaluation procedure of Clausman, van Wijngaarden and den Held (1984).

Variable/criterion	Short description	Daisy	Broad-leaved marsh orchid
GN	Number of grid units (100 m²) in the province	3269×10^4	3269×10^4
NG	Number of grid units in which the species is recorded	438×10^4	3000
MC	Mean cover per grid unit, defined as cover/$NG \times 100$	0.0087	0.0062
R_p	Provincial rarity of the species: $10 \times \log_{10} \{GN \times (1 - \log_{10}MC)/NG\}$ (Observed range of R_p is 6 to 81)	14	45
HFC	Hour-square frequency categories	9	4
URH	Unweighed national rarity based on HFC: $1.69 \times 10^7 \times 0.24^{HFC}$	44.65	56070
R_u	National rarity of the species: $10 \times \log_{10} \{URH \times (1 - \log_{10}MC)\}$ (Observed range of R_n is 11 to 72)	21	53
GA	Global area covered by the species (estimated percentages)	0.02	0.25
GC	Global area centre i.e. relative frequencies in rest of the global area as compared with the Netherlands	1	0.3
R_g	Global (mondial) rarity of the species: $R_n + \log_{10}(GC/GA)$ (Observed range of R_q is 14 to 112)	38	54
HD	Difference in hour-square frequency category between 1930 and 1980	0	−2
RT	Recent tendency (projection 1980–2000) of the species in classes 1 (sharp decrease) to 5 (increase)	3	1
VU	Vulnerability of the species, in classes 1 (very vulnerable) to 5 (not vulnerable)	4	1
TD	Tendency of the species: $4 \{HD + (RT - 3) + (VU - 3)\}/3$ (Observed range of TD is +8 to −10)	+1	−8
IV	Integral value of the species: $(R_p + R_n + R_g)/3 - TD$. (Note that the observed range of IV is 10 to 84)	23	59

Five comments can be made about the evaluation procedures.

1. Evaluations generally use one or two out of three different scale units recognized in the Netherlands. At the larger scale, 1 km² or 25 km², evaluation relates to the national topographical grid. This measure is good for statistical treatment of data, but it does not allow for spatial ecological relationships. At the smaller scale, the evaluation can be based on an ecotope, a spatially limited, homogeneous ecological unit (ecotopes can, therefore, be considered the spatial representation of ecosystems). Evaluation at this scale refers to ecological

entities but not to human use or to property rights. At an intermediate scale, the evaluation can be based on an ecochore, a spatially limited, heterogeneous ecological unit. This spatial unit is comparable to a 'land unit', and it mostly refers to areas whose boundaries are determined either by property rights or by artefacts like roads, canals and land use forms. The topographical approach is used in studies using rarity as the main criterion. In the approaches at smaller scales, other criteria are used as well: often ecotope values are combined to give an ecochore score.

2. Many studies have used a combination of some area measure (either the actual area or the relative surface area of an ecotope within an ecochore) with another criterion, mainly rarity and naturalness. This combination has been referred to in equation (7.5), which was developed by E. van der Maarel. Another form of this equation, used by GRIM (1974) for botanical evaluation, is

$$V_j = \sum_{i=1}^{n} Ap_i \times w_i^3, \qquad (7.7)$$

where V_j is the value of the jth ecochore, Ap_i is the relative surface area of the ith ecotope, within the jth ecochore, w_i is the national rarity value of ith ecotope (expressed in 5 classes), based on vegetation complexes, and n is the number of ecotopes present in the jth ecochore.

3. Many studies simply rank data on different criteria and then classify them ordinally or cardinally. After this classification, integration for the botanical or zoological aspects is done by adding class 'values' for each criterion per land unit. Sometimes sum values are not shown in tables but only on an integration map.

4. Even the study with the most detailed information (Clausman, van Wijngaarden and den Held, 1984) almost completely relies on 'rarity' as the most decisive criterion. Obviously, no profound study has yet been done so as to synthesize the information from field records into an ecological framework including pattern and process of ecosystems and landscapes.

5. Many studies adjust average results upwards for high values. This is a remarkable exercise, because no study appears to adjust downwards for low values. In other words, it seems that only the high values really count. One could wonder why people pay lip service to such a variety of criteria if they are only looking for Rembrandts!

7.4.2 Integrated evaluations

Table 7.6 shows some of the integration procedures (after van de Ploeg and Vlijm, 1978). Two comments can be made here. First, most studies

Table 7.6 Integrated evaluations (after Van der Ploeg and Vlijm, 1978)

Study-authors	Aspects used	Integration procedure	Final classes of value
1. Volthe-De Lutte Land Use Planning (Anon. 1971)	botany/fauna	Summation of separate botany (4) and fauna (1) criteria values (class 1–3), rearrangement into 3 classes.	highly valuable } nature reserve valuable valuable ecotope or ecochore
2. Green Space Arnhem Nijmegen (erkgroep GRAN 1973)	vegetation/fauna	Summation of ecochore aspects (vegetation plus 3 categories of fauna). Rearrangement into 5 classes.	low-high (I – V)
3. Kromme Rijn Projekt (1974)	soil/vegetation/ avifauna	Soil types on map of 'potential value' (3 classes). Maps of botanical values for 'actual value' (3 classes). Integration of maps into landscape evaluation map ($3 \times 3 = 9$ colours), with indication of important bird (abundance) and high rarity (plants) value areas.	very high, high, less high
4. Forelands (de Soet, 1976)	botany/avifauna/ geomorphology/ landscape quality	Summation of values (1–4) of the 4 aspects per foreland area. Rearrangement into 4 value classes.	very high, high, rather high, moderate to small
5. National Environmental Survey (Kalkhoven, Stumpel and Stumpel-Rienks, 1976)	vegetation/fauna	Based on 5 botanical value classes. Highest ranking fauna value puts ecochore in next higher value class.	5 classes ranking presence to absence of ecochores of national importance

(including ones not mentioned here) rearranged values into final value classes, probably to avoid non-perceivable results. Secondly, most studies do not mention 'low' values but only use indications like 'less high'. This is probably done to avoid misuse (for example for physical planning) of the results.

Since 1976, almost no integration procedures have been attempted. From the ecological point of view, this can be seen as an improvement. If ecosystems are what they are supposed to be, working structures containing interrelated components, then a few single indicators must be sufficient. In the final report of the regional study of Midden-Brabant (Werkgroep Methodologie, 1983), the changes in vegetation and fauna resulting from alternative land use plans have been indicated separately on computerized colour maps. For the vegetation, replaceability was used as the criterion, whereas for bird species the criterion was rarity on the national scale. After classifying land units into replaceability and rarity classes respectively, changes in value classes were indicated on the maps. This rather simple method has proved very effective once the information has been gathered and stored.

In some Environmental Impact Statements (for example Anonymous, 1981a) integration of effect scores has been done. The vegetation, flora and fauna are put into several effect classes, and, by using weight factors, these effect classes are integrated into a final classification (mostly qualitatively, ranking from $+++$ to $---$) for each alternative plan. To a certain extent such evaluations of impacts are as questionable as the integration of wildlife values. On the other hand, these impact evaluations mimick overall effects of land use plans on ecosystems and are useful if alternative plans are to be compared.

Another approach might be to look for complexity as an indicator of wildlife values. In that case, richness of different structures (for example geomorphological and vegetation types, historical landscape features) may simply be counted per unit area, supposing that wildlife values almost always are larger in areas with a larger variety of structures. This statement is certainly true for large parts of the Netherlands.

7.5   GENERAL DISCUSSION OF CRITERIA

Wildlife evaluation criteria can be grouped into the following three categories (Burggraaff et al., 1979)

(1) criteria used in an assessment of vulnerability (to human activities)
(2) criteria used in an assessment of suitability (for land use types)
(3) criteria used in an assessment of values (i.e. suitability assessment for wildlife conservation).

Most studies referred to in this chapter fall into either the first or the third category. Land use suitability assessment usually only requires information on soil properties and geomorphology. Most studies also indicate areas with the highest wildlife values, and therefore it can be concluded that nature conservation objectives always play a role. On the other hand, the reasons for doing these studies have always been the possible functions of the natural environment for man and society.

Thus, the general picture is that of a non-monetary cost-benefit or cost-effectiveness analysis, weighing land use (in the broadest sense) against wildlife interests. This is reflected in the choice of criteria: scarcity, richness and maturity are invariably valued highly. Only occasionally is scientific (ecological) knowledge used for the selection of evaluation criteria.

All criteria discussed in this chapter – rarity, representativeness, replaceability, naturalness, vulnerability, species richness and abundance – are suitable for wildlife value assessment. The most suitable, within the environment of a country such as the Netherlands with a very dense human population, seem to be rarity, species richness and abundance: they can be quantified easily. Ecological knowledge can be used to fit these three criteria into one 'quick-and-dirty' mathematical model which can be used to provide values, as well as projections of values in the future. Representativeness, replaceability and naturalness seem most suitable for vulnerability assessment. However, there is insufficient ecological knowledge to integrate these criteria into one model which would yield a value. Rarity is more difficult to use with regard to vulnerability assessment, as the reasons for rarity of habitats or of species may be manifold. Species richness, and abundance per species, are more suitable.

Recent trends in the Netherlands indicate a shift towards impact assessments, and biological monitoring (i.e. monitoring species because they are indicators of particular impacts) is gaining in importance. This, however, can only be done with relatively abundant species. Biological monitoring is almost impossible for ecosystems (or ecotopes) as the number of variables to be measured is too large. Therefore, it seems important that research on rarity, replaceability and representativeness should continue because such research may add to the scientific basis for nature conservation in the Netherlands. In the past, mistakes have been made, for example by developing mystical pseudomathematic expressions which include all possible criteria. Although this will undoubtedly happen again, there is a reasonable hope for a scientific approach to conservation objectives. In future, however, more attention should be paid to the fact that in ecological reality patterns change over time, which forces wildlife conservation to take a dynamic, rather than a static, viewpoint. Such may be the challenge for both conservation and research for the next decade.

7.6 SUMMARY

Wildlife conservation evaluation in the Netherlands has gained importance since the 1960s, as land use planning in this small country came more and more into conflict with objectives of environmental health and nature conservation. Many evaluation studies in the 1970s have been criticized because of their low scientific content and because of various political/strategical implications. Attention has gradually shifted to Environmental Impact Assessment methods in which some evaluation criteria might be incorporated.

In the Netherlands the most commonly used criterion is rarity at the national scale, either of species or of groups of species or of habitat types (ecotopes). Representativeness of species, or of species groups, and replaceability, are used as criteria for ecotope evaluation. Both replaceability and naturalness are based on expert judgements about succession and the degree of human influence.

Evaluation procedures may use different scales: topographical grids (commonly 1 km or 25 km squares), areas of an ecotope, or ecochores (usually with artificial boundaries such as roads or canals). Many studies combine such area measures with criteria such as rarity and naturalness so that ranking and integration is achieved by adding class values per land unit. There is, however, a tendency for presenting scores for vegetation and for fauna separately, as they may indicate different effects.

Most studies are concerned either with suitability assessment, for wildlife conservation, or with the assessment of vulnerability to human activities. The use of combinations of evaluation criteria, for either of these forms of assessment, requires further ecological research. Particular attention should be given to a dynamic approach, i.e. considering changes over time.

CHAPTER 8

Evaluation at the local scale: a region in Scotland

E. T. IDLE

8.1 Introduction
8.2 The criteria
    8.2.1 Area
    8.2.2 Diversity
    8.2.3 Rarity
    8.2.4 Naturalness
    8.2.5 Typicalness
    8.2.6 Fragility
    8.2.7 Other criteria
8.3 Other background considerations
    8.3.1 Scale
    8.3.2 Classification
    8.3.3 Data collection
8.4 An example from the South East Region of Scotland
    8.4.1 Introduction to the Region
    8.4.2 An example of site evaluation
8.5 Conclusion
8.6 Summary

Wildlife Conservation Evaluation. Edited by Michael B. Usher.
Published in 1986 by Chapman and Hall Ltd, 11 New Fetter Lane,
  London EC4P 4EE
© 1986 Chapman and Hall

8.1 INTRODUCTION

In Chapter 6 Ratcliffe discusses criteria for the evaluation of nature conservation importance in Great Britain, particularly in relation to the selection of a series of key, nationally important sites. At the local scale the problem is to apply these same criteria to information on sites which may be of local or regional importance, but which, in Great Britain, contribute to a network which is an important element in national nature conservation practice. The identification of the individual sites (known as Sites of Special Scientific Interest, abbreviated as SSSIs) within this network, and the subsequent influence of their land use and management on nature conservation objectives, is a crucial step in maintaining the present range of ecosystems and plant and animal species. Even though vegetation which has not been totally transformed from its original state occupies not more than 30% of the land area (in more intensively farmed areas in eastern and southern Britain the figure is less than 10%), some form of evaluation of potential nature conservation sites is necessary. Green (1981) points out that 'since all countryside cannot be protected, the setting aside of some land (for nature conservation) means that there has to be a process of selection based on an assessment of worthiness . . .'. Furthermore, because the identification of an area of land as an SSSI has economic, social and political implications, especially for its owners and occupiers, the process of selection needs to be more than a matter of personal preferences on the part of the individuals involved. Bunce (1981) emphasized the need for 'an explicit system of evaluation that is reproduceable', and Goode (1981) pointed out the importance of this approach in presenting arguments for nature conservation in the face of the competing claims of other land uses. In the last 35 years developments in the legislation for, and practice of, nature conservation have increased the need for a systematic approach to the assessment and selection of SSSIs.

Since the passage of the National Parks and Access to the Countryside Act in 1949, and planning legislation related to it, the Nature Conservancy Council (NCC) must be consulted by local planning authorities about applications for developments within or around SSSIs. Increasingly local and central Government have refused permission for proposals that would be harmful to these sites, and structure and local plans now frequently contain policy statements aimed at safeguarding SSSIs and sometimes other areas of wildlife or nature conservation importance. This close relationship, between the local authority areas and SSSIs resulting from the planning consultative process, led, in the 1960s and 1970s, to the production of county schedules of SSSIs (district schedules in Scotland following the reorganization of local government in 1973). Thus the 'area of search' within which SSSIs should be selected, or alternatively the extent or size of the local scale, is influenced by the planning/administrative boundaries of local government.

Land use activities are generally not defined as 'development' in terms of planning legislation, and therefore were not subject to consultation under the 1949 Act and subsequent Planning Orders. However, under the Wildlife and Countryside Act 1981 land use activities on SSSIs are subject to consultation between land owners/managers and the Nature Conservancy Council. The NCC must provide each owner or occupier of an SSSI with a list of the land use activities that are potentially damaging to the nature conservation interest of the site in question. If any owner or occupier plans or proposes to carry out any of these operations he must give the NCC written notice of doing so. Where the proposed activity is harmful to all or part of the SSSI, the 1981 Act empowers the NCC to negotiate a management agreement involving compensation payments, to the owner/occupier, for the profit foregone through abandoning or modifying the proposed activity. Where owners/occupiers refuse to negotiate management agreements, arbitration, compulsory purchase and 'stop-order' powers are available under the 1981 Act. Owners/occupiers failing to consult on potentially damaging operations are liable to prosecution under Section 28(7) of the 1981 Act.

Thus the more important nature conservation sites are covered by a set of statutory consultative procedures for both development and land use, backed up by planning, financial and legal provisions. Clearly a systematic approach must be adopted in the assessment and selection of SSSIs, the identification of potentially harmful land uses and development impacts, and the negotiation of management agreements that sometimes involve considerable sums of money.

8.2 THE CRITERIA

Although the criteria for evaluating the nature conservation importance of sites may have been embodied in the report of the Huxley Committee (1947), the main exposition of the criteria was by Ratcliffe (1977) in the Nature Conservation Review (see also Chapter 6). The criteria (size, diversity, naturalness, rarity, fragility, typicalness, recorded history, position in an ecological/geographical unit, potential value and intrinsic appeal) have been reviewed by several authors. Goldsmith (1983) distinguished between 'ecological criteria', such as size, diversity or richness and rarity, which could be more or less precisely measured, and 'conservation criteria', such as potential value and intrinsic appeal, which are much more value judgements. Margules and Usher (1984) developed this separation of criteria further in an investigation of the way that nine experienced conservationists evaluated the potential nature conservation interest of sites in north-east Yorkshire. They concluded that, for small sites, ecological fragility, threat, and both species and habitat rarity were the most important criteria, but representativeness, size, naturalness and position in an ecological/geographical unit were the most important for large sites. The 'conservation criteria' identified by Goldsmith (1983) were regarded as applying only at a later stage of the evaluation process, if at all. Margules and Usher also concluded that fixed weightings for individual criteria could not be made and that the flexible approach used by Ratcliffe should be endorsed, in contrast to the indexing of criteria outlined by Goldsmith (1975). Spellerberg (1981) and Hooper (1981) pointed out the potential conflict between criteria such as diversity and naturalness, or rarity, represented as the rare species associated with successional stages in grasslands or coppice woodland, and naturalness. Bunce (1981) and Goode (1981) indicated the way in which the criteria applied differentially to different habitats using as examples the natural species poverty and low diversity of the high plateau of the Cairngorms, and the relative importance of naturalness over species richness in the case of peat bogs.

The conclusions drawn from these studies meant that at the local scale in south east Scotland, and more generally where the Nature Conservancy Council assesses and selects SSSIs, each criterion is dealt with separately and applied systematically to information about sites. Criteria are not amalgamated, though, as Ratcliffe (1977) has pointed out, a single site might qualify as an SSSI on several criteria. Each site is therefore selected as the 'best' or 'one of the best' of its type, on the basis of at least one, and usually several, criteria. Not all criteria apply equally to all habitats, but in all cases the size of both the site and the habitats within it are recorded, and measures of diversity of species and habitats, and of rarity of species and habitats, are made. Judgements are required to assess the naturalness and typicalness of sites, but potential value and intrinsic appeal are used only occasionally.

8.2.1 Area

Although as a general guideline larger areas are more important for nature conservation than smaller areas, care is required in interpreting this. Large sites may be more likely to sustain populations of large predators and herbivores and be more robust to damage and threat, but Usher (1980), for particular plants, and Hopkins and Webb (1984), for invertebrates, have shown that larger sites do not necessarily support a larger total of species number. Furthermore, few sites are occupied by a single habitat, and, therefore, measurements of site area are always accompanied by measurements or estimates of the area of habitats.

8.2.2 Diversity

Much has been written about the relationship between the number of species and area, and mathematically formulated measures of diversity which combine species richness and relative abundance. It is generally accepted that the maintenance of species is one major nature conservation objective and that sites that support more species are more important than those with few. However, as discussed in Section 8.2.1, care must be taken with the interpretation of this criterion. Habitats that are naturally species poor, such as peat bogs or the Dorset heathlands (Webb and Hopkins, 1984), can be 'diversified', i..e enriched in species, either as a result of edge effects from surrounding land or due to past or present modification of habitats within the site. Thus the importance of diversity, however it is measured, in relation to other criteria needs to be considered in each type of habitat.

Generally, diversity is less important as a criterion in habitats which are naturally species poor. Peterken (1974) used 50 woodland 'indicator species' as a measure of species diversity to assess the conservation value of 85 Lincolnshire woods and he concluded that the ranking obtained was likely to be different from the ranking using the total number of species of vascular plants. Usher (1980) reached similar conclusions for limestone pavements in Yorkshire, as did Webb and Hopkins (1984) for invertebrate populations of Dorset heathlands. Thus, care is required not only about how diversity is used in relation to different habitats, but also about which groups of species it should be applied to. As Peterken (1974) states, his Lincolnshire example 'emphasises the need to develop separate lists of primary woodland species for each natural region'. The same can be said for other types of habitats.

8.2.3 Rarity

Rarity, of both species and habitats, is perhaps the single most readily appreciated nature conservation criterion, though the concentration on it in evaluation procedures has been criticized (Adams and Rose, 1978). Mabey (1980) has contrasted the use of 'common' and 'ordinary' with 'rarity' in arguing that nature conservation must have a broader appeal than to a narrow specialist 'scientific' constituency.

Ratcliffe (1977) has pointed out that rare species tend to be either recent arrivals or dwindling relics, while Margules and Usher (1981) have drawn attention to the 'peculiar' ecological requirements of rare species and have identified at least three different kinds of rarity – species of 'stressed' or changing environments, widespread but locally infrequent species, and species of few localities but with large numbers at each site (see also Chapter 1). Spellerberg (1981) and others have

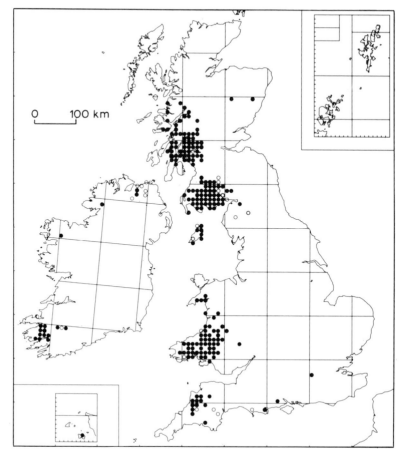

Fig. 8.1 The distribution of *Carum verticillatum*, a species showing a clear westerly distribution in Great Britain. The distribution map has been prepared by the Biological Records Centre.

highlighted the need to define the level of rarity – international, national, regional and local – and the need to take account of species at the edge of their geographical range. As Margules and Usher (1981) state, 'rare species and the areas that contain them can provide valuable natural laboratories for studies of biogeography and population regulation'.

Detailed information on species distribution, such as that given in the *Atlas of the British Flora* (Perring and Walters, 1962) (see Figs 8.1 to 8.3), and population records such as those collected by the Botanical Society of the British Isles, provide a basis for measuring rarity. Thorough site survey is required to produce species lists that can be compared with international, national and regional lists.

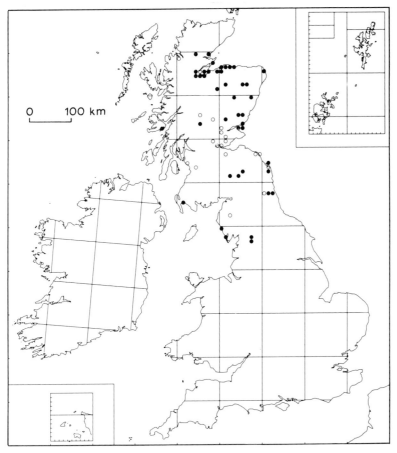

Fig. 8.2 The distribution of *Corallorhiza trifida*, a species that occurs abundantly at only a few sites. The distribution map has been prepared by the Biological Records Centre.

8.2.4 Naturalness

Little, if any, unmodified vegetation remains in Great Britain, and the only areas that remain relatively unaffected by direct human activity are the remote mountain tops or cliff coasts. Because of this, degrees of naturalness, involving judgements of the various categories, have to be included in nature conservation assessments. Margules and Usher (1981) described four categories of naturalness – undisturbed natural, disturbed natural, degraded natural and cultural. Few examples of the first two of these categories remain in any type of habitat in Great Britain. Most sites are in the 'degraded natural' category and judgements are required on the extent of modification. This is more important

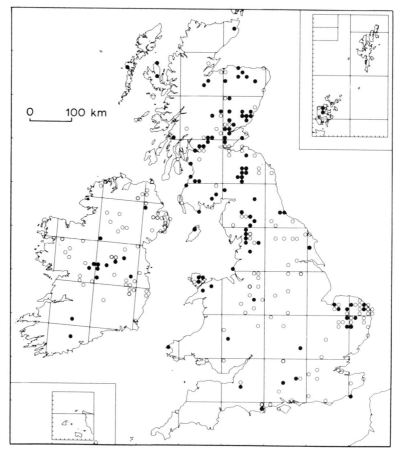

Fig. 8.3   The distribution of *Carex diandra*, a species that occurs in low numbers but which is scattered widely throughout the British Isles. The distribution map has been prepared by the Biological Records Centre, Institute of Terrestrial Ecology.

in habitats such as peatlands where 'naturalness' is relatively more important.

8.2.5 Typicalness

Goldsmith (1983) regards typicalness (and naturalness) as unmeasurable and of limited value in site assessment. However, as Ratcliffe (1977) points out, though the criterion is likely to be covered in the use of other criteria, there will be occasions when its use will identify sites which fulfil none of the other criteria but which fill a gap in the habitat coverage of SSSIs. The judgements required on whether one site is more typical than another are difficult and involve personal knowledge and assessment of all the other examples of that habitat type within the area of search.

8.2.6 Fragility

As Goldsmith (1983) states, the distinction between 'fragility' and 'vulnerability' is confusing and the two terms seem to be used almost as alternatives. In Britain the rate and scale of change of land use and development is so great that virtually all nature conservation sites are vulnerable to external pressures; in addition some habitats, such as permanent grasslands or heathlands, are vulnerable or threatened by internal successional change. In practice this criterion is used more in selecting the number of sites that should be notified as SSSIs and, as Ratcliffe (1977) states, this criterion is related to habitat rarity. All examples of very rare habitats, that are confined to five or less examples in Great Britain, are identified as SSSIs as are all examples of rare habitats that are under threat. Two or three examples of rare, but less threatened habitats, are selected as SSSIs in each area of search.

8.2.7 Other criteria

The other criteria, which are described by Ratcliffe (1977), include recorded history, potential value, position in a geographical unit, and intrinsic appeal. They are all less frequently used, and each involves judgement rather than 'measurement'. However, systematic consideration of each can be made for each site, and occasional examples of sites fulfilling one or other of these criteria do arise.

8.3   OTHER BACKGROUND CONSIDERATIONS

The criteria themselves, as discussed in Section 8.2, are clearly important background considerations when attempting an evaluation or series of evaluations. However, there are three other considerations, which are dealt with in the following subsections.

8.3.1 Scale

The extent of the area within which the comparative assessment of potential SSSIs is made was determined originally by the consultative arrangements between the Nature Conservancy (the forerunner to the NCC) and local planning authorities, arising from the National Parks and Access to the Countryside Act 1949. Thus, SSSIs were selected from within counties in England and Wales, and within districts in Scotland. This also ensured that a geographical spread of nature conservation sites was obtained, representing the present Great Britain distribution and variation of habitat types and their constituent species. In practice this means that areas of search of between 60 000 ha and 400 000 ha are used for site selection. Amalgamations are made when individual counties or districts are too small, and divisions are made when these administrative areas are too large. Similar habitats in adjoining counties or districts are taken into account in the evaluation process.

8.3.2   Classification

To make effective comparisons between potential SSSIs, a classification based on plant communities is employed, using the formations, sub-formations, and facies described in the *Nature Conservation Review*. The main formations are coastlands, woodlands, grasslands, wetlands, peatlands, and uplands, with sub-formations based on geographical, climatic and edaphic factors. Thus, coastlands are divided into flats, salt marshes, sand dunes, shingle beaches, hard and soft cliffs, lagoons and maritime heaths. Where necessary regional variants are identified. The large number of woodland categories used for SSSI evaluation is described by Peterken (1981): the classification is based on composition and structure as well as other factors. Thus, several upland or northern categories of woodland are used, for example upland birch–sessile oak woods, upland hazel–sessile oak woods, acid oak–pine woods, rowan–birch woods. Each of these categories is termed a 'habitat'. Sites may contain several habitats, but each example within a site should be compared with others in the same category from other sites within the area of search.

The habitats defined in this manner will be augmented in the future by the phytosociological categories identified by the National Vegetation Classification, as and when the parts of this new classification become available.

8.3.3   Data collection

The data required for site evaluation vary from habitat to habitat because of the relative and varying importance of the criteria to each habitat. Although some data, such as number of vascular plants and the presence of rare species, are common to all habitats, others vary with the habitat and the judgements of criteria that have to be made.

In all cases decisions on the habitat categories present are necessary, but it is rare that data on all groups of plants and animals are available, and in such cases comparisons are more difficult. Thus, in the case of invertebrates, it is often not possible to compare a series of sites because in only some are species lists available, though potential methods for collecting comparative data are outlined in Chapter 12. The data required for the six main habitat formations recognized in the *Nature Conservation Review* are discussed below.

Coastlands. The area occupied by the site and the habitats within it are estimated. A vascular plant species list, including rarities, is obtained by systematic search of the site, and data on both the invertebrates and birds are collected either directly or, where possible, from records. An estimate is made of the presence of various successional elements such as in the case of sand dunes, mobile and fixed dunes, dune grassland and dune heath, and an estimate of typicalness is made, i.e. the extent to which the habitats within the site represent other examples elsewhere. Other information such as the degree of disturbance, modification of the site, and use for educational or research purposes, is added where possible.

Woodlands. In addition to the estimates of size, the number of habitats present and the degree to which the site is typical of other similar examples are estimated. An estimate of structural variety is made by assessing the extent of the shrub layer, canopy and sub-canopy and the presence of overmature trees. The age of the woodland is investigated using old maps or literature searches. In Scotland the earliest maps of value for this purpose are those of Blaeu dated around 1650, but General Roy's maps made 100 years later are more accurate and can be used to assess whether a woodland should be regarded as 'primary' (*sensu* Peterken, 1981). When available, further information on other groups, including insects, birds and cryptograms, is added, though incomplete coverage makes comparisons more difficult.

Grasslands. As with coastlines and woodlands, estimates of site size, habitat type and range, number of species and rarity, are all made. Past modifications of the site by ploughing, draining or fertilizing are all estimated by literature search, or often by discussion with local farmers or landowners. Present grazing patterns, including different forms of use of the site, are measured either by direct observation or more frequently from discussion with the landowner.

Wetlands. In the case of wetlands, the site's size and the type and extent of the habitats can be readily assessed using air photographs for submerged macrophyte communities. The number of species and the presence of rarities, are much more difficult to measure and require either dredging or grab techniques. A species list can be derived by examination of plant material washed up on shores, but in view of the importance of *Potamogeton* species and doubts about identification of washed up fragments, the more thorough method using a boat is preferred. Information on the degree of modification of water bodies is collected, for example impoundment and fluctuation of water levels, and an assessment of mineral input from the catchment is made, particularly for oligotrophic lakes where unmodified catchments are more important. Assessments of modification of the animal communities, including both fish and invertebrates, are made from discussions with landowners and anglers.

Peatlands. The list of *Sphagnum* species and their extent is important in peatlands, where the total number of vascular plants may be no more than 25, except as a result of habitat modification on the periphery of the site. The presence and extent of *Sphagnum* lawns, as well as pool and hummock complexes, are important for between-site comparisons as a means of identifying the more active mire systems. Estimates of past disturbance, i.e. drainage, mowing and grazing, are made either from direct observation, from the use of air photographs, or from local discussion.

Uplands. As with other habitats, the number of species and habitat types are assessed, but in addition the altitudinal range of semi-natural vegetation is estimated. This means excluding improved grasslands and areas of conifer afforestation. The site's size is of importance in relation to predatory birds, but nest sites and feeding range must be considered in relation to site boundaries and adjacent land.

The comparison of sites on the basis of habitat data may not adequately cover the requirements of rare species or sites that support unusually large numbers of either common or uncommon species. Such examples are not selected on the basis of comparisons between sites, but rather on species distribution information from literature or local sources.

8.4   AN EXAMPLE FROM THE SOUTH EAST REGION OF SCOTLAND

8.4.1   Introduction to the Region

South East Scotland is one of four Nature Conservancy Council regions in Scotland (there are eight in England and three in Wales), and it covers about 800 000 ha. The region includes the five local authority Regions of Borders, Lothian, Central, Fife and Tayside which are divided into seventeen local authority Districts (Fig. 8.4). All of the local authority Districts fall within the size criterion of the area of search, except Perth and Kinross District which for SSSI assessment purposes was divided into two, but which, for administrative purposes such as local authority notification, was retained as a single District.

Borders Region covers the majority of the northern half of the catchment of the River Tweed stretching from the intensively farmed areas of the Berwickshire Merse in the east, to the round topped silurian hills of the Southern Uplands in Tweedale to the west, where conifer affores-

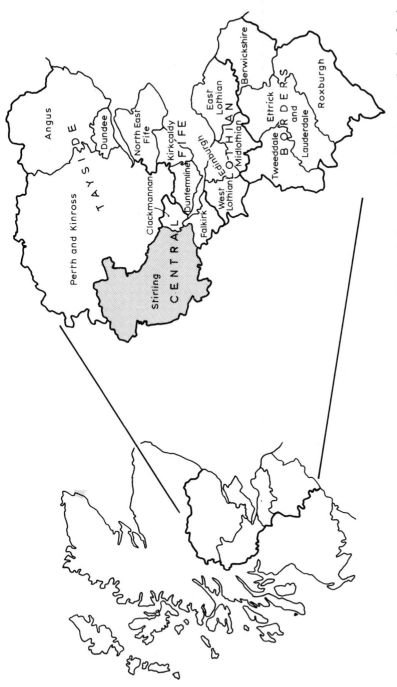

Fig. 8.4 A map of the Nature Conservancy Council's South East Region of Scotland showing the 5 administrative (local authority) Regions and the 17 administrative Districts. The district in which the woodlands, discussed in Section 8.4.2, are located is shaded on the map.

tation and sheep grazing are the two main land uses. The mid-Borders area, which is characterized by mixed farming, rises to an altitude of about 500 m. A history of intensive agricultural activity associated with monastic settlements in the 12th century has led to a reduction in potential SSSIs, particularly in woodlands and grasslands. However, the 'corrugated' topography of the mid-Borders area has meant that a large number of basin mires have formed and have survived, despite mineral enrichment from surrounding agricultural land.

Lothian Region lies on the southern side of the River Forth, east and west of Edinburgh. The southern edge of the Region is bounded by the northern slopes of the Lammermuir and Pentland Hills and the western edge by the millstone grits and carboniferous shales of the central valley of Scotland. East Lothian is intensively arable with potential SSSIs restricted to the uplands or coast or small areas of land which for historical or land-form reasons have remained relatively unmodified. Farming in West Lothian is subject to greater limitations. Altitudes are on average higher, with a correspondingly more exposed climate, and soils are less productive. Land use is concentrated on mixed farming, with relatively little arable, and on forestry. Industrial developments in West Lothian from 150 years ago have left characteristic spoil heaps, many of which have semi-natural vegetation, birch woodland or mires, associated with them. Areas of grazed blanket bog occur to the west where they can be compared with similar sites in the adjacent Districts of Strathclyde Region.

Fife Region represents a mirror image of Lothians, on the northern side of the Forth. Potential SSSIs are confined to the coastal strip and uplands, but the latter are generally lower, less exposed and are subject to more intensive grazing and disturbance than their Lothian counterparts. As in Lothian, mineral poor wetlands are confined to the uplands; in the lowlands unmodified wetlands are rare and all are subject to inputs of nitrate and phosphate from adjacent agricultural land.

The Highland Boundary Fault running north-eastwards from Loch Lomond, divides Central Region into an upland area to the north and west, and a lowland area to the south and east. The lowland area is characterized by intensive arable farming in the Forth Valley and sheep or cattle grazing and conifer afforestation on the peatlands west of Stirling or the plateau uplands to the south. Potential woodland, grassland and wetland SSSIs are all infrequent and small, but peatlands and estuarine mudflats are more common. The upland part of Central Region is underlain by rocks of Dalradian age and is dominated by mountains separated by narrow valleys, some of which are filled by large oligotrophic lakes. Land use is confined to sheep grazing, as well as some cattle grazing and forestry. Oak dominated woodlands, often undergrazed by sheep are relatively common and in some areas the calcareous schists of the Breadalbane hill range outcrop, though their species representation is never as large as at Ben Lui to the west or Ben Lawers or Glen Clova to the north east. The oligotrophic lakes often have well developed mire or fen habitats: unimproved lowland grasslands are scarce.

Tayside Region is large and is divided between three Districts, Perth and Kinross, Angus, and Dundee. As with Central Region, the main ecological division is produced by the Highland Boundary Fault giving an upland and a lowland area each of which may be compared in part with its neighbour in Central Region. In the upland area acid and base rich grasslands are common, as are deciduous woodlands. In addition, native conifer woodlands, dominated by Scots Pine, occur in a few well-known situations. Oligotrophic/mesotrophic lakes, with their associated mires and fens, are frequent, and species rich calcareous grasslands occur in the wider valleys on either side of the River Tay. The lowland area of Tayside is characterized by intensive arable farming with grazing in places and some afforestation. Loch Leven, a large eutrophic lake, lies to the south of this area but there are few other examples of this habitat. Deciduous woodlands are less common than in the north and west and usually occupy inaccessible or steep sites where past management or clearance would have been difficult. Species rich wet grassland is more common, particularly to the west where grazing by grey geese is a feature of winter use.

8.4.2 An example of site evaluation

The list of sites, on which data are collected, is usually derived from a large number of sources, including map and literature searches, air photographs, reports from amateur and professional naturalists and scientists, and museum records. Each habitat within a site is compared with similar habitats on other sites within the area of search. As a result of this process, some sites may rank highly for several habitats, while others for only one. The presentation of the information available for similar habitats on different sites in tabular form facilitates a systematic approach to the assessment of criteria. However, a mechanistic and over simple approach should be avoided.

The eight examples of upland birch–sessile oakwoods in Table 8.1 are taken from a District in Central Region, where a total of thirty such woodlands were visited, surveyed and assessed. All the woodlands listed are primary woodlands and have existed as woodlands for at least 350 years, as confirmed by map and literature searches. Non-primary woodlands were excluded from the table. In preparing the table, assessments of typicalness were excluded because each site seemed to represent adequately the range of habitat variation in all the others so that this

Table 8.1 An example of the evaluation of upland birch and sessile oak woods in the Central Region of Scotland. Diversity is ranked from 1 (= high) to 3 (= low). Local rarities occur in less than ten of the 10 km squares in Scotland, or are known to be rare in the area of search. National rarities occur in less than 15 of the 10 km squares in U.K. Other groups of known rarity are recorded only by group name.

Site	Area of whole site (ha)	Area of habitat (ha)	Structural diversity	Habitat diversity	Number of higher plant species	Rarity			Disturbance	Remarks
						National rarities (higher plants)	Local rarities	Other groups		
WA	71	53	2	3	70+	0	0	Bryophytes Birds	Undisturbed	Public body ownership
WB	13	13	2	3	68	0	0	Birds	Grazed	Public body ownership
WC	9	6	2	3	75	0	3	0	Sporting/ undisturbed	Private ownership
WD	14	14	1	2	109	0	3	Bryophytes	Grazed	Extensive farming
WE	20	18	2	2	43	0	0	0	Undisturbed	Public body ownership
WF	24	22	1/2	1/2	49+	0	0	Bryophytes Birds	Grazed	Private ownership
WG	18	16	2	2	87	1	6	Bryophytes Birds	Slightly grazed by goats	Public body ownership
WH	57	18	1	1/2	58+	0	0	0	Grazed and managed for timber	Private ownership

criterion could not be used in ranking one site more importantly than the others.

Examination of Table 8.1 shows that site WA is the largest (including the largest area of woodland) and contains an average number of higher plant species. Although it has no nationally or locally rare higher plant species, it supports notable populations of bryophytes and birds. It is undisturbed, in sharp contrast to the heavy grazing by sheep and cattle in most of the other woodlands, partly as a result of it being in public ownership. Although site WD is slightly smaller than the average, it supports more higher plant species than any other site, probably because of the habitat diversity introduced by extensive farming. It supports three locally rare higher plant species and notable bryophyte populations. It is privately owned and grazed by sheep particularly in winter. Site WG is of average size but supports an above average number of higher plants including one national and six local rarities, and contains notable populations of bryophytes and birds.

On the basis of this evaluation, amplified by further discussions of the criteria, three woodlands were selected as SSSIs. Site WA is 'the largest, least disturbed example of this type of woodland' in the District. Site WD is 'the most diverse, species rich woodland of this type supporting a number of locally rare vascular and lower plants'. Site WG 'supports several nationally rare vascular and lower plants and nesting rare birds'. In each case, the criteria, for which the site is important, are included in the description of the SSSI and can be used at a later stage when management options are considered.

8.5 CONCLUSION

The evaluation of potential SSSIs involves both a knowledge and understanding of sites and species and their context, and a systematic approach in comparing one with another. Data collection and tabular comparison made in relation to criteria are not straightforward, but do aid the step-by-step approach that is necessary if a reproduceable selection process is to be achieved. Other methods of applying nature conservation criteria to sites may be equally successful in site selection, but in each case criteria must be applied systematically in a way that the layman can see and understand. Once the important sites – SSSIs in the Scottish example – are selected, and the criteria they fulfil are known, decisions on important management practices can be pursued in a more rational framework of land use.

Perhaps the most important conclusion to come from the results of evaluating the birch–oak woodlands in the example is the demonstration of how the criteria are used in practice. In Chapters 3 and 5 attempts were made to integrate various attributes of the site into some kind of an index, after which the sites could be ranked according to the index value. In the example in Table 8.1 the woodland sites are not ranked, but the selection of three of them for scheduling as SSSIs shows that three criteria have been given prominence. Thus, WA was scheduled on the basis of the criterion of 'area' (although lack of disturbance also featured), WD on the basis of 'diversity' and WG on the basis of 'rarity'. The selection of these three sites has, therefore, depended on a multi-factor approach rather than being based on, say, the top three in terms of some arbitrarily defined criterion.

An approach such as this, where many of the attributes valued by society are included in the sites that are safeguarded, does allow for the values, placed on the criteria by society, to change without seriously changing the selection of areas. Such an advantage is of considerable benefit in the long-term use of land for nature conservation.

8.6 SUMMARY

This chapter is concerned with the process of evaluation within a small area of one nation. Although the criteria available to the evaluator are the same as those for national appraisals, it is likely that only a subset of the criteria will prove useful.

Several of the criteria are briefly reviewed: these include area, diversity, rarity, naturalness, typicalness and fragility. After a classification of Scotland's semi-natural communities into 6 broad categories, the criteria useful in each habitat type are considered. For coastlands the most important criteria are likely to be area, species diversity, rarity and successional state; for woodlands they are area, habitat diversity and naturalness; for grasslands they are area, both habitat and species diversity, rarity and naturalness; for wetlands they are the same as grasslands although more emphasis is placed on modification of the animal communities; for peatlands habitat diversity, diversity of *Sphagnum* moss species and lack of disturbance are the most important; whilst for uplands area, species and habitat diversity and altitudinal range are important.

The concepts are illustrated by reference to birch and sessile oak woods in the Central Region of Scotland. Data for 10 (out of 30) candidate woods are presented in tabular form, and a selection of three woods is made on the basis of the criteria of area, species diversity and rarity, together with lack of disturbance.

PART THREE

Specific habitats and groups of organisms

CHAPTER 9

Forest and woodland evaluation

KEITH KIRBY

9.1 Introduction
    9.1.1 Who uses the evaluation?
    9.1.2 Defining both the population and the evaluation units
    9.1.3 The effects of forest dynamics
9.2 The basis for many woodland evaluations
    9.2.1 Naturalness
    9.2.2 Species richness
    9.2.3 Rare or 'special' species
    9.2.4 Past records and future prospects
9.3 Examples of woodland and forest evaluation schemes
    9.3.1 Tropical forest evaluation
    9.3.2 Evaluations in mixed landscapes
    9.3.3 Between wood comparisons
    9.3.4 Within wood comparisons
    9.3.5 Assessing long-term change
9.4 Evaluation of British semi-natural woodlands
    9.4.1 Scoring systems
    9.4.2 Survey methodology in relation to evaluation
    9.4.3 Integrating and interpreting the data
    9.4.4 Achieving a representative selection
    9.4.5 How many sites should be protected?
    9.4.6 Judging the evaluation
9.5 Conclusions
9.6 Summary

Wildlife Conservation Evaluation. Edited by Michael B. Usher.
Published in 1986 by Chapman and Hall Ltd, 11 New Fetter Lane,
  London EC4P 4EE
© 1986 Chapman and Hall

9.1   INTRODUCTION

Woodland wildlife conservation can be approached in two ways. First, through the protection of the individual species and ecosystems that are characteristic of natural and semi-natural woodlands. Secondly, through the identification of practices and features within forest production systems that benefit wildlife generally – not necessarily just the wildlife of the natural forest. Wildlife conservation is distinct from, though related to, other aspects of conservation. Plantations of introduced tree species may maintain the water and nutrient balance of an area (good resource conservation) and make an attractive landscape (good landscape conservation) but reduce populations of the native flora and fauna (poor wildlife conservation).

In this chapter various factors which affect the type of woodland evaluation used are considered together with some of the assumptions that lie behind many evaluation systems. Among the examples, methods for assessing the value of semi-natural woodland in Britain are examined in detail.

9.1.1   Who uses the evaluation?

Different groups of people seek different results from woodland evaluations. Planning authorities may wish for the grouping of all woodlands into a few distinct categories which range from those that are very valuable, to those of little importance, for wildlife (Selman, 1981). Conservation organizations may wish to have a finer ranking of sites which are potential nature reserves and to lump together the other sites. In deciding what constitutes 'important for wildlife' the general public may wish to give most weight to impressive or attractive species (Bull, 1981).

Naturalists are often most concerned with the protection of rare species or the protection of natural areas, whilst specialists may assess a site according to its value for a particular taxonomic group (for example see British Lichen Society, 1982) or for some particular feature such as undisturbed soils (Ball and Stevens, 1981) or the genetic resource contained in the native pinewoods of Scotland (Steven and Carlisle, 1959). The evaluation results are usually expressed on some internally defined scale of conservation value, but there have been attempts to convert these values to monetary terms.

9.1.2   Defining both the population and the evaluation units

Before evaluation can proceed the following need to be defined: (i) the total population or area of sites within which the evaluation is to take place, and (ii) the basic unit for evaluation. In Britain, for example, some systems evaluate large stretches of countryside using map grid squares as the unit for comparison (Tubbs and Blackwood, 1971), while others compare individual sites (Goodfellow and Peterken, 1981).

Site-by-site comparison methods break down in areas of more or less continuous, fairly uniform woodland, because the boundaries used to define sites may be purely arbitrary. In such regions the conservation need is to identify 'special localities' that contain features or species absent from the bulk of the woodland, as well as the boundaries of the area that contains all the woodland of roughly comparable quality, any part of which could be selected as representative of the whole (Peterken, 1981).

Even where woodland is more discrete, the precise boundary of a site may be ambiguous, if for example, a wood is subdivided by clearance (Fig. 9.1a), or previously separate sites become linked by recent woodland development (Fig. 9.1b), or where a wood merges into some other habitat (Fig. 9.1c). Very different results might be obtained in terms of the number of sites involved in each case, their area and their species content depending on the rules used in the survey and evaluation.

Priorities within an evaluation vary according to the region, as well as the scale, on which it takes place. In Lincolnshire, woods dominated by

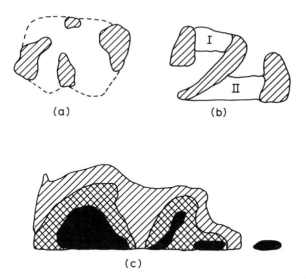

(a)   (b)

(c)

Fig. 9.1 Problems of site definition caused by (a) the creation of four small woods from one large woodland block by recent agricultural clearance (‒ ‒ ‒ former woodland boundary, ⎯ modern boundary); (b) linkage of three formerly separate woods by new plantations of introduced tree species (I) and naturally regenerated woodland (II); and (c) gradual transitions from woodland to non-woodland habitat without clear-cut boundaries (▬ tree cover more than 60%, ▨ tree cover 26–60%, ▨ tree cover 6–25%).

lime (*Tilia cordata*) are fairly common and a county evaluation might give more weight to other woodland types (Game and Peterken, 1984). In Britain as a whole, such woods are rare and, in a national selection of important sites, are well represented (Ratcliffe, 1977).

An evaluation may initially compare all sites independently to produce a ranking, but often one aim is to identify a group of woods which between them contain the main features of conservation value. In such circumstances, the potential value of two woods as additions to a group of existing reserves may differ from their individual conservation value when considered in isolation (Hooper, 1981; Kirkpatrick, 1983). A species poor wood may be important as a possible reserve if it contains features not represented in the group, whereas a much richer wood may only duplicate conditions found already in existing reserves. Several small woods together may be more valuable than one large wood (Higgs and Usher, 1980). A degree of opportunism, acquiring as reserves whichever woods become available, may be desirable (Game and Peterken, 1984).

9.1.3 The effects of forest dynamics

Woodland evaluation must take account of the dynamics of the forest system which includes successional stages and regeneration cycles spread over many years (Grubb, 1977; Pickett and Thompson, 1978). The patterns formed may be either natural or man-induced. As a result, the present state of a wood is not always a true reflection of its long-term value.

A recently felled and restocked conifer plantation may temporarily hold more species than a semi-natural coppice wood whereas over the whole rotation the latter tends to be more species rich. If all tree age-classes are not represented equally on a site, the populations of the flora and fauna associated with individual age classes will fluctuate correspondingly (Ash and Barkham, 1976; Hill, 1979). Continuity of habitat is important for some species, especially invertebrates with annual life cycles. Absence of their required food plant for even a few years, or the shading over of glades where the species normally occurs, may eliminate the species from the site. Plants, on the other hand, may survive unfavourable periods as buried seed (Brown and Oosterhuis, 1981) or regeneration may occur only at infrequent intervals (Peterken and Tubbs, 1965). Woodland successional stages pose particular evaluation problems as do all successional sites (see Chapter 3). Regeneration of yew (*Taxus baccata*) at Kingley Vale Reserve (Sussex) occurs not within the wood but in adjacent scrub and grassland (Tittensor, 1980). These areas are an essential part of the woodland system, but might be difficult to fit into a woodland evaluation scheme.

9.2   THE BASIS FOR MANY WOODLAND EVALUATIONS

9.2.1   Naturalness

Large areas of natural forest remain in parts of the tropics, but in Europe most woodland has been cut-over or burnt by man, or it has been grazed by domestic stock. The larger forest mammals (bison, bear, wolf, beaver) have been eliminated from most forests. In Britain, after many centuries of management, woods may at best be classed as partially semi-natural.

One problem in determining the naturalness of a forest is that both management and natural processes may sometimes produce the same stand structure (Whitehead, 1982). Virgin stands in the northern temperate zone are often partly even-aged (Jones, 1945), a feature more commonly associated with plantations. Similarly, mixed-aged, apparently natural stands may result from changes in the grazing patterns of domestic stock in managed woods (Tittensor, 1981). Nevertheless some sites and features within them are clearly more natural than others. The distribution of trees and shrubs in the coppice layer of British woods, that can often be traced to underlying variations in soil, and where successive generations of trees are derived from previous ones by stump regrowth or natural regeneration, is more natural than the tree layer in a plantation. Within a plantation the ground flora is usually more natural than the tree layer.

Forestry systems whose woodland structure mimics natural forests tend to be valuable for wildlife, but so also may be other, more artificial systems such as the 5–20 year coppice system, formerly widespread in Britain. The latter has allowed a wide range of the native woodland flora and fauna to survive within quite small forest blocks.

Peterken (1977) distinguishes 'past-natural' features that survive from the original natural woodland, from 'future-natural' ones that will become part of natural woodland allowed to develop from this point onward. Old stools of lime (*Tilia cordata*), possible descendants from the extensive lime forests that once covered the English Midlands, are a 'past-natural' feature of some old woods. The grey squirrel (*Sciurus carolinensis*), an introduction in the nineteenth century, will become a 'future-natural' feature of British woods. 'Past-natural' features, and sites which contain them, are generally more important for nature conservation than 'future-natural' ones (Peterken, 1977, 1981).

9.2.2   Species richness

Wildlife conservation is ultimately about species survival. Sites which contain self-supporting populations of many species are more valuable than those with few. The simplest measure of species richness for a site

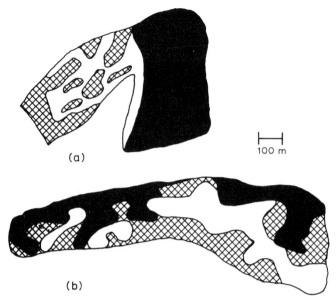

Fig. 9.2 Variations in species richness within a wood; (a) total number of vascular plants recorded from ash–hazel area (■■) 71, from alder areas (▭) 82, from the oak–birch areas (▦) 47; (b) number of species found in a 200 m² quadrat, ■■ more than 30, ▦ 21–30, ▭ 11–21 vascular plants.

is the number of species on a species list, although diversity indices that take some account of the population size for each species can sometimes be more useful. Single estimates of species richness become less valuable as the area surveyed gets larger or more variable. Species are seldom distributed evenly, so that separate lists for different areas, and maps either of plant communities or of particularly species rich areas can aid evaluation (Fig. 9.2).

Many evaluations use vascular plant richness as a measure of conservation interest, because plants are often the easiest group to record, and they provide the food and habitat for other species. Lower plants are more difficult to identify so that their enumeration often relies on visits to an area by specialists. Direct observations on birds (see Chapter 11) mammals and invertebrates (see Chapter 12) require several visits to a site, either during particular conditions (for example warm, sunny days for butterflies) or during a restricted time of year (such as the breeding season), to produce adequate species lists. Standard recording methods are, however, available, for example see Ralph and Scott (1981).

Direct measures of the number of species may be replaced by indirect assessments based on the relation between species richness and features such as woodland structure. The different layers of vegetation, the presence of dead trees, the occurrence and distribution of open areas and water bodies, all greatly affect the range of species present, whether invertebrates, birds or mammals are considered (for example: MacArthur, 1965; Elton, 1966; Stubbs, 1972; Massey, 1974; Geier and Best, 1980).

Species richness may be misleading if the evaluation is directed specifically towards woodland conservation: a site may be rich because of the range of non-woodland habitats it contains. This effect may be reduced by either keeping records from non-woodland areas separate during the survey, or by only counting in the evaluation those species that are associated with, though not necessarily confined to, woodland. Woods may also be rich in one group of woodland species such as bryophytes, but poor in others. Different types of site may need to be favoured in the evaluation for different groups of species.

Assessment may be further refined to concentrate on those species characteristic or indicative of long-established or undisturbed woodland conditions. In Britain potential indicator species have been found among vascular plants (Pigott, 1969; Peterken, 1974), lichens (Rose, 1976), bryophytes (Ratcliffe, 1968), molluscs (Boycott, 1934; Cameron and Down, 1980) and beetles (Hunter, 1977; Harding, 1981). The behaviour of such species, however, varies from region to region; vascular plants with a strong affinity for ancient woodlands in Lincolnshire (Peterken, 1974) are not always the same in Essex (Rackham, 1980) or Sussex (Tittensor, 1981). The presence of only one or two 'indicator' species is not as useful as the presence of a whole suite of such species.

9.2.3 Rare or 'special' species

Evaluation schemes often give extra weight to rare or threatened species because they may be less likely than common ones to survive without special conservation measures. Selection of sites based on their rare species content may be as effective an evaluation method as selection methods based on total plant lists (Game and Peterken, 1984). Species may be rare in the region concerned but common elsewhere, or fairly common in the region but rare in a wider context. For example, the may lily (*Maianthemum bifolium*) is rare in Britain but common in Europe. Species may be threatened because they have poor colonizing abilities; new populations are unlikely to develop and replace those lost when a site is destroyed. Species may also be vulnerable if a large proportion of the total population depends on particular small areas, as black hairstreak butterflies (*Strymonidia pruni*) tend to concentrate around only a few blackthorn bushes (*Prunus spinosa*). Other examples are given in Peterken (1981).

Decisions as to which species should be classed as rare or threatened can be made more or less objectively from a knowledge of their distribution and life history. Other species may be given special weight in an evaluation according to the interests of the user (Adamus and Clough, 1978; Helliwell, 1973). Orchids, however common, tend to excite more interest than rare slugs. Provided it is accepted that the decision to give certain species additional weight is subjective, their treatment in the subsequent evaluation can be the same as for rare species.

9.2.4 Past records and future prospects

Evaluations are usually based on surveys carried out at one particular time and therefore, they reflect only the current interest of the site. Major external threats such as proposed motorways or agricultural improvement schemes may be envisaged, which mean that whatever the current state, the future value of a site for woodland conservation is very low unless such threats can be averted. Equally there may be internal indications of future change, either natural or man-induced, which suggest that the present value is unlikely to be maintained. Interpretation of an evaluation exercise needs to allow for such factors.

Historical aspects of woodland nature conservation in Britain have recently received much attention (for example Rackham, 1976, 1980). Woods may contain features of historical and archaeological interest such as old banks or settlement remains, or they may be examples of former management systems (deer parks, working coppices). Woods may be associated with famous events or people; other sites are notable as the sites used in earlier research work. These do not directly alter the conservation value of a wood, but may increase the interest of an area in a broader ecological sense and the potential for future research may be greater than in sites whose history is unknown.

9.3 EXAMPLES OF WOODLAND AND FOREST EVALUATION SCHEMES

Evaluation schemes may include comparisons at a variety of levels. From the widest to the narrowest they may include comparisons of broad tracts of country so as to identify potential national parks in the tropics, to comparisons within one particular woodland. At any of these levels there can be a need to assess changes over time (and hence the requirement for monitoring).

9.3.1 Tropical forest evaluation

Ashton (1976) suggests that 2000 ha might be the minimum area needed to support viable populations of the main forest trees in Sarawak, and for protection of the major types in tropical rain forest even larger areas will be needed (Myers, 1981, and see Chapter 4). The principles used in selecting such areas – the need for large forest blocks to maintain species of special interest (Lovejoy et al., 1983), the inclusion of examples of different communities and particularly species rich areas or rare populations (Ashton, 1981) – may not be dissimilar to those used for British woodland reserves, but the scale of working is very different. Detailed ecological knowledge of the areas is usually lacking, and the problems of setting up the reserve areas are very much greater.

9.3.2 Evaluations in mixed landscapes

Comparisons of mixed landscapes including woodland are usually based on some form of habitat mapping (Tubbs and Blackwood, 1971; Buse, 1974; Goldsmith, 1975; Buckley and Forbes, 1979; Nichol, 1982) involving direct field survey of the areas, remote sensing techniques, including the use of satellite imagery, or data extracted from existing maps (Bunce and Smith, 1978). In mixed landscape comparisons the presence of woodland, even plantations, is usually rated highly because its potential wildlife interest is much greater than the surrounding agricultural land. Another type of comparison is involved in the debate in Britain over the conversion of upland moorland to coniferous plantation. The relative conservation value of the two types of habitat has been assessed using the species found in each (Helliwell, 1972). Other studies have examined the changing bird populations during and after the period when an area is afforested (Lack and Lack, 1951; Moss, 1979). However, the main point at issue is not whether one habitat is in some way richer than another, but concerns the effects of replacing semi-natural moorland habitats by plantations and the consequent loss of particular, individual species.

9.3.3 Between wood comparisons

Between wood evaluations may simply group sites into broad categories of importance for wildlife using, for example, information on old maps and aerial photographs (Fig. 9.3) (Kirby et al., 1984). Since in Britain long-established semi-natural woodland tends to have the greatest importance for wildlife conservation (Peterken, 1977) a simple evaluation of all sites is produced. Evans (1984) similarly uses a dichotomous key, based largely on the management and on the tree species present, to group woods for conservation purposes. More detailed evaluations usually involve some form of field survey of the area; features thought to be of importance for wildlife are assessed, often in a semi-quantitative way, and the woods are ranked on the basis of these results. Most of the

Fig 9.3 ■ sites which have been continuously wooded since AD *c.*1600 and retain a semi-naural tree layer; ▦ sites which have been continuously wooded since *c.*1600, but which have now been replanted; ☐ sites which have developed as woodland only in the last 400 years. The first category is the most important for nature conservation as a rule. (From an unpublished report on Essex woodlands by J. W. Spencer.)

features used in evaluations of semi-natural woodland are direct or indirect measures of the diversity of the site, and hence of its probable species richness, or are indications of relative naturalness or rarity (see Table 9.1). Other examples can be found in Edwards (undated), Forbes (1979), Keymer (1980), MacDonald (1981) and Smith (1981), while Sparrowe and Sparrowe (1978) illustrate an equivalent approach used in North American forests.

9.3.4 Within wood comparisons

Comparisons of areas within the same block of woodland tend to employ similar methods to between site comparisons, but there is the important ecological difference that the areas concerned are not isolated from each other. Mount (1981) and Anonymous (1981b) used vegetation sampling to assess the wildlife interest of different areas within conifer plantations in southern Scotland. They underline the importance of unplanted areas, streamsides, broadleaf fringes, etc., in increasing the

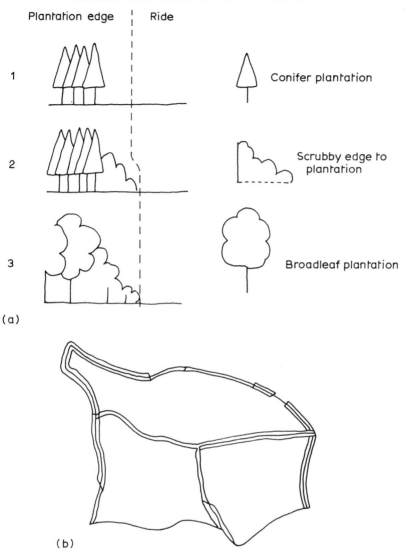

Fig. 9.4 Assessment of forest rides as habitats for invertebrates, using (a) plantation edge structure types (likely value for invertebrates 1<2<3); and (b) relative abundance of nectar-supplying plant species per length of ride (≡ high,═ medium,— low).

range of plants found in large blocks of relatively uniform age and species composition. Open areas can attract butterflies (Jenkyn, 1968), and also birds which would not otherwise be present (Moss, 1979). Maps of habitat features help identify the main areas likely to be of higher than average interest in the plantation. Elsewhere, the value for wildlife of retaining native woodland areas within plantations is generally recognized (for example Roche, 1979; Friend, 1980).

In southern England many broadleaf woodland blocks have been converted to coniferous plantations. Much of the former flora and fauna of the individual compartments has been lost, but the management of the rides has sometimes enabled rich invertebrate populations to survive (Peachey, 1980). Rides likely to be rich in invertebrates may be identified using either ride edge structure, or food plant distribution (Fig. 9.4) (England Field Unit, 1982). Peterken (1981) used an index of the number of plant species found in different compartments to pick out a species rich block which might form the basis of a reserve. Alternatively, a map of vegetation types could be used to identify the areas of highest variability within the wood. In studies of the flora under different tree crops by Ovington (1955) and Anderson (1979), frequency of species occurrence in quadrats was used as the basis for within site evaluation.

9.3.5 Assessing long-term change

Changes in the wildlife value of a site may be detected by comparing surveys from two different dates provided that (i) the original records are stored in such a way that they are not lost, (ii) the areas to which the records refer are relocatable, and (iii) the methods used at the two times are similar.

Examples of this type of study include work on the changes in the flora of a developing conifer plantation (Hill and Jones, 1978) or the effects of reductions in grazing levels within a wood (Pigott, 1983). Other monitoring systems may involve either permanent sample plots (Dawkins and Field, 1978; Sykes, 1981) or systems based on a series of well documented photographs (Proctor, Spooner and Spooner, 1970; Thalen, 1979). Monitoring year to year changes may be particularly valuable for some animal groups, for example the 'butterfly transect' recording scheme (Pollard, 1982).

9.4   EVALUATION OF BRITISH SEMI-NATURAL WOODLANDS

Woodland evaluations try to extrapolate the overall conservation value of a wood from a limited number of quantitative and qualitative observations. An analogous problem confronts soil scientists who must make broad land capability classifications from a limited number of field measurements (Davidson, 1980). There is usually good agreement between different people about the merits of very good or very bad sites (Margules and Usher, 1984), so that formal evaluation schemes are most useful in assessing sites of intermediate quality. Subjective judgements are involved in all stages of the evaluation, including the choice of the features used in the evaluation, the way they are recorded (see Table 9.1), and the means of combining the scores for the different features in the final assessment.

Most of the surveys discussed here involved a single visit of between three and eight hours for a wood of about 30 ha. It was generally impractical to make direct detailed records for groups other than vascular plants, although for some sites additional records were available from previous surveys. Such extra information must be considered in the evaluation, since to know from past survey that a wood has a rich bird community is preferable to presuming that the bird life should be rich because the wood has a diverse structure. Past records may also help to interpret the current state of the wood. However, some records may be out of date: for example the species found ten years ago are no longer there. Also, the evaluation may become biassed towards sites which are already well known because, for example, the extra time spent recording in the wood will result in more species being found.

9.4.1 Scoring systems

Table 9.1 includes a variety of scoring systems for features such as size, number of plants, habitat diversity. Little emphasis should be placed on small score differences between woods, particularly with respect to features such as woodland structure where ambiguity in the definition of categories, or differences due to recording location within the wood, may lead to high variability between observers. Even wood size may not always be easy to define precisely (see Fig. 9.1). Grouping woods into size classes of increasing range (0–2 ha, 2–5 ha, 5–25 ha, etc.) to some extent reduces this problem. The relatively small size differences between sites in the same size class are then ignored for evaluation purposes.

The importance, in nature conservation terms, of differences between evaluation scores is usually harder to interpret the more complicated the score used. Comparing woods on the basis of the number of plant species in each, the number of rare species, or the number of introduced species, is easier to interpret than the use of a composite score where, for example, most species are scored at 1 point each, rarities at 3 points each and introductions at −1 point each. Graphs of the number of species in a site plotted against area show which woods are richer than average for their size. However, there is no general rule that a small species rich wood is necessarily more important than a large, species poor one; nor is a wood with an average number of species for its size necessarily 'typical' of that woodland type.

In-built biasses towards particular types of woodland occur in the use

Table 9.1 Features and scoring systems used in British woodland evaluation schemes.

Feature	Method of scoring	Reference*
Size	1 point per 0.4 ha of wood (up to 100 points maximum)	(g)
	1 point per 1.0 ha of wood (no upper limit)	(f)
	1 point per 4.0 ha of wood (no upper limit)	(d)
	1 point per 10.0 ha of wood (no upper limit)	(c)
	Grading system (in ha), I, >125; II, 41–125; III, 13–40; IV, 5–12; V, 1–4; VI, <1	(a)
	Each vegetation type in wood assigned to one of the following categories: <0.5, 0.5–2.0, 2.1–10.0, >10.0 ha	(b)
Vascular plant species richness	1 point for each species (up to 100 points maximum)	(g)
	1 point for each 'woodland' species (no upper limit)	(b)
	As above + extra points for rare species	(d)
	As above + extra points to rare species, negative points for introduced species	(e)
	1 point per 5 tree or shrub species + 1 point for each 'indicator' species present	(c)
	1 point for each species + species area index calculated (I = species no./log area)	(f)
	I–VI grading of woods according to number of 'woodland' species relative to the mean number for woods of a similar size	(a)
Structure	Points given for presence in the wood of 15 structural categories, including trees, coppice, pollards, young trees, saplings, seedlings	(c)
	Points for presence in wood of well-developed shrub layer, mixed-age structure, mature trees	(d)
	Points given according to % tree and shrub layer in sample quadrats and presence in quadrats of over mature trees, regeneration, creepers	(f)(g)
	I–VI grading of the woods based on overall assessment of structure by the surveyor	(a)
Other habitats in the wood	Points given for presence in wood of the following habitats: pond, stream, ditch, wet flush, glade, ride, bank	(c)
	As above but with the following habitats: pond, stream, flush, glade or ride, dead wood, rocks	(d)
	As above but with the following habitats: pond, stream, river, marsh, ride, glade, field, track, path, boulder outcrop, cliff, scree, gorge, hedge, and dead wood abundance estimated from sample quadrats	(f)

	As above but with over 60 distinct categories within the following habitat types: water bodies, open areas, rock, dead wood, human artefacts (old buildings, roads etc.)	(g)
	I–VI grading of the woods based on surveyor's assessment of the state of the boundary present in the wood	
Adjacent land	Points given for adjacent semi-natural vegetation	(a)
	Points given for adjacent broadleaf woodland	(d)
	I–VI grading of the woods based on surveyor's assessment of the state of the boundary and adjacent land	(c)
Management	Points for quality of past management and assessment of the woods future prospects	(a)
	Points for lack of grazing	(c)
	% cover of native trees and shrubs used as indication of good management	(d)(g)
	I–VI grading of the woods based on surveyor's assessment of the quality of the management	(f)
Other	Quality of historical records for wood	(a)
	Vegetation types present	(a)(b)
		(a)(b)(c)(d)
		(e)(f)(g)
	Geology of the wood	(d)
	Presence and abundance of dense bryophyte carpets on ground	(d)(f)

*References

(a) East Anglian woodland evaluation (Dr O. Rackham, Cambridge University, unpublished)
(b) Norfolk woodland evaluation (Goodfellow and Peterken, 1981)
(c) Hertfordshire woodland evaluation (Hinton, 1978)
(d) Dyfed-Powys woodland evaluation (Massey, Peterken and Woods, 1977)
(e) Tayside woodland evaluation (Dr R. Smith, Nature Conservancy Council, unpublished)
(f) Gwynedd woodland evaluation (Smith, 1981)
(g) Yorkshire Dales woodland evaluation (Pilling, Gibson and Crawley, 1978)

of some of the features in Table 9.1. Vascular plant scores favour naturally species rich types or sites with a mixture of types against those which are more uniform or of a species poor type. Ranking woods according to their size is likely to bias the evaluation towards woodland on infertile or difficult soils since large woods are less likely to survive on good agricultural land.

Simple presence/absence scoring for features such as streams or over-mature trees becomes less useful in large sites because most sites contain at least one such feature. Scores based on the frequency with which the feature occurs in a sample series of quadrats in the wood are then more appropriate. Alternatively presence/absence scores may be supple-mented by descriptions of the nature, extent or abundance of that feature in a given site.

The importance of some features varies so that a particular score for, say grazing, must be assessed in the context of a given site. A given level of grazing may be beneficial in some woods, because it encourages the growth of a dense bryophyte mat, largely irrelevant to the wildlife interest in others, and detrimental in another wood where regeneration is suppressed.

9.4.2 Survey methodology in relation to evaluation

The smallest difference in any score that can be accepted as significant in the evaluation must be larger than the level of error inherent in the survey method. Equally there is little point in striving for greater preci-sion in the survey method than can actually be used in the evaluation,

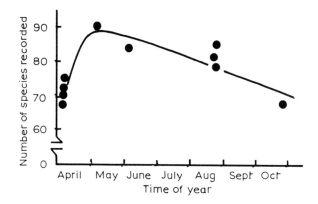

Fig. 9.5 Number of vascular plant species (out of 148 known to be in the area) recorded on a three hour walk through 30 ha of broadleaf woodland at different seasons of the year. The lines are drawn through the points by eye. (Kirby *et al.*, 1986.)

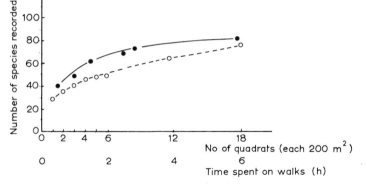

Fig. 9.6 The relation between survey effort and the number of vascular plant species recorded for quadrat surveys (o---o) and walk surveys (●—●) in a broadleaved wood in April. Each quadrat took about ¾ hour to record. The lines are drawn through the points by eye. (Kirby *et al.*, 1986.)

particularly if the increase in precision results in fewer sites being visited (Goode, 1981). As far as possible the intensity of survey between sites, and certainly the survey methods used, should remain constant from wood to wood.

Most woodland survey for nature conservation purposes in Britain involves plant lists and descriptions made either during a walk through a wood, or from a series of sample points in the wood, or a combination of the two approaches (Peterken, 1981; Bunce and Shaw, 1973; Kirby, 1982).

The need to consider the survey methodology in interpreting the results of an evaluation is illustrated in Figs 9.5, 9.6 and Table 9.2. Variations in the number of plant species recorded from the one wood were found between survey methods, survey intensities, and the time of

Table 9.2 Variations in the number of vascular plant species recorded by two observers (A and B) using two survey methods in the same wood in April. Each value is the number of vascular plant species found either on one of six different walk routes (each 3.0–3.5 km in length) or in one of six sets of randomly placed, 200 m² quadrats (six quadrats per set) (Kirby *et al.*, 1986).

	Observers	
Method	*A*	*B*
Walk survey	80	78
	83	88
	86	89
Quadrat survey	51	55
	59	56
	59	62

year, and also between the results of two independent observers using the same method at the same time of year (Kirby *et al.*, 1986). Anything which makes survey work more difficult (bad weather, dense undergrowth, steep slopes) can affect results. If the differences between the survey intensity in different woods are large it may be worth considering 'correction factors' (Goodfellow and Peterken, 1981). Ralph and Scott (1981) discuss how differences in survey methodology may affect bird survey results.

9.4.3   Integrating and interpreting the data

In some evaluation schemes the numerical scores for different features (size, structure, number of plants, etc.) are combined, by addition or multiplication, to produce a single value for each wood. This simplifies subsequent comparisons of the scores for different woods, but masks the real complexities of the differences between sites. The features are not independent (size and number of species for example) and both positive and negative correlations may exist between them. It is difficult to define levels of equivalence between the feature scores. As a result, a small, bryophyte rich wood with a well developed structure might produce the same overall score as a larger wood with fewer bryophytes, a poor structure, but more vascular plants.

Sites can be set out in a series of 'league tables' according to their separate scores for each feature or according to some combined score. Any qualitative information about the sites (such as 'contains rich butterfly populations', or 'has good historical records back to 1500') is set alongside their positions in these rankings. This qualitative information may be sufficiently important to outweigh some of the smaller numerical differences between woods so that some re-ordering of the tables is needed. A knowledge of the woods and the survey methods may suggest at this stage that one of the rankings is likely to be a better indicator of conservation values than the others. For example, the scoring for vascular plants may have been generally more consistent between observers than the scoring for woodland structure; conversely the vascular plant scores may be thought unreliable because not all records could be made at the most favourable season. Alternatively, all the rankings may be regarded as equally important.

If there are positive correlations between the scores for the different features, sites can be selected which are highly ranked in several tables. If rankings in some tables are inversely correlated, then one option is to identify sites that do not have any particularly low rankings across the board. Equally, however, they are unlikely to have any very high rankings and this option could produce a series of generally mediocre sites. Alternatively, distinct groups of sites can be chosen, each of which ranks highly for one set of features. In Britain, for example, two separate groups of sites may be selected: parklands, which are rich in mature timber invertebrates and lichens but generally poor in vascular plants, and coppice woods where the reverse is often true.

9.4.4 Achieving a representative selection

One object of some woodland evaluations is to produce a selection of sites from across the spectrum of natural and semi-natural woodland, whether the spectrum of variation is defined in terms of vegetation types, geology or woodland structure (Massey, Peterken and Woods, 1977; Ratcliffe, 1977; Goodfellow and Peterken, 1981). Making a representative selection is a separate exercise from the ranking of woods discussed earlier, because it may be necessary to choose woods low down the main ranking to represent certain types. Woodland characteristic of acid soils, for example, tends to have few vascular plants, and hence is correspondingly low in a league table of plant richness scores. Where woods contain more than one type (however defined) there are advantages in evaluating the area of each type independently of the wood as a whole.

9.4.5 How many sites should be protected?

In any region there are usually a small number of very high value sites and a large number of sites with a value that is slightly below average.

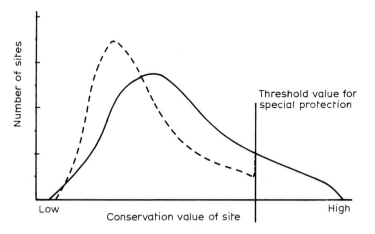

Fig. 9.7 Existing and probable future distribution of conservation value between sites if there is complete protection above a certain threshold level but generally unsympathetic management on the majority of sites (——— existing distribution of value,----future distribution of value).

As a result of the evaluation the high value sites may receive special protection (Fig. 9.7). However, serious reduction of conservation interest through unsympathetic management in sites which fall just below the threshold level could be just as damaging to overall wildlife conservation as damage to one or two sites just above the threshold. The more likely it is that sites are to be damaged, the lower the threshold level needs to be set.

9.4.6 Judging the evaluation

There is no absolute standard against which any evaluation can be judged. However, once any sort of ranking of sites has been produced, there should be a final check against the full survey records to ensure the following three points.

(1) Sites towards the top of the list individually, and collectively, should contain the features that were identified as important in the original survey and in the evaluation objectives.
(2) Sites towards the bottom of the list should not contain any important features that are not found in the top sites.
(3) There should be no borderline sites whose position must be changed due to the omission of some features from the scoring system.

9.5 CONCLUSIONS

Woodlands and forests are complex habitats in terms of their structure and species composition, both of which are heavily influenced by the past history and management; perhaps more than in any other habitat, the need to define precisely what the evaluation should achieve is crucial. No single evaluation method is appropriate for all occasions, although three criteria – naturalness, species richness and rarity – form the basis for most methods. There is considerable variability in the way in which these concepts can be translated into woodland attributes which can be recorded in the field, and conversely the methods used to record the attributes, whether they are area measurements or estimates of the number of vascular plant species present, introduce biases or sources of error. Subjective decisions and judgements cannot be totally eliminated, but a measure of objectivity is introduced into woodland evaluations by treating the evaluation process in a systematic and preferably quantitative, manner – as far as this is possible.

9.6 SUMMARY

Woodland evaluations aim to compare both sites and forestry practices so as to determine which contribute most to woodland conservation objectives. These objectives, and hence the evaluation method, vary according to the interests of the users and in relation to whether a broad grouping of sites or a detailed ranking of them is required. The region within which the evaluation takes place, and the size of the sites themselves, also affect the result. The evaluation procedures must allow for the effect of regeneration cycles and successional patterns within the woods.

Most evaluation schemes are based on the assumptions that natural and semi-natural areas are more important than plantations; that species rich areas tend to contribute more than species poor sites; and that particular attention needs to be given to rare, threatened or otherwise special species. Historical features are also highly valued within British woods.

Examples of different levels of evaluation are given: these range from comparisons of broad tracts of country to surveys of differences within a wood. The evaluation of British semi-natural woodlands is discussed in detail. The features which have been used to characterize the conservation value of a wood, and some methods of quantifying these features, are described. The influence of survey methods on the results, ways of combining the observations made on different features, and the selection of representative woods, are mentioned.

CHAPTER 10

Evaluating the wildlife of agricultural environments: an aid to conservation

RALPH COBHAM

and

JANET ROWE

10.1   Introduction: the agricultural enviroment
    10.1.1   Agricultural dominance
    10.1.2   Principal habitats and agricultural impacts
    10.1.3   Wildlife conservation strategies and evaluation
10.2   The evaluation task
    10.2.1   Objectives of survey
    10.2.2   Documentary material
    10.2.3   Preliminary qualitative survey
10.3   More detailed survey and monitoring methods
10.4   The results achieved
    10.4.1   A background to the assessment process
    10.4.2   The single-purpose plan
    10.4.3   The multi-purpose plan
10.5   Overall evaluation
    10.5.1   Wildlife and husbandry relationships
    10.5.2   The impact of change
10.6   The economic and political environment
10.7   Summary

Wildlife Conservation Evaluation. Edited by Michael B. Usher.
Published in 1986 by Chapman and Hall Ltd, 11 New Fetter Lane,
  London EC4P 4EE
© Chapman and Hall

10.1 INTRODUCTION: THE AGRICULTURAL ENVIRONMENT

10.1.1 Agricultural dominance

In most countries of the world agriculture dominates the use of cultivatable land. Man has been of major importance in shaping the natural history of Britain and other countries of lowland Europe. The same is true of the farmlands of North America and Australasia, and it is increasingly true in Third World countries as they adopt western technology. In Britain, one of the most densely populated countries of the world, 75% of the land surface is farmed. Today just under half of this is included in arable rotation of different kinds. In contrast, Sweden has less than 10% of the total land area devoted to arable cropping, with forest and woodland occupying about 50%. Other comparisons are shown in Table 10.1.

With one of the most intensive farming systems in Europe, Britain can be viewed largely as an agricultural environment: pasture and arable land generally predominate with woodland or other habitats occupying areas less suitable for cultivation. The present British landscape has been described (Hoskins, 1970; Tittensor, 1981) as a mosaic of habitats with individual units dating from known phases of man's past activities: only a tiny number of sites can have escaped man's influence at some time (Evans, 1975). Indeed, until the recent increase in arable cropping, traditional farming practices largely sustained what has come to be regarded as the 'natural heritage'.

The term 'agricultural land' is a description of use rather than of biological characteristics, ranging from land which is subject to the frequent and far-reaching disturbances of arable rotation (which often militate against the development of habitats suitable for wildlife) to that which is often regarded as 'semi-natural' (Duffey, 1971). The latter includes moors and downlands where for centuries there may have been only low density grazing. On more marginal land, human use may not only have preserved the continuity of habitat, but it may have actually increased the variety of species present.

In general, the proportions of national land areas dominated by nature conservation interests are small, ranging from 0.03 ha per person in Great Britain, West Germany and Trinidad and Tobago, to 2 ha per person in Australia and Canada. There is, thus, a close interdependence between wildlife and agriculture. The willingness of those who own and farm the reserves to co-operate with national wildlife interests is of critical importance. In countries, such as the United States, where the land surface is greater and population density lower, it has been possible to set aside large, continuous tracts of land as reserves and refuges. The interdependence is especially marked in countries, such as Britain,

| Country | Population[b] (million) | Total land area (millions of ha) | Population density (ha/person) | Percentage of | | | | | | |
				Forest and woodlands	Other[c]	Arable (1)	Pasture (2)	Extensive grazing (3)	Percentage of total agricultural area (1) + (2) + (3)
North America									
Canada	24.3	997.5	41.0	44	50[d]	4		2	6
USA	226.1	916.9	4.1	32	23	18	4	26	48
Europe									
West Germany	61.3	24.7	0.4	29.5	24	29		19	48
Sweden	8.3	45.0	5.4	50	40[e]	7		3	10
UK	55.8	24.1	0.4	9	16	29	21	25	75
Australasia									
Australia	15.3	769.3	50.2	6	26	3	3	62	68
New Zealand	3.2	26.9	8.4	27	19	2		52	54
Asia									
Nepal	16.1	14.5	0.9	29	39[f]	19		13	32
Carribean									
Trinidad and Tobago	1.2	0.5	0.4	50	25[g]	20		5	25
Africa									
Zambia	6.2	74.3	12.0	48	23[h]	16		13	29

[a] Based on the most recent land use figures available.

[b] Latest census results: North America, Europe and Australasia from *Whitaker's Almanac* (1984), 116th edition; Nepal (1982) from the *National conservation Strategy for Nepal* (1983), a prospectus, IUCN, Trinidad and Tobago (1980) from *Trinidad and Tobago Agricultural Census* (preliminary report no. 1), published by the Central Statistical Office, Port of Spain, and Zambia (1982) from the *National Conservation Strategy for Zambia* (1984), background papers, statistics compiled by J. B. Mutelo; *Statistical Year Book for Agriculture of the Federal German Republic* (1984).

[c] 'Other' generally includes: urban areas, roads, national parks and nature reserves, mountain and wasteland.

[d] Includes water 8%.

[e] Includes lakes 9%.

[f] Other habitats 33%, nature reserves 5%, urban 1%.

[g] Includes nature reserves 6%.

[h] Swamp, escarpment, etc. 13%, nature reserves 8%, urban 1%.

where only about 1% of the land surface is contained within nature reserves managed by national, local or voluntary organizations. Even if protection were to be given to all the sites described as being of key importance to the conservation of the British flora and fauna (Ratcliffe, 1977), such sites would still account for less than 4% of the total land area.

10.1.2 Principal habitats and agricultural impacts

In the lowland regions of Britain, the majority of the native flora and fauna has survived largely in the uncropped areas: traditionally the hedgerows, copses, woodlands, wetlands and old pastures. It is difficult to assess the precise extent of these semi-natural habitats on a national level (Cobham, 1983). Areas of natural and semi-natural vegetation may account for up to a quarter of the total area of Britain, but the proportion of these in agricultural areas is much lower and has been more or less influenced within different regions by agricultural policies, systems and management. An example of the variation in the incidence of wildlife habitats between different farms is borne out by figures for the nine lowland commercial farms which were selected by the Countryside Commission to participate in the Demonstration Farms Project (see Table 10.2).

Historically, some species, whose natural habitats have been lost, have survived in extensively farmed areas: for example, elements of the chalk downland flora are thought to have come from such diverse natural communities as cliff margins, sand dunes, molehills in forest glades and anthills in salt marshes (Tittensor, 1981). Chalk downland itself requires careful maintenance (Jones, 1973), the chief agents being close grazing, ideally by sheep or rabbits, and burning. Thus, many wildlife species have become dependent on certain methods of husbandry and do not readily survive changes.

The last 40 years have seen sweeping changes in agricultural methods aimed at maximizing yields. Crop breeding has played a significant role, chemicals have been used increasingly to achieve this crop dominance, and land farmed extensively for centuries has been brought into the arable rotation. Furthermore, many uncropped areas and wildlife features of the lowlands have been eroded or removed, having come to be regarded as incompatible with 'modern' farming methods (Westmacott and Worthington, 1974, 1984; Nature Conservancy Council, 1977, 1984). As an example, in Britain more than 90% of the herb rich meadows and 25% of hedgerows have disappeared (Wildlife Link, 1983). Those wildlife features which survive do so without the labour intensive management systems (and in some cases sheep) which are necessary to

Table 10.2 The main wildlife conservation features on the nine lowland Demonstration Farms at the outset of the project. All of these farms also contained features of visual and historical interest.

	Total area of farm (ha)	Woodland (ha)	Scrub (ha)	Hedges (km)	River/stream (km)	Wetland (ha)	Amenity spinneys (ha)	Green lanes and tracks (km)	Herb rich grassland (ha)
Kingston Hill	294	16.9	3.7	14.5	1.5	2.5	—	0.8	—
Bovingdon Hall	691	93.0	—	41.9	2.0	1.5	1.8	8.0	—
Catsholme	263	5.0	—	1.4	3.0	6.9	—	1.3	—
Tynllan	104	9.1	1.5	6.5	1.0	0.5	—	1.9	—
Home Farm	296	6.5	0.4	19.2	4.0	0.1	0.2	1.5	—
Cwm Risca	213	17.5	—	7.6	1.4	9.4	—	6.5	—
Dobcross	187	49.6	0.5	12.1	1.0	0.4	—	2.5	0.1
Hopewell house	223	1.5	1.7	14.2	3.0	1.7	—	2.8	0.3
Manor Farm	1183	57.4	12.0	16.8	3.0	0.9	—	1.0	150.0

maintain their value. In a discussion of the causes of decline and extinction of rare plants since 1800, it has been suggested (Perring, 1970) that at least half can be attributed to changes in agricultural practices and drainage. Another 15% have resulted from changes in scrub and woodland management and from habitat destruction. These losses are paralleled in the reduced populations of more common species. Whilst the sensitivity of species to environmental conditions vary, the instances of species adapting to sudden changes are few: a rare example (Parrinder, 1964) is the little ringed plover (*Charadrius dubius*), which is now virtually dependent on the man-made habitats of wet gravel and sandpits.

Semi-natural habitats are under increasing pressure within the main agricultural areas. In Britain, great importance is attached to the extensively farmed uplands because it is there that much of the remaining wildlife exists. This is highlighted by the fact that of the 3000 areas (with a total coverage, in 1981, of 5.5% of the land surface), that are designated as biological Sites of Special Scientific Interest (SSSIs), 48% are located in the uplands. In the lowlands, the uncropped areas are becoming fewer, smaller and increasingly fragmented. Their populations have characteristics of true islands: as they become smaller and more isolated they tend to be more liable to accidental extinction and genetic change.

10.1.3 Wildlife conservation strategies and evaluation

Agriculture is a major factor influencing both the extent and quality of wildlife habitats in both lowlands and uplands. Due to limitations concerning the proportion of a country which can effectively be designated for nature reserves, it is essential that agricultural areas should feature in any national wildlife conservation strategy. Indeed, reference to the latter is an essential starting point when undertaking an evaluation of wildlife in agricultural environments. The general objectives of any conservation strategy must include: the prevention of further losses of scientific and genetic resources; the maintenance and improvement of those habitats and features which the public, rural landowners, tenants, managers and workers regard, both nationally and locally, to be part of their 'natural' heritage; and the improvement of resources which are available for public education and enjoyment. Where no national strategy exists, accurate wildlife evaluation is itself an essential starting point for the preparation of a conservation strategy.

The British strategy (Nature Conservancy Council, 1977) consisted of three main elements: establishing the most important sites as nature reserves and managing them primarily for conservation; helping owners to manage Sites of Special Scientific Interest so that they retain their scientific interest; and collaborating with other land users in the wider countryside, so that as much wildlife conservation is achieved in the better habitats of unscheduled land as is compatible with their primary land use (this objective refers primarily to agricultural land).

Following on from the general strategy considerations, the scientific evaluation of wildlife in agricultural environments is required for the fulfilment of one or more functions, which include (i) the establishment of the facts about existing species, populations and communities and the changes which have taken place over time; (ii) the assessment of the potential for conservation of habitats and species, and for improvement through management; (iii) the determination of the scope for increasing diversity through the creation of new habitats and features; (iv) the provision of a basis for formulating national, regional and local land use policies and plans, together with the various statutory instruments and measures (reserve designations, grants, tax reliefs, access controls, etc.) required for their achievement; (v) communication to the public, so as to secure support and co-operation for conservation measures and funding at national levels. Although there has been a tendency for evaluation work to focus upon losses of wildlife habitats and species, it is important that the more positive and creative aspects of the task are not overlooked. For example, during the Demonstration Farms Project, the 'gains' as well as the 'losses' were identified. Although the responsibility of agricultural practice for the 'losses' has been correctly emphasized, it must be recognized that the 'gains' are the direct result of an increasing awareness amongst landowners and managers of the importance of wildlife conservation within an agricultural environment.

10.2   THE EVALUATION TASK

10.2.1   Objectives of survey

In Britain, the Nature Conservancy Council (1977) recognized three categories of habitat in terms of wildlife value (Table 10.3). Category 2 includes the majority of wildlife habitats, which occur within agricultural environments. Wildlife surveys of these agricultural environments pose special problems, since they need to be conducted concurrently at two distinct levels. At the first level, the species that are typical of farmland, are identified and their relative abundance is assessed. At the second level, the composition of the habitats that harbour these species is investigated. The habitats are usually small in area (for example ponds) and are often widely dispersed (for example hedges, copses, and track verges). Many of the birds and other animals may range widely over neighbouring areas, thereby extending the scale of survey work.

The objectives of a survey should be carefully defined at the outset if meaningful results are to be obtained. Typical survey situations include the following three categories.

(1) General surveys, where the attention of the landowner or farmer must be directed to those areas of land that are assessed as being of the greatest biological interest, so that advice can be given on management practice.

(2) Specific surveys, where advice is needed to resolve land use conflicts, perhaps between farming and forestry, recreation or conservation, or where difficult decisions are to be made on the conservation of particular features.

Table 10.3 The nature conservation value of various habitats, arranged in three categories from the most important for wildlife (category 1), through those of moderate importance (category 2) to those of little importance (category 3). The information is adapted from the Natural Conservancy Council (1977).

	Categories		
Type of Land	1	2	3
Woodland	Primary woodlands	Broadleafed plantations Mature conifer plantations Recently planted conifer plantations Copses and field corner plantations Hedges	Conifer plantations with no ground cover
Heathland	Lowland heaths High mountain tops	Moorland and rough grazing	———
Wetland	Unpolluted and untreated rivers lakes, canals, permanent dykes, large marshes and bogs	Gravel pits, clay pits, farm ponds, small marshes and bogs	Polluted water of all kinds Temporary water bodies
Grassland	Permanent pastures and meadows untreated with fertilizers and herbicides	Road and railway verges Disused quarries	Grass leys Improved pasture Playing fields Airports
Other	Coastal habitats (cliffs, dunes, salt marsh, etc.)	Large gardens Golf courses Arable land with rich weed flora Neglected orchards	Small gardens Allotments Arable land with poor weed flora Horticultural crops Commercial orchards Industrial and urban land

(3) Specific surveys, where information is required on areas which surround existing or potential nature reserves, so that management plans can be formulated.

The objectives of the survey will largely determine the degree of detail required. For example, a farmer may be interested only in a few 'indi-

Table 10.4 Examples of the wide ranging objectives established for three different ecological surveys undertaken on agricultural land in England and Wales.

Royal Agricultural College Cirencester, Eysey Farm[a]	Countryside Commission Demonstration Farms[b]	West Dean Estate Sussex[c]
To establish the qualitative and quantitative state of wildlife at the present time	To collect baseline data for determining: the wildlife assets of the farms; the existing management practices which directly or indirectly affect those assets; and the potential for improving the wildlife habitats and populations through capital works and/or changes in management practices	To determine the ecological status of the estate
To provide some measure of change in wildlife and its relation to agricultural practice on the farm		To list the sites of ecological importance and to classify them according to habitat and degree of importance
To provide information which may be considered by the college when formulating its farming policies		To determine the developmental status of the major habitats
	To assist the preparation of 'complete farm' plans which enable commercial and conservation interests to be combined	To prescribe conservation management principles for major habitat types in general and the sites of ecological importance in particular
To expand both the knowledge and interest within the agricultural education function of the college and its student body	To monitor and evaluate the actual changes in wildlife habitats and species resulting from the implementation of multiple land use plans	To list the species of major groups and to map the distribution of rare or interesting species
To provide information which may help towards a cost effectiveness study of wildlife conservation on farmland	To demonstrate the results achieved and their financial implications to visiting landowners, farmers, students, politicians, etc.	To provide a basis for running residential courses for amateur and specialist naturalists

[a] Royal Agricultural College (1979). *Eysey Farm – Farming and Wildlife Study.*
[b] Cobham *et al.* (1984). *Agricultural Landscapes: Demonstration Farms*, Countryside Commission.
[c] Heymann, M. and Tittensor, R. (1981) Ecological Appraisal of an Agricultural Estate, *Chartered Surveyor*, **113**, No. 6, 424–25.

cator' species which can be visually appreciated, these species representing the tip of the 'conservation iceberg'. At the other end of the scale is the study designed to further scientific understanding of the ways in which management practices can be adjusted to increase the diversity and richness of species. Surveys may have to cater to some extent for both these extremes. Examples of projects which entailed conducting ecological surveys, with mixed objectives, are described in Table 10.4. In all cases the results have been vital in the formulation and implementation of multiple land use plans.

10.2.2 Documentary material

From a comparison of early topographical maps (often originating from the first half of the nineteenth century) with modern maps, it is possible to identify areas and features of probable biological interest, such as remnants of old woodland, parish boundaries, wetlands, old pastures and common lands. Geological and soil survey maps indicate where edaphic factors are likely to influence vegetation type, particularly the location of 'lower grade' soils which probably will have been subjected to less intensive farming practices. Aerial photographs may show habitats and features such as hedgerows, scrub, woodland rides and recent plantations. Details of 'existing' reserves and any related management agreements should be available from nature conservation organizations or from farmers and landowners. Local government authorities, libraries and museums may have relevant information, and other likely sources include Community Councils, Educational Associations, Naturalists' Trusts, local societies, clubs (for example the Sierra Club in the USA) and amenity groups.

10.2.3 Preliminary qualitative survey

Several organizations, which in Britain include the Agricultural Development Advisory Service (ADAS) and Farming and Wildlife Advisory Groups (FWAGs) provide 'layman's guides', summarizing the likely wildlife value of existing uncropped areas on farm land, as in Table 10.5. It will be seen that the guidelines reflect the size and age of the habitats, as well as both the variety of plants present and the management practices adopted. Predictably there is a close relationship between this specific classification and the more generalized one shown in Table 10.3. Size is important in relation to the likely degree of disturbance from activities on adjacent land. Age is relevant because in temperate zones old habitats are generally, though not always (Tittensor, 1981), richer in flora and fauna. They also contain species more sensitive to changes of management than do more recent or newly-

Table 10.5 Values of existing habitat: a guide to assessing priorities for wildlife on the farm (from the FWAG *Guide to Priorities for Wildlife on the Farm*, 1982, The Lodge, Sandy, Beds).

Habitat	Outstanding value	Great value	Lower value
Woodland	Old (shown in first ordinance survey map) deciduous woodlands of more than 1 ha, especially those with trees of different age classes	Smaller (less than 1 ha) old woodlands Secondary woodland i.e. woodland that has developed on grassland or moorland. Plantations of native broadleaf trees of more than 1 ha	Plantations of conifers, and small plantations of broadleaf trees
Hedges	Thick old hedges (i.e. those with four or more species per 30 m)	Poorly managed old hedges. Thick younger hedges (less than four woody species)	Poorly managed young hedges
Heaths and moorlands	Large (more than 8 ha) with both dry and wet areas, not intensively grazed	Smaller heaths and moors. Large heaths very heavily grazed	
Grasslands	Unimproved pastures and meadows with numerous plant species. Flood meadows	Unimproved pastures and meadows with few plant species	Grasslands improved by fertilizers, older grass leys
Wetlands	Large, well-established ponds and lakes with at least one edge free of trees. Unpolluted rivers, streams and ditches containing permanent water (in the lowlands this is shown by the presence of numerous species of submerged water plants)	Smaller ponds, totally shaded larger ponds. Permanent rivers, etc., showing signs of pollution (algae, etc.) or which are treated chemically	Small polluted and/or shaded ponds. Ponds and ditches which are dry for some part of the year

formed habitats (Duffey, 1974; Peterken, 1974; Peterken and Harding, 1975; Rose, 1976).

The qualitative study of vegetation rather than fauna will almost certainly prove to be the best basis for a preliminary survey, since animals are generally mobile and elusive. Vegetation is the most conspicuous element in the agricultural environment: plants are good site indicators. The 'indicator' approach does not demand detailed knowledge of community floristics: rather it concentrates on identifying the predominant species present and classifying them, by a relatively quick inspection, on the basis of an established set of main morphological categories, i.e. growth form, stratification, seasonality, density, etc. (Pears, 1977). However, during the 1960 Silsoe exercise (Barber, 1970) on 162 ha of lowland mixed farmland in Hertfordshire, it was suggested that bird communities would provide the most rapid and accurate indication of other flora and fauna. The expertise of the survey team will be relevant in choosing suitable methods. Observations of birds and other conspicuous groups such as the Lepidoptera and Odonata, amphibians, reptiles and mammals can also provide useful breadth to preliminary surveys. The use of birds and invertebrates in evaluation is described in Chapters 11 and 12 respectively.

The main wildlife habitats, both existing and potential, can be mapped. Factors such as the diversity of ground and herb layers (if present), the heights of shrubs and the diversity of headlands and banks provide useful information about present management. The shapes of, and relationships between, semi-natural and uncropped areas may be significant. Thus, for instance, a thick belt of scrub may buffer a remnant strip of chalk downland turf on a scarp slope against spray drift from adjacent arable land. The effects of disturbance on old woodlands, meadows and marshes of 20 ha or less are likely to be minimized if the length of the edge or perimeter is minimal in relation to the total area, as occurs when the shape is close to a circle or a square (Duffey, 1974). In contrast, a long boundary in relation to area may be advantageous in intensively managed systems such as plantations or arable fields where wildlife is largely confined to the periphery. Special attention should be paid to fringes and margins, which often support particularly diverse communities.

10.3   MORE DETAILED SURVEY AND MONITORING METHODS

The methods described below are those which have been used successfully in the course of the Demonstration Farms Project on ten farms ranging in size from 100 ha to 1370 ha at locations dispersed throughout England and Wales. The methods, thought to be the minimum necessary to demonstrate to mixed audiences the principles involved, can generally be applied only to the main wildlife habitats identified through preliminary survey, both because they are labour intensive and because the availability of suitable labour has proved a major problem. The following five distinct classes of wildlife usually require investigation.

Plants. Species lists are prepared to identify areas of particular floristic interest. These areas are then monitored by systematic techniques, such as temporary or permanent quadrats and transects, in situations where environmental change is likely to occur as a result of the adoption of new management practices. 'Before and after' photography is useful, although the initial choice of photographic location may be critical.

Birds. Breeding birds are usually monitored by the 'Common Bird Census' technique devised by the British Trust for Ornithology (Marchant, 1983). This is a mapping method in which an observer makes between 10 and 12 visits to the farm during the breeding season (between April and June). The size of the area which can be covered by one recorder depends on the number of hedges and copses: providing the wildlife habitats account for less than 10% of the total area, one observer can cover a farm of 60 to 80 ha. Non-breeding birds and migrants are observed on a less systematic basis.

Invertebrates. Butterflies are monitored using the 'standard walk' technique (Pollard *et al.*, 1975). A route, of about 3 km, which passes through representative habitats or parts of habitats is chosen and walked at least once a week from April to September inclusive: each butterfly sighting is recorded by species. Although moth populations can be monitored using a light trap, the practical difficulties mean that the method is not recommended for general use. Surface-active invertebrates can be sampled in pitfall traps, and soil invertebrates can be studied using soil cores followed by extraction in Tullgren funnels. These methods can be successfully used only by competent entomologists over several seasons. Flying invertebrates can be monitored by systematic netting in selected habitats, although populations of Odonata and other more noticeable groups are adequately observed by less systematic means. The methods described in Chapter 12 have yet to be applied to farmland.

Mammals. Small mammals can be monitored by systematic live trapping using Longworth traps. However, this method poses problems of both a practical and interpretative nature, since some species exhibit complex cycles of population increase and decrease that are not necessarily directly related to known environmental conditions. Qualitative observation is usually sufficient to assess the status of species within each habitat: such observation is also used for bats, hares, badgers, etc., and for aquatic species, such as water voles.

Amphibians and reptiles. Qualitative observation is the only practicable method of recording and monitoring.

Table 10.6   Methods found useful in recording and monitoring the wildlife on the ten farms included in the Demonstration Farms Project.

Survey method	Frequency of use
Plants	
Permanent plots used for monitoring vegetation change	9
Birds	
Complete common bird census	7
Monitoring of plots established in key habitats	2
General surveys	4*
Mammals	
Live trapping carried out systematically	4
General surveys	10
Invertebrates	
'Butterfly walk' undertaken within key habitats	4
General surveys	10

* used in the three farms for which there was no Common Bird Census.

Practical experience. The systematic methods, which were found to be in use after at least two years' evaluation on the ten Demonstration Farms, are shown in Table 10.6. The recording, which was carried out mostly by local amateur naturalists under the guidance of a professional biologist, suffered from a lack of sufficient labour to achieve consistent results. The most successful monitoring was carried out by closely-knit self-motivating local naturalists' associations, whose recording was part of their annual programme of activities.

Difficulties were also experienced at Eysey Farm (Table 10.4; Royal Agricultural College, 1979): these emphasize the need for advance planning, budgeting and close supervision. An initial conservation plan for the 162 ha farm was drawn up in the summer of 1971 by three experts. This stimulated changes in management, although not always in the desired direction. Subsequently, a team of 36 professional and amateur naturalists conducted a six-year survey using the above methods. Although there was no shortage of labour, two problems were encountered. First, there was the absence of an overall co-ordinator capable of integrating the various contributions and of offering timely, clear-cut advice. Secondly, there was the lack of dialogue and understanding between members of the survey team and those responsible for the overall management of the farm. Two technical problems also emerged. First, since botanical survey methods varied between habitat unit, recorder and year, quantitative comparisons could not be made. However, variations within habitats, associated with seasonal climatic conditions and time of recording, were noted. Secondly, in attempting a complete Common Bird Census of the farm, difficulties were experienced in delineating the areas surveyed, and it was thought possible to compare only numbers and not densities of breeding species in different habitats.

10.4   THE RESULTS ACHIEVED

10.4.1   A background to the assessment process

The results achieved are best illustrated by describing the wildlife survey and evaluation exercise undertaken on one of the Demonstration Farms, Hopewell House Farm in Yorkshire. The farm, situated on the western side of the Vale of York, and typical of many farms in the area, is composed of three separate units (Hopewell House Farm, Cockstone Farm, Haugh's Farm) which together cover 223 ha. The main unit, mostly arable land, includes a rich diversity of pond, pasture, hedge and woodland habitats. It contrasts sharply with the large area of featureless, flat arable land that forms the second unit (Cockstone Farm). Reclaimed gravel land, now under permanent pasture, which forms part of the third unit (Haugh's Farm), provides further contrast, especially since it is bordered by the River Nidd, a tributary of the River Ouse.

The wildlife surveys were conducted as part of the preparation of an overall management plan for the three units. Detailed surveys, relating to each of the land use and conservation interests, preceded the preparation of 'single-purpose' plans, which sought to propose and achieve optimum results for each interest (for example the best schemes for the development of the farming enterprises, or for maximizing the conser-

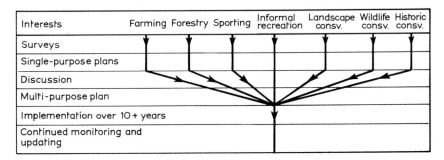

Interests	Farming	Forestry	Sporting	Informal recreation	Landscape consv.	Wildlife consv.	Historic consv.
Surveys							
Single-purpose plans							
Discussion							
Multi-purpose plan							
Implementation over 10+ years							
Continued monitoring and updating							

Fig. 10.1   The overall approach used in the Demonstration Farms Project. This shows the process adopted on each of the farms from the individual survey by each interested party to the final preparation and implementation of the multipurpose management plan for the farm. The exercise is a dynamic one. Continued monitoring and experience produce the need for both periodic (annual or biennial) review and updating of the management plan.

vation interest). The plans were reviewed so as to identify agriculturally compatible, conflicting and unacceptable proposals. The costs and benefits of all the proposals were carefully considered. Whilst the costs of implementing most of the elements of the 'single-purpose' plan could be determined relatively simply, the benefits were often less easy to quantify. In these cases, decisions were based on what was practicable as well as what had scientific or amenity value. This overall approach or process is shown in Fig. 10.1. The 'multi-purpose' (management) plan, arrived at by compromise between the different demands on land use, took into account the priorities and wishes of the landowner.

The starting point for the wildlife evaluation involved the formulation of the specific objective, namely to maximize the abundance and diversity of the farm's wildlife, without involving the loss of a significant area of agriculturally productive land.

10.4.2   The single-purpose plan

A quick reconnaissance of the three units indicated that there was considerable scope for improving the management of existing habitats and for creating new conservation features. Against this background, and the general demonstration objectives of the project, it was relatively simple to establish the required type and extent of the surveys, which are described below.

Botanical survey. The aim of the botanical survey was to record as many wild plant species as could be found during the survey period (late May to the end of September 1979). The recording unit was a landscape feature (i.e. a hedgerow, a track, a field, etc.), and within that unit a crude measure of the abundance of each species was scored, as shown in Table 10.7. Problems were of two kinds. First, there was a seasonal effect since the survey did not start in early spring. This meant that many spring-flowering plants like bluebell (*Endymion non-scriptus*), lords-and-ladies (*Arum maculatum*) and primrose (*Primula vulgaris*) were

Table 10.7 Hopewell House Farm botanical survey: the abundance classes, and their definition, used in recording the plant species data. Frequencies of either plants or clumps apply to the whole feature (a hedgerow, copse, etc.) being surveyed.

Abundance	Abbreviation	Number of plants	Number of groups (clumps)
Abundant	a	more than 100	More than 10
Frequent	f	31–100	4–10
Occasional	o	11–30	2–3
Rare	r	1–10	1

Table 10.8 Hopewell House Farm botanical survey: the definition of the plant species diversity classes and the number of features (hedges, fences, fields, etc.) included in each of those classes. The percentage shows the percentage of features in that class measured over all three farm units.

Assessment diversity class	Score	Number of features	Percentage
Very diverse	⩽60	3	2
Diverse	40–59	19	13
Moderately diverse	25–39	42	28
Low diversity	⩾24	86	57
Total		150	100

likely to be under-recorded. Secondly, there were taxonomically difficult groups that amateur naturalists often found difficult: these included the rushes (*Juncus* spp.) and the majority of grasses. Two thresholds for species richness were subjectively chosen. Features with a score of 40 or more were assessed as diverse (see Table 10.8), whilst those with a score of 24 or less were assessed as being of low diversity. The levels were chosen so that approximately 15% of features were in the diverse group, 25% in the moderate group, and 60% in the low group. Rare plant species were assessed as those that occur in less than 1% of the 1 km grid squares of the Harrogate District (see Jowsey, 1978); uncommon species were assessed as those which occur in less than 3% of the squares. In total, 12 local rarities, and 16 locally uncommon species, were recorded on the farm.

Mammal survey. Four weekends were spent trapping on the farm by a group of interested amateurs (the Yorkshire Mammal Group). They used about 100 Longworth traps, weasel traps, and made a number of observations of mammals, amphibians and tracks.

Ornithological survey. Like the botanical survey, the ornithological survey aimed to cover the whole area of the Demonstration Farm. The survey was based on the Common Bird Census. Recorders were asked to make at least ten visits to the farm, mapping the location of common birds seen and recording behaviour such as singing, fighting, etc. Recorders were also asked to prepare notes on non-breeding species, such as migrants, wintering flocks, etc. A quantitative method of analysis was used (North, 1978).

Invertebrate survey. Butterfly walks, which were undertaken on each farm unit, involved recording all butterflies seen within a 5 m band of the predetermined route. A brief survey of spiders was undertaken (37

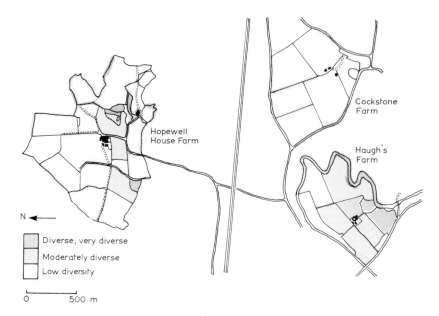

Fig. 10.2 The results of a wildlife habitat survey on the three farms which together compose the Hopewell House Demonstration Farm in the Vale of York, northern England. Although the survey collected data on higher plants, birds, mammals and several invertebrate groups, diversity in this diagram is based only on the higher plants (see Table 10.8 for a definition of the diversity classes).

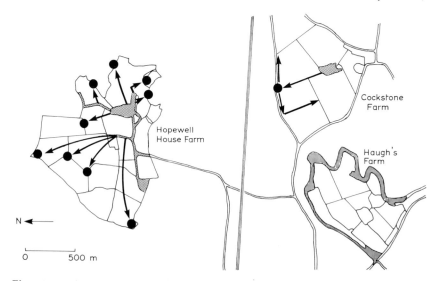

Fig. 10.3 The planned wildlife corridors and centres (stippled) and the linkages (shown by arrows and circles, the so-called 'stepping stones') planned for Hopewell House Demonstration Farm.

species being collected). In addition, spring-tails (Collembola) inhabiting the litter and soil surface were surveyed (21 species were identified, including one rarity).

1979 survey results. 'With a list of about 280 species of higher plants and 56 species of birds, as well as other bird species seen, such as the hobby and golden plover but not included in the Common Bird Census returns, the farm contains a reasonable diversity of wildlife . . . which can be considered typical of farmland in Yorkshire' (quotation from the single-purpose plan). When the results were classified on the basis of the diversity criteria (Table 10.8), it was noted that within the farm significant differences existed both between the individual units and between particular features (see Fig. 10.2).

Good distribution of wildlife cover, together with strong links between the areas containing the most species, were considered desirable, as shown in Fig. 10.3. Whereas two of the units possess a strong spine or corridor of wildlife habitats, the wildlife structure on the arable unit is relatively weak. In addition, the differing nature of the types of corridor is interesting: at one unit the hedges, wet areas and copses

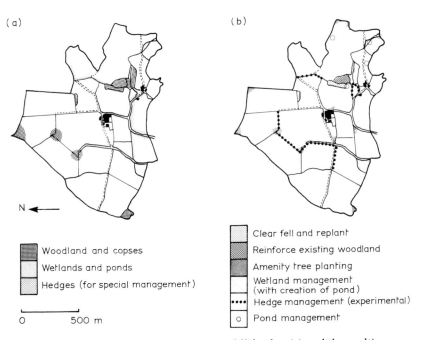

(a)

(b)

N ←——

Woodland and copses

Wetlands and ponds

Hedges (for special management)

0   500 m

Clear fell and replant

Reinforce existing woodland

Amenity tree planting

Wetland management
(with creation of pond)

•••• Hedge management (experimental)

o Pond management

Fig. 10.4 Features of the single-purpose wildlife plan (a) and the multi-purpose plan (b) on one of the three farms in the Hopewell House Demonstration Farm Project. In drawing these plans some features, especially in (a), have been exaggerated in order to draw attention to them. The agreed wildlife features account for approximately 3.5% of the land area of the farm.

provide a strong internal structure, whereas at the other the river provides a distinct boundary corridor.

In preparing the 'single-purpose' plan for Hopewell House, the three part evaluation of the diversity of features continued to be helpful. In the areas and features supporting a diversity of plants (which correlated positively with other groups of wildlife), the proposals were geared primarily towards protective management. For the areas with low diversity, it was recognized that wildlife interest would remain virtually insignificant. The areas with moderate diversity were, however, recognized to be important as providing continuity of habitat throughout the farm and as possessing potential for improvement. The main features of the single-purpose plan are shown in Fig. 10.4a. Specific proposals for woodland and copses included planting, development and management, increasing such wildlife habitats from 0.7% to 2.3% of the farm area. For wetlands, through developing new ponds, cleaning existing ponds and creating small areas of alder carr, it was proposed to increase the total wetland from 0.7% to 0.8% of the farm area. Particular attention was paid to a small area of botanically rich Magnesian Limestone grassland; conservation of this, together with all small areas of permanent pasture, accounted for 1% of the farm area. Proposals for additional planting of hedgerows (410 m), their laying (820 m) and cutting to particular heights featured prominently in the plan. In summary, the plan called for a total of 12 ha, representing just over 5% of the total farm area, to be used primarily for wildlife interests.

10.4.3   The multi-purpose plan

As Fig. 10.4b indicates, a very high proportion of the wildlife proposals proved acceptable both to the other interests and to the owner, albeit as a result of compromise in some cases. The reasons for the rejection of the unsuccessful proposals are interesting in that they provide insights into the factors, other than wildlife, which form part of the decision-making process. Often they reflect the rights or privileges of other people or parties over the agricultural land – a resident's right to unobstructed views, or the responsibility of an Internal Drainage Board – but sometimes they indicate an unwillingness to surrender good arable land to wildlife conservation.

10.5   OVERALL EVALUATION

10.5.1   Wildlife and husbandry relationships

A knowledge of farm management methods must, to a greater or lesser extent, form an integral part of any evaluation of the wildlife of agricultural land. On grazing land, for example, the density and rotation of stock, the use of fertilizers and the length and treatment of leys need to be considered. The management of hedges, ponds, etc. is likely to affect values; for example, are all hedges trimmed, to what shape, at what time of the year, and are they fenced from stock? In woodlands and shelter belts, access by grazing stock will affect the herbaceous and shrub layers. In plantations, the choice of species and planting patterns, the presence of rides and glades, and the methods of weeding, brashing, and thinning, will be important. Good examples of the interaction between husbandry methods and wildlife conservation achievements are to be found in the Demonstration Farms Project (Cobham *et al.*, 1984). These include observations on all the main habitats and features: woodlands, copses, scrub, hedgerows, grassland and wetlands, as well as historic and landscape features.

Nationally, the adoption and pursuit of multiple land use policies and practices can contribute significantly to the wildlife resources of agricultural environments (Barber, 1983; Cobham Resource Consultants, 1983; British Association of Shooting and Conservation, 1983; Coles, 1984). Although not exclusively beneficial, the provision and management of sporting habitats (for fishing, shooting, stalking and hunting) have been shown to benefit wildlife (Standing Conference on Countryside Sports, 1983).

10.5.2   The impact of change

The evaluation process, in the highly managed conditions prevailing in most agricultural environments, must include estimates of potential value to wildlife (see also Chapter 3). For example, evaluation must include an estimate of the survival chances of the habitat in the long-term, an assessment of the extent to which its species and communities are isolated, and an appraisal of whether the agricultural practices on neighbouring land might affect it. In the case of the large wetland system of international importance, the Somerset Levels, it has recently been recognized that small areas cannot be conserved if neighbouring land is drained.

Sometimes a change in the overall management of a farm would materially increase its value to wildlife. An example might be in the time and extent of hedge trimming, so that both winter food and summer nest sites for birds become available. Part of the evaluation exercise also involves an assessment of the scope for creating new wildlife habitats. The conventional response might be the digging of a pond or the planting of field corners with trees and shrubs, although as the Demonstration Farms Project shows, rather more imaginative responses are possible.

10.6   THE ECONOMIC AND POLITICAL ENVIRONMENT

Agricultural environments are part of much larger systems dominated by political and financial forces. These systems have hitherto shown little regard for the wildlife of individual farms, regions or countries. Throughout the second half of this century, nations and international agencies have been preoccupied with maximizing food production. This has generated a growing need for, and interest in, conservation action throughout the world. However, in comparison with semi-natural areas, little has been done to evaluate the wildlife resources of farmlands. The *World Conservation Strategy* (IUCN 1980) emphasizes their importance in global conservation: national working examples of the Strategy are helping to create a more favourable climate for the survival of wildlife with agriculture. Particularly encouraging are the changing attitudes of landowners and farmers toward their 'food producing' roles. Increasingly there is a demand for 'optimizing' rather than 'maximizing' food production which, coupled with a genuine concern for wildlife, has positive implications for the development of integrated multiple land use policies.

In 'developing' countries, the World Conservation Strategy has a more creative role to play: it can serve as a platform upon which surveys of wildlife resources and their conservation can be more effectively organized (IUCN 1980). In contrast, in 'developed' countries, the national conservation strategies largely seek to reinforce structures and policies which provide for a balance between conservation and development. This usually involves highlighting the need for conservation considerations to receive greater attention whenever economic policy and resource management issues are at stake.

Newby (1980), Shoard (1980) and Cobham *et al.* (1984) all voice the opinion that farming and conservation interests are reconcilable. However, in Britain no system currently exists for harnessing the goodwill, understanding and commercial motives of landowners and farmers. Rather there are what Cobham *et al.* described as 'fragmented land use policies, few incentives, an inadequate advisory network, no effective controls . . . a government previously pre-occupied with production . . . in short, a recipe for continued erosion of valuable countryside features'. As Wibberley (1982) points out, 'a chauvinistic fundamentalism spreads through the agencies, pressure groups and even to civil servants, making them push strongly for their own particular interests'. Three pressing needs can be identified. First, there is the need for central government, with the help of different interest groups, to formulate clearly-stated land use policies. Secondly, the need for all interests to assist government in devising ways of rewarding landowners and farmers who contribute to wildlife conservation is important. Thirdly, there is a need for all advisers and students of land use to be well trained and practised in the skills of multi-purpose management.

The need for education is paramount, and conservation is now on the syllabus of most courses concerned with aspects of land management. Farming and Wildlife Advisory Groups provide an independent forum where all interests in the countryside can meet for discussion and advice, and some local authorities also employ advisory officers. The Countryside Commission, both through its Project Officers in urban fringe and lowland areas and Demonstration Farms Project, attempts to lead by education and example. Voluntary bodies such as the County Naturalists' Trusts and the Royal Society for the Protection of Birds lend evaluation expertise, and other bodies, such as the British Trust for Conservation Volunteers, offer practical assistance with conservation management. The Country Landowners' Association and others make annual awards to farmers who successfully combine conservation with farming, contributing to a positive climate of public opinion. All of these bodies, however, function within the constraints of existing agricultural policy, and in the absence of a framework for integrated land use.

In the face of mounting public pressure it will be difficult for any government to avoid some restructuring of its approach to countryside management. In this the evaluation and conservation of wildlife in agricultural environments is likely to play an increasing role. However, those involved in planning and undertaking wildlife surveys must appreciate their wider responsibilities and opportunities. These include: the importance of understanding the agricultural ecosystem and the various forces (constraints, incentives and pressures) operating both within it and upon it; the need to communicate the benefits of wildlife, especially sound 'conservation-cum-commercial' management, to all those landowners, managers and organizations involved (persuasion and demonstration are likely to prove more effective than perpetual 'chastisement'); and the education of landowners, farmers, advisers, etc., so that they are better able to recognize important wildlife habitats and thereby derive pleasure from them. Undoubtedly an increase in wildlife surveying and monitoring, which is undertaken in the full knowledge of these wider responsibilities, will do much towards counteracting the mismanagement of wildlife resources throughout the agricultural lands of the world.

10.7 SUMMARY

With the modernization and intensification of agriculture, the survival of the flora and fauna in agricultural environments depends upon reconciling the interests of food production and habitat conservation. Wildlife conservation evaluation is a necessary pre-requisite in the drawing up of plans for integrated land management at the farm level. Surveys, which may be general or specific, are demonstrated by reference to the examples of Eysey Farm for the Royal Agricultural College and the Demonstration Farms Project of the Countryside Commission which involves ten farms, one of which is described in detail. Qualitative and quantitative methods used in these surveys are described, along with the practical experience gained in the execution of them. Diversity (species richness), especially of plants, is found to be generally the most useful criterion in assessing the sites of greatest value for wildlife. The all-important relationships between wildlife and husbandry are discussed.

CHAPTER 11

Ornithological evaluation for wildlife conservation

ROBERT J. FULLER

and

DEREK R. LANGSLOW

11.1   Introduction
11.2   The objectives of evaluation
11.3   Attributes for ornithological evaluation
11.4   Examples of ornithological evaluation
11.5   Ornithological evaluation of wetlands for non-breeding birds: the 1% principle
11.6   Problem areas in ornithological evaluation in Britain
       11.6.1   Woodland birds
       11.6.2   River systems
       11.6.3   Moorland and mountain birds
       11.6.4   Farmland birds
       11.6.5   Breeding waders of lowland habitats
11.7   Summary

Wildlife Conservation Evaluation. Edited by Michael B. Usher.
Published in 1986 by Chapman and Hall Ltd, 11 New Fetter Lane,
   London EC4P 4EE
© 1986 Chapman and Hall

11.1   INTRODUCTION

Within the field of wildlife conservation birds enjoy great popular appeal. This popularity is partly because birds are conspicuous and often brightly coloured organisms, and partly because there is a long tradition of ornithology. Conservation organizations are increasingly using ornithological evaluations as one component of wildlife assessment, partly as a result of the huge interest in birds but also perhaps in recognition that some sites with interesting bird populations are otherwise of limited wildlife interest. Birds are also valued for other than aesthetic or cultural reasons. Certain species may act as indicators of environmental pollution when heavy metal and pesticide residues can be measured from tissue analysis; birds of prey and seabirds are notable examples. In the most extreme cases severe population declines have provided warning of pervasive and insidious environmental deterioration, as with the peregrine, *Falco peregrinus*, virtually world-wide (Ratcliffe, 1981). Because of their conspicuousness, ubiquity and ecological diversity birds have also been used as indicators of broad environmental quality in extensive land use planning (for example Graber and Graber, 1976; Svensson, 1977; van der Ploeg and Vlijm, 1978; Blana, 1980).

Two major components of conservation action are, first, the safeguarding of sites and/or habitats of particular interest and secondly, the protection of individual species through legislation. Ornithological evaluation is relevant to both these aspects of conservation action. In the first case the importance of sites is assessed on a limited geographical scale to define key localities; in the second, individual species must be assessed in order to establish priorities for protection. This chapter considers both these elements of evaluation. The objectives of evaluations are discussed and the selection of appropriate attributes, with which to judge ornithological interest, is considered. The range of objectives and methods of evaluation is illustrated by published examples and some particular problems of ornithological evaluation and possible solutions are outlined. Western Europe is taken as the focus for this review, with references to some of the relevant studies elsewhere.

11.2   THE OBJECTIVES OF EVALUATION

There is no universal approach to ornithological evaluation: the choice depends on the objectives of the evaluation exercise and is constrained by the availability of appropriate information. It is just as valid to ask which localities are important for a specified species in a local area as it is to identify sites of ornithological value throughout an entire country or international region. Between these extremes lies a continuum of scales and complexities for evaluation. A system of evaluation designed for one purpose can seldom be applied directly to another. The first step in undertaking any evaluation must therefore be to define the objectives and context as precisely as possible. This can involve asking the following seven questions.

(1) Who will use the evaluation(s)?
(2) What is the purpose of the evaluation(s) (for example reserve selection, evaluation of a specific threatened site, environmental planning, species protection legislation)?
(3) What is the geographical context (for example town or city, political or administrative region, country, biogeographical region, continent, planet)?
(4) Is the evaluation restricted to one or a few habitat types or does it interpret ornithological interest independently from any habitat framework?
(5) If the evaluation aims to compare sites, how are the sites to be defined? Boundaries should be defensible on biological and practical grounds.
(6) Which birds are the object of the evaluation (for example a single species, other taxonomic group, ecological guild, community)?
(7) At which time of year is the evaluation relevant (i.e. breeding, migrant or wintering populations)?

At the core of an evaluation system are the *attributes* (such as species richness, rarity) into which wildlife conservation interest can be classified. The final choice, and any possible weighting, of attributes in an evaluation scheme is entirely dependent upon the objectives and context. No individual attribute has a greater intrinsic merit than any other. The following section considers the range of potential attributes and their relevance to ornithological evaluation. Within each attribute the extent of interest can be judged by the use of *criteria*. Whilst it is sometimes possible to quantify criteria (for example the number of species recorded may be used as a measure of species richness), quantification should not be confused with objectivity. Quantification merely ensures a standard, or consistent, application of the criteria. It does not imply that the evaluation is objective, for the choice of attributes and criteria is essentially subjective. Furthermore, the choice of attributes is rarely determined by absolute biological factors but rather by human perceptions and interpretation of wildlife. Finally, once the level of conservation interest has been measured within each attribute, the problem of relating this measurement to a scale of values remains. Occasionally it may be sufficient simply to rank the sites. Usually, however, some form of classification is required, entailing grouping sites according to defined levels of importance which are often related to

Honey buzzard
(*Pernis apivorus*)

Red kite
(*Milvus milvus*)

Osprey
(*Pandion haliaetus*)

Woodlark
(*Lullula arborea*)

Redwing
(*Turdus iliacus*)

Marsh warbler
(*Acrocephalus palustris*)

Dartford warbler
(*Sylvia undata*)

Firecrest
(*Regulus ignicapillus*)

Cirl bunting
(*Emberiza cirlus*)

Fig. 11.1 Examples of species which are on the edge of their breeding range in Britain. The maps show the approximate breeding distributions in the mid-1970s.

geographical areas within which the site has special significance (for example national or regional). The choice of threshold values separating such levels of importance is subjective and depends on the context of the evaluation. The black-tailed godwit, *Limosa limosa*, in Europe, provides a good illustration. The Dutch breeding population is about 100 000 pairs (van Dijk, 1983) and the species is widely distributed throughout the Netherlands (Teixeira, 1979). By contrast the British population is about 100 pairs and all breeding sites could be regarded as nationally significant which would not be the case in The Netherlands. Other examples of species which are on the fringe of their range in Britain are shown in Fig. 11.1. Such situations raise the question of whether a wider international perspective should be sought in evaluations of the importance of individual species.

So far we have stressed the importance of defining the spatial context (geographical region, habitat type, bird community type), but there remains the question of temporal context. Concepts behind both the attributes and the criteria adopted may change as human perceptions of wildlife alter and new numerical techniques become available. Some birds may also show major changes in status during a short period. Hence, criteria established at one time may not be appropriate at a later date.

11.3   ATTRIBUTES FOR ORNITHOLOGICAL EVALUATION

Not all the attributes used in wildlife conservation evaluation can easily be applied to ornithological interest and the purpose of this section is to assess which are the most relevant. General discussions and reviews of attributes may be found in this book (see especially in Chapters 1 and 6) and in Ratcliffe (1976, 1977), van der Ploeg and Vlijm (1978), Adams and Rose (1978) and Margules and Usher (1981). It is assumed, therefore, that the reader is familiar with the range of attributes and the philosophy underlying each. Attributes can be divided into two categories. The first consists of those attributes whose values can be expressed in terms of criteria derived from biological research. The second comprises attributes which recognize cultural values; these cannot be measured by biological survey and reflect some of the motives for which wildlife conservation is practised. It includes intrinsic appeal, recorded history, educational value, scientific (research) value and recreational value. Such attributes are not considered further here because we can add little to Ratcliffe's (1976) stimulating discussion of these topics. The following is a discussion of the more ornithologically relevant attributes that can be assessed by survey.

Size (extent)

This is an important attribute for ornithological evaluation. The question: 'is the site large enough to meet the needs of the species?', must be asked. For example, golden eagles, *Aquila chrysaetos*, have home ranges of 5000–10 000ha and any evaluation based on breeding sites must include a consideration of hunting ranges. Within estuaries, the requirements for both roosting and feeding areas for different species of shorebirds at different seasons must be allowed for. Generally a larger area will contain a larger population of each species and this has practical implications since it is desirable that potential nature reserves have as large a population as possible of desirable species.

Diversity and species richness

Measures of bird species diversity or of species richness have often been used in evaluation systems. Most diversity indices have both biological and statistical problems (for example May, 1976; Southwood, 1978b; James and Rathbun, 1982). Several methods (see Section 11.4) of ornithological evaluation have been proposed which place very heavy emphasis on the use of diversity measures. Too much reliance on diversity (or species richness) can lead to other important attributes being ignored, for example population size, rarity, potential value for wildlife. Indeed, Väisänen and Järvinen (1977) have shown how diversity may decline in some habitats as conservation interest increases. The use of indices of bird species diversity in evaluation systems cannot be recommended and estimates of species richness are preferable. Theoretically it is easy to record the number of species present on a site. In practice, comparable methods and sampling effort have often been used to obtain a relative measure of species richness at sites of similar habitat. Birds are very mobile and isolated (in time) breeding attempts or transient visits by individual birds may be recorded. Hence some requirement as to the frequency of occurrence should preferably be included. As well as considering the total number of species, this attribute can be applied to particular taxa, guilds or other ecological groups (for example Fuller, 1982).

Population size

Rather surprisingly, this attribute has only been applied in limited circumstances. The greater the proportion of the total population present on one site, the greater its potential for human interest and for biological importance. Population density, as distinct from absolute numbers, has rarely been used in evaluation systems although it has potential for comparing sites supporting communities of solitary (i.e. non-colonial) birds. Population size is primarily useful as an attribute where an accurate total population estimate (within the chosen geo-

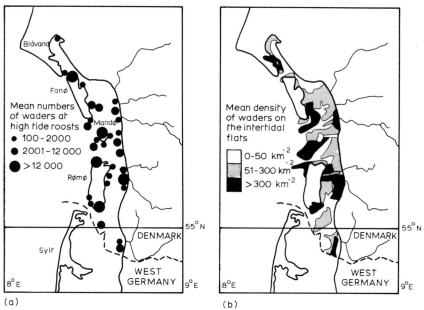

Fig. 11.2 The distribution of waders in the Danish Wadden Sea in August and September 1980. (a) mean numbers at high tide roosts. (b) mean numbers on the inter-tidal flats. (Redrawn from Laursen and Frikke, 1984.)

graphical scale) is available and where the birds are concentrated into a limited number of discrete sites. Non-breeding waterfowl on coastal estuaries and lakes meet these criteria particularly well. It is therefore not surprising that the most widely used evaluation system incorporating population size is the 1% criterion for assessing the importance of sites for non-breeding waterfowl. This system treats any site as of international importance if it supports 1% or more of an entire population of a species, or of a recognizable migratory or distinct geographical population. The criterion is considered in greater detail in Section 11.5. The 1% criterion has also been extended to colonial breeding birds, especially to seabirds (Fuller, 1980; Lloyd, 1984). For scarce and dispersed breeding species, comparison of sites on the basis of population density is often more appropriate than the use of the 1% criterion.

Although frequently used to compare different sites, the population size attribute can also be used to assess the relative significance of the different parts of a single site. An example is given in Fig. 11.2, which shows the usage of roost sites and feeding areas by waders within the Danish Wadden Sea. Such detailed information can be useful in assessing the significance of threats to sites and in constructing site management plans.

The frequency and type of use made of a site by birds needs to be considered when applying the population size attribute. A site which holds large numbers of wildfowl every winter is likely to be of greater significance than one which does so only periodically. Such cases need to be interpreted with care, however, for some sites may be extremely critical for large numbers of birds for very short periods. For example, in occasional periods of severe weather, estuarine birds may congregate in a small number of sites where their chances of survival are best. Another problem is that population counts of migrant or wintering birds do not give a true reflection of the numbers of birds using a site if there is a turnover of individuals in the population (for example Moser and Carrier, 1983; Pienkowski, 1983). In such cases the number of birds using the site during a defined period is higher than the number present on any one day during that period. For conservation purposes, a site should be judged on the real numbers, perhaps measured as bird-days or total population, rather than the apparent numbers of birds using it.

While a large population at a site may usually be desirable, this is not always so. Green (1984) has urged caution in using total counts of individuals to assess the value of migration and wintering sites for birds because factors such as the reliability of the food resources and safety of moulting grounds for relatively few birds may be more important. Where the population size attribute is being applied to breeding popula- tions, one should ideally take productivity into account. The largest or most dense populations do not necessarily produce the greatest net number of fledged juveniles as Pienkowski and Evans (1982) demon- strated for shelducks, *Tadorna tadorna*, nesting on the Firth of Forth in Scotland. In assessing habitat quality one should also take into account temporal changes in habitat preferences which may be related to the overall population of a species (for example O'Connor and Fuller, 1985). Some species occupy a series of habitats when their population is large, but occur in a smaller number of optimal habitats when numbers are low. Breeding success (and/or survival) in such optimal habitats can be particularly high, and these habitats may therefore be of critical impor- tance in enabling a species to recover from a population decline. Whilst it is important to be aware of these potential limitations, in the absence of detailed biological information, population size can be a practicable attribute for evaluating sites for gregarious species, or bird populations which are confined to a limited number of discrete patches of habitat.

Rarity

Most people ascribe a high value to rare birds and this attribute is found in virtually all evaluation schemes for wildlife conservation. It is difficult to define rarity since it encompasses both biological and human aspects.

Four major types of geographical distribution may be identified for rare birds (Drury, 1974; Margules and Usher, 1981): (i) species which occur at widely scattered localities within a large area of apparently suitable habitat but have few individuals; (ii) species which are always found in small numbers but which are widely dispersed in each community and occur in many areas within their range; (iii) species which occur in large numbers but in a restricted number of localities; and (iv) species which inhabit a changing or transitional environment.

The application of rarity within an evaluation scheme requires a clear definition of geographical context. Rarity is occasionally self-evident at all scales such as for the Seychelles magpie robin, *Copsychus sechellarum*. However, a species may be rare in one part of its range but common elsewhere and in such cases a decision must be taken about the region within which the status of the species is to be examined. Rarity can be assessed for evaluation purposes either by using an atlas of bird distribution or by defining the size of a viable population. The former method is usually the most practical since Atlases have been produced in several countries (for example Britain and Ireland (Sharrock, 1976), Natal (Cyrus and Robson, 1980), Netherlands (Teixeira, 1979), Switzerland (Schweizerische Vogelwarte Sempach, 1980)) whereas the latter method requires a far more detailed knowledge of the species (Frankel and Soulé, 1981).

Endangered/Vulnerable

These terms are special aspects of rarity. Not all rare species are endangered or vulnerable; for example the golden eagle in Britain may be considered rare but is neither endangered nor vulnerable. *Endangered Birds of the World: the ICBP Bird Red Data Book* (King, 1981) includes endangered and vulnerable species as its two most critical categories and defines them as follows. Endangered species are those: 'in danger of extinction and whose survival is unlikely if the causal factors continue operating'. Vulnerable species are defined as: 'Taxa believed likely to move into the endangered category in the near future if the causal factors continue operating'.

Fragility

This assesses the sensitivity of wildlife to environmental changes such as direct human exploitation, changes in land use around a site, or birds favouring a particular stage in a natural vegetational succession. Fragility is closely related to the endangered/vulnerable attribute. Many birds are opportunists and this attribute is frequently more appropriate to evaluations of plant and insect, rather than bird, communities. However, insular species which are confined to one or a few small islands could be dependent upon a fragile habitat (such as Lord Howe wood rail, *Tricholimnas sylvestris*, which is confined to the moss forest plateaux of Lord Howe Island).

Potential value for wildlife
Like fragility, this attribute depends upon the judgement of the surveyor. It can be an important consideration in site selection for reserves whose main function is educational and where ornithological interest can be developed by active management. However, there are habitats in an early stage of succession, such as abandoned gravel workings (see Chapter 3), where one can predict with some certainty that an interesting bird community will develop (for example Milne, 1974; Catchpole and Tydeman, 1975).

Habitat diversity
When either the ornithological survey information is inadequate or when a rapid assessment of ornithological interest is needed, the number of habitats and ecotones present within a known area can be used to forecast potential bird species richness and density. This approach has been used infrequently in ornithological evaluations, but it may become more widely applied as ornithological knowledge increases.

Naturalness
This is considered by reference to a habitat type unmodified by man. Such situations are so rare in western Europe that its usefulness as an attribute in that region is questionable; however it may assume greater relevance elsewhere. The mobility of birds often results in their spending part of their time in one habitat and part in another, for example, geese nesting in the high Arctic are in natural environments but they often spend the winter on intensively cultivated land. For birds, a natural community is difficult to define but research conducted in such environments (for example Tomiałojć, Walankiewicz and Wesołowski, 1977) can provide an important background to other studies.

Representativeness
Conservation of bird species and bird communities that are representative examples within the range of ecological variation is often a fundamental aim of biological conservation (UNESCO, 1974; Ratcliffe, 1977). This attribute can be applied through a comparative measure of the completeness of the bird community within a particular habitat and geographical context. In practice this attribute can be difficult to apply because of the problems of classifying the bird communities, although multivariate techniques of analysis may help in this area (see Chapter 2).

An example of such an approach is that of Opdam and Retel Helmrich (1984) who classified Dutch heathland bird communities into eight species rich and five species poor types.

Fragmentation

For a particular species, a minimum area is required to support a breeding pair. To function as a viable population, several such areas must be functionally connected so that a chance extinction in one area allows recolonization from another. Although many species of birds disperse readily, some species, notably in tropical forests, are reluctant to cross even small areas of non-forest habitat. This is an important attribute to consider for rare species and those found in scarce habitats.

Ecological function in the life cycle

The long term conservation of any bird species demands that all requirements of its life cycle are met. This includes breeding, migrating, wintering and roosting areas. Ideally, therefore, a sound ecological knowledge of the species is desirable but is frequently not available. For few migrant species are the exact relationships between breeding grounds, migration stop-overs and winter sites known. There has been a tendency in many ornithological evaluation systems to concentrate on sites supporting very large numbers and/or a great variety of birds at one time of the year. There is a need to take more frequent account of the significance of certain sites/habitats to the population dynamics of the species, i.e. which sites have the greatest breeding productivity and which are most critical for individual survival.

11.4   EXAMPLES OF ORNITHOLOGICAL EVALUATION

This section presents a selective review of published approaches to ornithological evaluation. The examples chosen embrace a variety of objectives in terms of geographical context and types of bird communities, and it is hoped that they illustrate the need to tailor methods to fit objectives. The first three examples deal with strictly defined bird communities at specific localities in particular geographical areas: non-breeding wildfowl in southern England (Williams, 1980), breeding seabirds in the Republic of Ireland (Lloyd, 1984) and breeding bird communities of lakes and mires in southern Sweden (Nilsson and Nilsson, 1976). These are followed by two studies which have attempted to define the ornithological value of different habitat types within defined regions (Järvinen and Väisänen, 1978) for northern Norway and Blana (1980) for an area in West Germany). The sixth method, from Fuller (1980 and 1982) classifies the ornithological interest of a large number of British sites, differing in habitats and bird communities. Finally, a system, developed by Adamus and Clough (1978), for evaluating individual species is described. Section 11.5 looks in some detail at the history of evaluating wetlands as winter habitats for birds.

Williams (1980) ranked wildfowl sites in west Surrey, England, using an index which might have more general application. The index for site value (SV) was

$$SV = \frac{D^2}{S} \sum_{i=1}^{n} (\bar{d}_i w_i)$$

where $\bar{d}$ is the average number of days usage by the ith species for x seasons; w_i is the rarity value weight for the ith species; D is the average number of species present for x seasons; S is the scale value (Williams used 1000), and n is the number of species. Thus, Williams used the attributes of population size (days usage), rarity and number of species and the criteria he applied were clearly stated. Williams considered peak wildfowl counts misleading and applied days usage instead which is a practical approach to overcoming the problems of regularity of usage by birds and population turnover (see above). He calculated a rarity value for each species in the context of the sites in his local sample: similar weightings could be calculated for any defined collection of wetland sites. A criticism of the method is that it may give too much emphasis to the number of species recorded because this factor is not only squared in his index but is also multiplied by the summed weighted values of days usage. A general problem with indices which summate or, as in this case, multiply scores assigned to different attributes is that they obscure potentially useful information.

Lloyd (1984) developed a method for evaluating colonies of breeding seabirds. Although her method was developed for colonies in the Republic of Ireland, it could be applied to breeding seabirds elsewhere. The attributes used were population size, diversity and rarity. The simplest way to describe the method is to give an expanded version of Lloyd's own summary of how to set about using the method. It involves the following nine steps.

(1) For each species calculate 1% of the national and international populations.
(2) Select those colonies which hold one or more species in numbers exceeding the 1% national population level.
(3) Obtain the approximate annual total number of birds (breeding pairs) which have regularly/recently bred (all species combined) at each colony.
(4) Obtain the number of species which have regularly/recently bred at each colony.

(5) Devise scoring systems for the rarity of each species on both national and international scales.

(6) For each colony calculate a national rarity score and an international rarity score for those species which breed in nationally important numbers (see 2 above).

(7) For each colony list additional features of interest or of conservation value.

(8) For the attributes in 3, 4 and 6 determine suitable thresholds in terms of quantitative criteria for three levels of importance (A, B, C) where A is the most important category of site.

(9) Select the key seabird colonies of special importance. Lloyd described three methods of doing this, using three categories (A, B, C), but she favoured the simple approach of treating any colony with at least one category A as a key site.

Lloyd applied her method to more than 400 colonies of sea birds in the Republic of Ireland and identified 133 as nationally important, 11 as internationally important and selected 31 colonies as key sites.

Nilsson and Nilsson (1976) wanted to rank the conservation value of lakes and mires in southern Sweden for breeding birds. Their method incorporated species richness, rarity and population size as the attributes. Each species was given a conservation value inversely proportional to its estimated breeding population in western Europe and its breeding frequency in southern Sweden. For each site the conservation values for all breeding species present were summed to give the conservation value of the site. The method was designed for use in a particular geographical context but the species weightings took into account the total population over a wide area. The method was similar to that used by Fuller (1980) (see below) to assess the quality of the breeding bird community at British sites. A similar approach could be readily applied to any series of sites for which lists of breeding bird species had been established by systematic survey work. Nilsson and Nilsson (1976) also suggested an alternative index which could be used when more exact information was available on the western European populations of birds.

Järvinen and Väisänen (1978) devised a scheme for assessing the conservation importance of bird species or habitats and applied this to land birds in northern Norway. They did not define the precise objectives of their scheme but recognized a need for objective criteria for choosing priority conservation areas. Nilsson and Nilsson's (1976) index was used as a basis but it was felt that this gave too much emphasis to abundant species. Järvinen and Väisänen (1978) considered that the more common species were generally not vulnerable and therefore gave a substantial weighting to rarity. Conservation value of the kth habitat

(V_k) was defined by an index which was calculated for each habitat

$$V_k = \sum_j V_{jk}$$

where V_{jk} is the conservation value of the jth species in habitat k, and is given by

$$V_{jk} = \frac{n_j A \, d_{jk}}{N_j^2}$$

where n_j is the size of population of species j in northern Norway, N_j is the size of population of species j in northern Europe, A is the area of northern Norway, and d_{jk} is the density of species j in habitat k.

This index had a clear geographical context but did give great emphasis to rarity. For example, Järvinen and Väisänen found that the snowy owl, *Nyctea scandiaca*, contributed 66% of the total conservation value of one habitat (stony ground and oceanic mountain heath) which supported 26 bird species. The properties of V_{jk} were such that it was high if a species bred in only one habitat or was very rare in northern Europe. The index was low for those species which were rare but widespread (for example the osprey, *Pandion haliaetus*), and for those species with only a small part of their total population in northern Norway. The index is an interesting and original approach for it permits evaluations of entire habitats, as opposed to sites. However, the authors themselves stress that the index should be applied with caution because it is mainly affected by just a few species in each habitat. Järvinen and Väisänen conclude that 'It seems to us that V combined *with common sense* might be a good strategy for many conservational purposes' (our italics). This method and the following one require density estimates which are available only in special circumstances.

Blana (1980) developed a novel approach to evaluating habitats, within a West German landscape near Cologne, in terms of their breeding birds. He combined results from extensive atlas work and breeding bird censuses on sample plots in order to define the ornithological value of the different habitats in the study area. The atlas results were used to establish the regional rarity of each species. For each habitat the censuses gave the number of species present and the relative abundance of each species. These results were used to calculate two indices of ornithological value for each habitat. The first index was a Shannon–Weaver index (H') of bird species diversity. The second index (S) was one of bird species rarity derived from the sum of weighted rarity values for each species occurring in the habitat. Finally, these two indices were combined to produce a single expression of the ornithological value (*OV*) of the habitat in the form

$$OV = Sexp(H').$$

Integration of information collected systematically by census and atlas work has considerable potential application in both landscape and site-orientated evaluations. However, the exact approach adopted by Blana has drawbacks, not least the strong emphasis placed on bird species diversity (see above). It is also uncertain to what extent Blana's two indices are independent measures of ornithological value. The index of bird species rarity alone, which incorporated the relative abundance of each species in the habitat with a measure of its regional rarity, may have been a sufficiently useful indicator of ornithological value. Blana made the important point that ideally such appraisals of habitats should be based not just on breeding birds, but on their populations of migrant and wintering birds as well.

Fuller (1980 and 1982) described a method to classify the ornithological interest of more than 3000 sites in Britain. The task was a substantial one because the sites were extremely heterogeneous in terms of size, internal ecological variation, and the recorded survey information. He selected three attributes as being equally desirable for the conservation of a species or bird community. These were: (i) population size (or size of aggregations); (ii) two aspects of diversity were considered (these were the number of species recorded (species richness) either breeding, wintering or on passage, and the quality of the breeding bird community at a site, which was indicated by a simple index based on the national scarcity of the species occurring at the site); (iii) rarity (the number of nationally rare species).

The attributes were carefully explained together with their quantitative criteria. It may have been more appropriate to have treated the index of breeding bird community quality as a special case of rarity and to have termed the diversity attribute 'species richness'. Each of the above attributes was assessed separately using quantitative criteria which related the attribute to five levels of conservation importance: international, national, regional, county and local. No attempt was made to combine the assessments derived from each attribute in a single score or index. Each site was simply assigned to the one highest level of conservation importance produced by applying the various sets of criteria to the available site data. Fuller commented that 'their (the attributes) ecological significance is not clear'. This reflects a general unease that the ecological requirements of species, and perhaps communities, are not properly considered within the attributes.

Adamus and Clough (1978) developed a system to evaluate the relative importance of different species (not just birds) as part of a project to identify natural areas for protection in the state of Maine, USA. The work of Adamus and Clough may be unique in that it considers which attributes are the most useful when attempting to assess the conservation importance of a species and hence the priorities for protection and/ or legislation. They distinguished between species which were 'suitable' for protection and management and those species 'desirable' for additional protection. The attributes of suitability were site tenacity, seasonal mobility, area size needs and spatial distribution, while those of desirability were relative scarcity, status changes, endemicity, peripherality, habitat specialization, habitat scarcity, susceptibility to immoderate human presence, unique scientific values and aesthetic amenities and use. The former group includes measurable features while the latter is a mixture of human perception and biological properties. These attributes may provide a useful checklist for evaluating species once the context and purpose of the evaluation system have been defined. Perhaps the main weakness of Adamus and Clough's scheme is that it is governed by species which are easy to conserve rather than those really in need of special protection. For example, they suggest that species with only local mobility are more suitable for protection 'because a greater proportion of their seasonal habitats can be protected'. Species which have large area requirements are difficult to incorporate into site evaluation systems and such species tend to be excluded by Adamus and Clough's attributes.

11.5   ORNITHOLOGICAL EVALUATION OF WETLANDS FOR NON-BREEDING BIRDS: THE 1% PRINCIPLE

The history of the development of criteria for the selection of wetlands of international importance in the past 20 years reflects a microcosm of the development of ornithological evaluation. Initially there was the mistaken belief that truly objective criteria could be found. Later cultural and other criteria were included whose application was impractical and whose rationale was suspect; the most recent stage was a reversion to explicitly defined features which represented widely acceptable standards by which to judge sites.

The discussion which eventually led to the Ramsar Convention on wetlands of international importance identified the need for criteria to assess the international importance of wetlands for birds. Szijj (1972) considered the criteria and standards for selecting sites on the MAR list of internationally important wetlands compiled by the IUCN. He suggested numerical criteria based on the proportion of the total flyway population of waterfowl or the population of one species using a site. Subsequently, this proportion – 1% of a defined population – has become widely accepted for evaluating wetlands as habitats for non-breeding waterfowl. Szijj also proposed additional criteria for sites used regularly by endangered species, sites important as staging posts for migrants, sites representative of a vanishing type of wetland, sites complementary to a wetland of international importance and sites having an educational value transcending national requirements. Hence Szijj's criteria subsumed a very wide range of attributes.

Atkinson-Willes (1976) discussed the problems and shortcomings of choosing 1% of a defined population as the criterion for international value, particularly its arbitrariness. However, he suggested that a strong justification for this criterion was that it produced 'a justifiable number of sites for the majority of species' and a list of potential reserves which was not too long and gave some choice in the national context. The 1% of the population of a species (or a total count of at least 10000 wildfowl or 20000 waders at one site) has proved a shrewd choice as a criterion since it has rapidly gained wide acceptance amongst scientists, conservationists, planners, politicians and the concerned public although this does not, of course, override the objection that it lacks a biological basis (see 'Population size' in Section 11.3). The 1% criterion plus additional criteria relating to the practicality of conserving and managing a site, representativeness and the research/educational/recreational values were drawn together to produce the Heiligenhafen criteria (Smart, 1976). Revised criteria – the Cagliari criteria – were agreed for incorporation in the Ramsar Convention (International Waterfowl Research Bureau, 1980): the aim was to include only aspects which could be measured using biological facts. Thus all the Heiligenhafen criteria dealing with human use were dropped and the numerical criteria were retained with the extension of the 1% criterion to colonially breeding waterfowl. The Cagliari criteria also included reference to rare, endemic, endangered or vulnerable animals and plants in wetland ecosystems. Indeed one of the striking features of these criteria is the shift in emphasis away from waterfowl to a more general recognition of wetland habitats. Thus, despite criticism (see Section 11.3), the basic philosophy behind the 1% criterion has stood the test of time and remains widely accepted as a practical means of assessing the importance of waterfowl habitats.

The work of Saeijs and Baptist (1977) provides a good example of the application of numerical criteria to one important wetland area: the Delta area in the south-west Netherlands. They used the criteria of Szijj (1972), but disagreed with the implication that ornithological interest was an indicator of general ecological value. Saeijs and Baptist (1977) raised questions about the need to define site boundaries and what to conserve: they suggested the philosophically difficult concept of a target optimal population for each species on the site. The problem was also recognized that direct application of the 1% principle to the peak count is likely to seriously underestimate the usage of a site because of population turnover.

The use by Saeijs and Baptist (1977) of 1% criteria on a wetland subjected to a major development raises an important point about evaluations. Criteria are designed merely for making judgements about the level of wildlife interest attained; they cannot be used to predict the impacts of an environmental change. Impact assessments must be based upon analogous case studies and/or relevant biological observations conducted as part of a specific research programme. Evaluations may be used to judge whether a threatened site is sufficiently important to justify such research and to ascertain whether the site remains of high conservation importance following change.

A different approach to waterfowl evaluation was taken by Bezzel (1976) and Bezzel and Reichholf (1974). They suggested the use of diversity or evenness indices in preference to other measures for evaluating waterfowl usage of wetlands and argued that such indices were more stable than direct counts. Reichholf (1981) suggested that, as waterfowl were an integral part of the wetland ecosystem, they could be used as indicators of the ecological state of wetland ecosystems. He proposed an evaluation based on the structure of the bird community to calculate a Wetland Index which incorporated not only a diversity index but also the mean numbers of species and of individuals (all species combined). A major problem with these approaches is that diversity indices have many unsuitable properties when applied to biological communities (see above) and their use in evaluation systems does not lend any greater biological credibility than does the use of the 1% criterion.

11.6 PROBLEM AREAS IN ORNITHOLOGICAL EVALUATION IN BRITAIN

Certain species, populations or communities pose particular problems of evaluation. This section discusses these problems, using British examples, and suggests approaches for defining important sites.

11.6.1 Woodland birds

For most woodland types in Britain it is difficult to identify sites of obvious ornithological importance since most woodland birds are widely distributed and occur in many fragments of woodland. Consequently individual woods rarely carry more than a tiny proportion of the national population of a species. Notable exceptions are the remaining extensive tracts of Caledonian Scots pine, *Pinus sylvestris*, forest which support a group of scarce and localized bird species including large proportions of the crested tit, *Parus cristatus*, and the Scottish crossbill, *Loxia scotica*, populations in Great Britain. The latter species is the only endemic species of bird in Britain. Conservationists are, however, often concerned to know which examples from a sequence of woods are the most interesting for birds. This is particularly true on a local scale where a conservation body may wish to know which is the 'best buy' of several sites. Several approaches can be used for such assessments. Firstly, sites of high species richness can be identified. Species richness depends partly on site area so comparisons must be based on a standardized area. Sometimes, plotting the number of species against log area may be adequate to identify the particularly rich sites as those lying well above the regression line, although rarefaction is an alternative method (James and Rathbun, 1982). Secondly, the 'completeness' of the community can be measured in terms of the proportion of potentially occurring species supported by the wood. This requires the initial definition of the species pool for the region concerned; atlases of bird distribution can provide an objective means of defining the regional species pool. Thirdly, density (breeding pairs/unit area) has rarely been used in ornithological evaluation, probably because a very large amount of fieldwork is required to obtain reasonable estimates of density for most woodland birds. Where density estimates are available for a sequence of sites, they can be compared directly. Two-dimensional ordinations of the numbers of species and total number of individuals, in equal sized plots, may be a valuable technique for comparing bird community structure (James and Rathbun, 1982).

These three approaches treat all species as of equal value. However, it would be possible to weight individual species in terms of any of the following qualities: scarcity/localized, special habitat requirements, representativeness of certain woodland types, human popularity. Comparisons of the relative or absolute abundance of such species could then be used to produce a ranking of sites.

In Britain, identification of nationally important woodland sites for birds remains difficult. With the exception of the Caledonian pine forest, one must question whether such an objective is even desirable. The Nature Conservancy Council's strategy for woodland nature conservation (selection of a range of woodlands of great historical/ecological interest, see Chapter 9) is likely to be adequate to conserve a representative range of British woodland bird communities. Furthermore, the present character of woodland avifaunas is often the direct result of recent forest management. An important implication is that, unless managed sympathetically, a site which holds a particularly rich bird community now may not do so in 10 or 20 years' time.

11.6.2 River systems

Rivers are vulnerable to many gross habitat changes. These include channel alterations aimed to improve drainage, flood control or navigation. In mountain areas, damming for hydro-electric schemes and changes in land use, especially afforestation, can affect rates of flow and water quality. The river systems of Britain are collectively important for their bird populations: those of kingfisher (*Alcedo atthis*), dipper (*Cinclus cinclus*) and grey wagtail (*Motacilla cinerea*) are largely associated with rivers and the total numbers of several other aquatic and wetland species breeding along rivers are also very large.

Despite this overall ornithological value of the river network, it is difficult to identify discrete sections of outstanding ornithological interest, because birds are often more or less evenly distributed along the length of the watercourse. Conventional site-based systems of wildlife evaluation are inappropriate here. A promising approach for mitigating the effects of modern river management is evaluation of the different physical features (for example glides, pools, riffles, vertical banks) which comprise the river environment, in terms of their wildlife interest. By identifying these, and liaising with river engineers, it may prove possible to retain varied habitats for wildlife within a managed river.

Analogies can be drawn between rivers and certain other linear habitats: canals, railway lines, hedgerows and rocky shores. The latter are increasingly recognized as major winter grounds for waders, complementing the estuarine populations (for example Summers and Buxton, 1983). Waders may be spread along several miles of coastline, nevertheless it is usually possible to identify stretches holding major concentrations.

11.6.3 Moorland and mountain birds

The extreme fragmentation of the British lowland landscape facilitates the identification of potential sites of wildlife conservation value. Individual wetlands, woodlands, heaths and unimproved grasslands are clearly etched against a backdrop of farmland. The moorlands and mountains are a different landscape, forming the last wilderness in Britain, greatly valued for both scenery and wildlife. Birds are distributed patchily across this landscape. Many species of birds are restricted to this land, several of which are nationally scarce and carry great human appeal. It is important that these bird communities should receive adequate protection from those land uses which have greatest impact on the prevailing character of the uplands. Such changes in land use include afforestation, drainage, burning, moorland reclamation and overgrazing. But, in the absence of obvious site boundaries, the difficulty lies in deciding which tracts of upland are worthy of special protection.

Many of the scarcer upland birds, especially raptors – golden eagle (*Aquila chrysaetos*), peregrine (*Falco peregrinus*), merlin (*Falco columbarius*) and hen harrier (*Circus cyaneus*) – are widely scattered at very low densities. Therefore, to include a substantial population of any one of these species within protected areas vast sites would be needed and conservation of these rare species cannot be achieved purely through site designation. Nevertheless, some upland areas hold much greater densities of birds than others; examples include those northern Scottish blanket bogs with large populations of golden plover (*Pluvialis apricaria*), dunlin (*Calidris alpina*) and greenshank (*Tringa nebularia*) (Reed, Langslow and Symonds, 1983). Extensive surveys of the distribution of upland breeding birds can provide an objective basis for identifying such high density areas.

11.6.4 Farmland birds

Lowland farmland is the most extensive habitat (see Chapter 10) used by birds in England, and as such carries a huge bird population. These birds are broadly distributed and the definition of important farmland sites seems an inappropriate concept although evaluation of this nature has been attempted by Helliwell (1978). Maintenance of a rich fauna and flora throughout the farmed countryside is an essential conservation objective which will be realized only through publicity and liaison with the farming community. Research which monitors and studies the effects of modern agricultural practices on farmland bird populations is also an important part of this conservation effort.

In summer, most of the farmland birds are common and widespread; breeding waders on wet meadows are an exception discussed below. In winter, the situation is different, for British farmland then carries huge aggregations of birds, notably lapwings (*Vanellus vanellus*), golden plovers (*Pluvialis apricaria*), fieldfares (*Turdus pilaris*), redwings (*Turdus iliacus*) and starlings (*Sturnus vulgaris*). Many of these birds are of Scandinavian and European origin, attracted to Britain by its comparatively mild winter climate. Specific sites of importance are hard to identify even though most of these species are gregarious. These birds feed mainly in fields and are strongly dependent on agricultural practices which may change annually with consequent shifts in distribution of the birds. The two grassland plovers show much site fidelity to particular pastures (Fuller and Youngman, 1979). Such fields may be suitable for management agreements with farmers, but the best prospect for the long-term conservation of these winter bird populations lies in discouraging practices which may damage the habitat requirements and food resources of the birds over large areas. Research is required to under-stand the full effects of loss of grassland and chemical applications on both the food of the birds and on the birds themselves.

11.6.5   Breeding waders of lowland habitats

Lowland waders in Britain provide an elegant example of the impor-tance of the geographical context in evaluation systems. The main species concerned are redshank (*Tringa totanus*), snipe (*Gallinago gallin-ago*) and lapwing (*Vanellus vanellus*) which are distributed throughout Britain but in greatly variable densities. Inland meadows in England and Wales, particularly those in the southern counties, now carry very small populations of these birds (Smith, 1983). By contrast, the populations are very much greater on certain saltmarshes (redshank), and especially in some northern Scottish marshes notably those of the machair of the Uists, Outer Hebrides (Galbraith, Furness and Fuller, 1984). Reasonable population estimates exist to enable these islands to be placed in a national context. In 1983, the redshank population of the Uists probably exceeded 2000 breeding pairs. The entire 1982 Redshank population of England and Wales was approximately 6000 pairs of which over 2000 were on estuarine marshes (Smith, 1983). The national significance of the Hebridean machair is beyond doubt, but where does this place the English and Welsh populations, particularly the inland ones which, although small, are nevertheless valued by local ornithologists and form part of the wildlife diversity of lowland England? This example illus-trates that what is typical in one part of the country may be highly valued elsewhere.

11.7   SUMMARY

Ornithological evaluations can be used to define sites or habitats of particular conservation value for their bird populations, or they can be used to identify bird species which are in need of special protection. There is no 'best method' for conducting an ornithological evaluation. The examples given in this chapter illustrate a variety of approaches that have been, and could be, used to tackle different problems.

   If it is decided that an evaluation of some ornithological resource(s) would provide worthwhile information for wildlife conservation, the following basic steps may help to clarify the sort of method to use.

(1) Define the objectives of the evaluation. Try to answer the questions posed in Section 11.2. Who is to use the results of the evaluation and what is its purpose? What is the geographical scale and which habitats are to be studied? How are 'sites' (if any) to be defined?

Which species are to be studied? To which season is the evaluation relevant?

(2) Once the objectives have been clarified it will be easier to draw up a list of potentially suitable attributes. Those most frequently used by ornithologists are population size, diversity or species richness, and rarity (including measures of regional occurrence). However, the full range of attributes, and the relevance of each to a particular study, should be considered.

(3) Can the chosen attributes be quantified using measurable criteria? If so, would any type of criteria used in previous studies be appropriate for your own evaluation?

(4) Is it adequate simply to rank the sites or species on a continuous quantitative scale (for example density of birds) or must the sites be classified into classes of conservation importance? If the latter, then suitable threshold criteria must be determined so that each site or species can be clearly allocated to a class.

Evaluations cannot be fundamentally objective. The application of an evaluation method can be a systematic and routine procedure, but all such methods have inbuilt value judgements. The choice of attributes and the criteria by which to measure them, and the determination of classes of importance, all involve some initial subjective decisions. Although the ideal is to base all conservation priorities on biological fact, sufficiently detailed knowledge of the ecological requirements of most species of birds is currently lacking.

CHAPTER 12

Assessments using invertebrates: posing the problem

R. HENRY L. DISNEY

12.1   Introduction
12.2   Criteria that can be used with invertebrates
    12.2.1   Naturalness
    12.2.2   Rarity
    12.2.3   Area
    12.2.4   Diversity
12.3   Comprehensive surveys
12.4   Sample surveys
    12.4.1   Sampling problems
    12.4.2   The terms 'collecting-success' and 'collecting-efficiency'
    12.4.3   Choice of collecting methods
    12.4.4   Achieving equivalence in terms of collecting-efficiency and habitat type
    12.4.5   The comparison of data obtained with sets of traps
    12.4.6   Ranking of sites
12.5   Conclusion
12.6   Summary

Wildlife Conservation Evaluation. Edited by Michael B. Usher.
Published in 1986 by Chapman and Hall Ltd, 11 New Fetter Lane,
  London EC4P 4EE
© 1986 Chapman and Hall

12.1 INTRODUCTION

From the viewpoint of anybody considering competing proposals for a particular parcel of land the aim of any conservation evaluation exercise must be to rank the particular site in relation to other sites in the region under consideration (for example see McHarg, 1969).

At present the use of invertebrates in the process of evaluating terrestrial sites for conservation purposes rarely moves beyond the anecdotal stage. Even when quantitative information is employed it is either difficult to interpret or else it is not employed to rank the particular site in relation to a set of comparable sites in the same region. Perusal of *A Nature Conservation Review* (Ratcliffe, 1977) indicates that even for the most widely recorded invertebrate taxa (for example spiders) data are available for only a limited number of sites; and even then the data are only indicative, at most, of species richness and/or the presence of rare species. For terrestrial habitats there is a dearth of quantitative data that could be used for ranking sites objectively, but better data are frequently available for freshwater habitats.

Two questions can be asked about the simplest measure of the total number of species recorded. How comprehensive was the survey which produced the species list for each particular site? How comparable, in terms of equivalent habitat type(s), are the different sites being compared?

For freshwater habitats the answers to these two questions may be satisfactory for selected taxa of macro-invertebrates. The sampling problems and taxonomic impediments, in principle, are not so formidable as for terrestrial habitats. This chapter, however, will focus primarily on the problems relating to the use of invertebrates in the ranking of terrestrial sites.

Assuming one is able to achieve comprehensive surveys of some invertebrate taxonomic groups for a series of comparable sites, there still remains the question of which criteria can be given a conservation value which can be meaningfully applied to data on invertebrate faunas. Furthermore it is evident that for many invertebrate groups the achievement of a comprehensive survey of even a single site is so expensive in terms of both time and resources that such an attempt is not an available option. It is essential, therefore, to consider the use of limited (or sample) surveys for such groups. However, the use of limited surveys will further restrict the possible criteria that could be meaningfully employed. The relevant criteria are reviewed below. Before proceeding, however, a brief consideration of the wider viewpoint a conservationist brings to his consideration of particular sites seems appropriate.

The aim of the wildlife conservation movement has been defined as being 'to convey the maximum diversity into the next century' (Disney,

1981); it being understood in this context that the notion of diversity refers to the native flora and fauna and refers not only to a diversity of species but also to a diversity of vegetation types, of animal communities, and of natural and semi-natural ecosystems. There are three strategies to achieve this aim.

(1) *Emergency action to save endangered species considered worthy of conservation.* The conservationist is not concerned with the perpetuation of all species, as has been argued by extremists, but his position is to give the benefit of the doubt to every species. Where there is no doubt that mankind would be better off without certain species (such as the human malaria parasites) conservationists endorse, or at least do not oppose, the efforts of those working for their extinction. A decision to save a particular species may only be executed by securing suitable pieces of habitat. The identification of suitable sites in this instance is derived directly from a detailed knowledge of the ecological requirements of the species concerned.

(2) *The establishment of an adequate series of protected nature reserves.* The philosophy and problems inolved in the identification of the relevant key sites in Great Britain have been discussed in detail by Ratcliffe (1977) and in Chapter 6 of this book.

(3) *The re-establishment and maintenance of a landscape embracing a high density of wildlife refuges.* Elsewhere (Disney, 1975) it was suggested that 'scheduled reserves can be regarded as the spectacular motorway system by which we hope to convey a diverse wildlife into the future', and that a 'motorway system is useless without a system of supporting roads feeding into it. Much the same can be said for scheduled reserves and the conservation of wildlife'. The analogue of the 'supporting roads' in conservation strategy is a high habitat diversity per km^2 in the landscape as a whole.

The very success of the efforts to establish protected reserves has tended to undervalue the importance of maintaining adequate wildlife refuges in the rest of the landscape. It is increasingly evident that habitat destruction is the principal threat to wildlife (for example Perring and Mellanby, 1977; Mellanby, 1981). It is equally clear that under the impact of man, landscapes are constantly changing and refuges for particular species are lost. Equally, we are able to create new refuge habitats. We can also deliberately re-introduce species that have disappeared from the region. The conservationist, therefore, is concerned with the guidance of change so that whilst particular pieces of habitat may go, or be created, the habitat diversity per km^2 remains high. The evaluation of any particular habitat parcel will therefore primarily be with reference to the immediate neighbourhood. A good rule of thumb in Britain is to consider the actual 10 km square of the national grid in which the site is located plus every adjoining 10 km square. The farmer sympathetic to conservation wishes to maximize his contribution to wildlife conservation while minimizing the financial penalties. He needs to know the potential for provision of valuable habitat on his land. He also needs to know how valuable, in conservation terms, is any particular existing habitat parcel on his farm. The value of a parcel of habitat will, first, depend on how rare that habitat type is in the neighbourhood. Secondly, if it is of a habitat type that is fairly frequent locally, is the native fauna and flora more or less valuable when compared with similar sites in the neighbourhood? The latter question can only be answered by the collection of quantitative data which will allow the ranking of a candidate site along with all the similar sites.

It is apparent that site evaluation in practice is mainly focused on relatively small, discrete, parcels of habitat. Larger complexes are composed of mosaics of different habitat parcels. While this greatly complicates the process of evaluating a site it is nevertheless desirable that the constituent habitat parcels should be individually assessed. Only then is one in a position to evaluate the complex as a whole.

12.2   CRITERIA THAT CAN BE USED WITH INVERTEBRATES

The choice of criteria to be applied in a particular evaluation exercise will be in part determined by whether comprehensive surveys or sample surveys are to be employed in the collection of the relevant data.

When one considers the criteria reviewed by Ratcliffe (1977) or Margules and Usher (1981) it is evident that the fragmentary state of our knowledge of invertebrate faunas, even in a country as well documented as Britain, precludes the use of most criteria. In view of our ignorance only naturalness, rarity, area and diversity would seem to merit consideration. These are briefly reviewed below.

12.2.1   Naturalness

At first sight the zoologist can only look with envy at the detailed data available to botanists on the history of the flora in a particular region. Much of this history has been pieced together from a study of sub-fossil pollen preserved in peat, and other recent deposits. The knowledge obtained allows reconstruction of the flora before the advent of habitat destruction by man. Along with reconstructions of climatic changes in the post-Pleistocene period this knowledge enables the assessment of the degree of naturalness of the present flora. Equivalent data for the fauna are impoverished by comparison. However, knowledge is steadily advancing. Molluscs are particularly well preserved in certain soils (for example Barrett and Chatfield, 1978) and this fact, combined with the

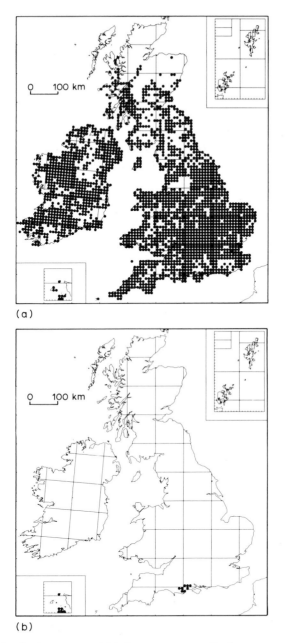

(a)

(b)

Fig. 12.1 Distribution maps of (a) the common snail, *Cepaea nemoralis*, and (b) the rare grasshopper, *Chorthippus vagans*. These maps are reproduced by permission of the Biological Records Centre, Institute of Terrestrial Ecology and Mr E. C. M. Haes (*C. vagans*) and Dr M. P. Kerney and the Conchological Society of Great Britain and Ireland (*C. nemoralis*).

relative immobility of terrestrial molluscs, means that one ought to be able to develop measures of the degree of naturalness of a particular molluscan fauna surveyed today.

Peat systems preserve more than just the pollen of past vegetation. Grospietsch (1965) has reported on the tests of testate rhizopods, and Bryce (1962) reported on the head capsules of larval Chironomid midges. The remains of beetles have been recovered from a range of soil types (see Crowson, 1981). The principal impediment to the use of such material is that one is identifying parts of organisms only. Furthermore, there are almost no keys that can be used with confidence by a non-specialist.

As knowledge advances we should be able to utilize the criterion of naturalness with growing confidence. At present we can only employ this criterion with a few taxa, in a limited range of habitat types, in only a few parts of the world.

12.2.2 Rarity

Rarity is a criterion that can be used providing two requirements are fulfilled. First, one must be dealing with data produced as a result of a comprehensive survey of a site. Secondly, the taxonomic groups being considered must be those for which detailed knowledge of distribution is available.

Rarity, as a criterion, can only be used with data derived from comprehensive surveys because it is now well established that the vast majority of sites habour at least a few rare species of invertebrate. Thus the assignment of a high rank value to a site will depend upon the number of rare species rather than the presence of rare species.

The second requirement is necessary due to the need for an objective definition of rarity. In Britain, the efforts of the recording schemes organized by the Biological Records Centre are primarily designed to provide detailed knowledge of distribution (Heath and Scott, 1977; Heath and Harding, 1981). Once a group has been mapped in sufficient detail one can define degrees of rarity. For example, species reported from less than 15 of the few thousand 10 km squares surveyed might be designated Category 1 rarities. Those reported from 15–50 squares might be designated Category 2 rarities, and so on. The distribution of a common species, which occurs in about three quarters of the 10 km squares, is shown in Fig. 12.1a, whilst the distribution of a rare species, occurring in only eight 10 km squares (and hence Category 1 above), is shown in Fig. 12.1b. Two forms of Category 2 rarity are shown in Fig. 12.2: one is distributed locally, whilst the other is widespread and isolated (Fig. 12.2a and b respectively). One can then score the frequency of such rarities for a series of sites that have been comprehensively surveyed. One could also score the proportion of rare species in one's species lists.

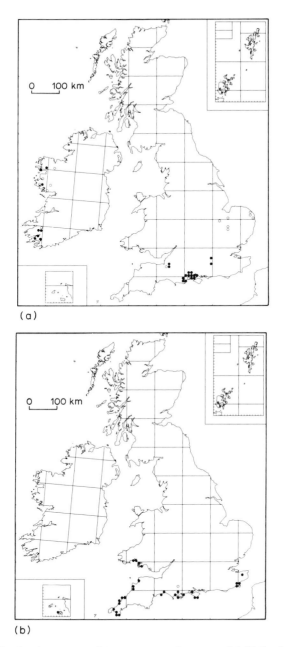

(a)

(b)

Fig. 12.2 Distribution maps of two rare grasshoppers, (a) *Stethophyma grossum*, which is now distributed only in a small part of the British Isles, and (b) *Tetrix ceperoi*, which is widely distributed in isolated areas in southern England and southern Wales. Open circles indicate records up to, and including, 1960; filled circles are used for records from 1961 to 1982 inclusive. These maps are reproduced by permission of the Biological Records Centre, Institute of Terrestrial Ecology, and Mr E. C. M. Haes.

For most groups of invertebrates we lack the distribution data to be in a position to define what are rare species. For most parts of the world there is nothing comparable to the Biological Records Centre mapping schemes of Britain. It is possible to obtain some subjective assessment of rarity by seeking the opinion of 'experts', or by examining museum collections. However, the former assessments are frequently distorted by a predilection for a particular collecting method. The latter tend to be distorted by over-representation of specimens of rare species.

12.2.3 Area

Insufficient attention has been given to the criterion of area by zoologists concerned with conservation evaluation. Such data as exist suggest 'that the traditional concept of species–area relationships and theories of island biogeography were not helpful in analysing the presence and abundance of invertebrate species' (Webb, 1982). However, these data also indicate that area is a factor that must be taken into consideration when comparing sites in order to rank them in terms of wildlife conservation value. Smaller sites tend to be richer in species than expected. In part this derives from the mobility of most invertebrates and the consequent influx of species from adjacent habitats. While such enrichment of the fauna is largely confined to the margins in a large site, in a small site the entire habitat may be affected. Moran and Southwood (1982), in sorting arthropods from trees into their various ecological guilds, found it necessary to add an extra guild of 'tourists'. It would be interesting to examine data procured in exercises designed to investigate the relationship between species richness and area with, and then without, the inclusion of the 'tourists' in the analysis. Certainly the anomalous patterns reported by Webb (1982) indicate the need for critical investigations before we can use the criterion of area. In particular an assessment of the diversity of the habitat mosaic surrounding the particular parcel of habitat under consideration is needed in order to be able to interpret species/area data (see Webb, Clarke and Nichols, 1984).

12.2.4 Diversity

This appears to be the one criterion that can be used in any part of the world, with any invertebrate group, whatever our state of knowledge of the local fauna. Furthermore one can use this criterion with data derived from either comprehensive surveys or from limited (sample) surveys, providing the latter are comparable in terms of their methodology. For most insects this last consideration demands surveys performed simultaneously at the different sites being considered; a condition that is difficult to achieve in practice.

In view of the fact that diversity is the one criterion that can be applied regardless of the level of knowledge of the faunas being surveyed, attention will be concentrated on this criterion. We can consider its use in relation to attempted comprehensive surveys and then in relation to intentionally limited surveys.

12.3 COMPREHENSIVE SURVEYS

An absolutely comprehensive survey of the total fauna of a site is an unattainable ideal for the majority of habitats, and possibly for all habitats. It is, however, possible to approximate a comprehensive survey for selected groups of invertebrates.

For relatively immobile organisms like molluscs direct searching and sorting techniques may suffice to produce a reasonably complete species list for a site (for example Cameron and Redfern, 1972). More mobile creatures which are relatively large and conspicuous, like butterflies or grasshoppers, require a little more effort. With the Orthoptera visual searching can, at the right time of year, be supplemented by listening for songs characteristic of particular species. Other organisms may be less conspicuous individually but may, because of their habits, be amenable to direct searching. Ants, by virtue of their nest building behaviour and social organization, are such a group. Likewise gall-forming or leaf-mining insects are readily recorded in direct-search surveys. Homoptera (hopper-bugs, leaf-hoppers, spittle bugs, etc.) are readily surveyed, at appropriate times of year, by sweep-netting or suction samplers, in certain types of habitat (for example grassland).

For most invertebrates any particular collecting method is very selective; some species are readily procured whilst others are seldom, if ever, caught by that method, even when they are common. Furthermore, seasonal changes in the availability of species militate against the achievement of a comprehensive survey. Even when collecting effort is extended over all seasons the selectivity of particular collecting methods will ensure failure to achieve a comprehensive inventory of the species present. For example, in year-long surveys of the invertebrates of 42 sites in the Pennines by means of pitfall traps, the results for Sepsidae (Randall, Coulson and Butterfield, 1981) and Phoridae (Disney, Coulson and Butterfield, 1981) are far from comprehensive in terms of species representation, as evidenced by the returns from other collecting methods employed in similar Pennine sites. The reason is that pitfall traps are a most inefficient means of collecting these two families of flies (Disney et al., 1982).

For most groups of invertebrates it would require an unacceptable effort in order to achieve even a crude approximation towards a comprehensive species list for a single small site. For example between 1954 and

1958 members of the Yorkshire Naturalists' Union carried out a survey of the insects of the Malham Tarn area in North Yorkshire, England. The survey listed 436 species of Diptera (Flint, 1963). Since 1971 attention has been directed towards Diptera in an effort to achieve a comprehensive survey. The total number of species of Diptera recorded now exceeds 1000.

The increase results from increased effort, a diversification of collecting methods, and the publication of improved identification keys for previously poorly understood families. Despite an expected diminishing return for increasing effort the list has more than doubled. Table 12.1 presents examples of the increases in the numbers of recorded species for selected families. The augmented figures for Chironomidae, Dolichopodidae and Muscidae are largely due to additional effort and better keys. For Heleomyzidae and Syrphidae increases are due to more effort and a wider range of collecting methods. Some families, for example Cecidomyiidae, are still largely unrecorded. It is clear that there is much more to do before one could even consider claiming to have listed more than 90% of the species of Diptera occurring on the Malham Tarn Estate Nature Reserve; and yet it should be noted that this is one of the best surveyed localities in the world as far as the listing of the total Diptera

Table 12.1 Number of species in an attempted comprehensive survey of the Diptera of the Malham Tarn Estate Nature Reserve in North Yorkshire. Only a selection of families are tabulated.

	Number of species	
Family	*Recorded before 1960 (Flint, 1963)*	*Recorded by 1983 (unpublished data)*
Chironomidae	45	96
Dixidae	2	8
Mycetophilidae	11	42
Psychodidae	0	16
Sciaridae	2	12
Tipulidae	93	96
Dolichopodidae	37	54
Empididae	42	66
Phoridae	0	88
Syrphidae	48	79
Anthomyiidae	8	28
Calliphoridae	5	9
Muscidae	35	83
Agromyzidae	0	42
Drosophilidae	0	9
Heleomyzidae	9	18
Total (all families)	436	>1000

fauna is concerned. One might ask whether such attempted inventory (or comprehensive) surveys are of any value in conservation evaluation. The answer is probably that as a tool for the initial recognition of high-ranking sites inventory surveys are only an available option for invertebrate taxa that can be comprehensively surveyed with relatively little effort. Nevertheless, preliminary inventory surveys of other groups may provide the basis for planning a programme of limited surveys. For example the results of the initial attempts at an inventory survey of the fauna of the Hortobágy National Park in Hungary (Mahunka, 1981) simultaneously highlight the interest of the area in general, and of constituent habitats in particular, as well as underlining the high level of ignorance.

Inventory surveys really come into their own when a site has been identified as being high-ranking on the basis of surveys not directed at the invertebrate fauna. In Britain, most biological Sites of Special Scientific Interest have been designated on the basis of vegetation surveys. There is, however, no necessary correlation between a high-rank for vegetation and a high-rank for the invertebrate fauna (for example see Brooker, 1982). Some, largely circumstantial, evidence indicates that a wood managed in a manner which allows accumulation of dead timber will support a richer insect fauna than a wood with a richer ground flora but from which dead timber is continually removed (for example see Chapter 14 in Elton, 1966). The results of many inventory surveys will be needed before the results of vegetation surveys can be extrapolated to make generalizations about associated faunas. Furthermore, the management of sites set aside for conservation will be more soundly based if results of inventory surveys of invertebrate faunas are available to supplement vegetation surveys.

It is evident that for most taxa and for most habitat types the effort required to achieve comprehensive surveys precludes their use for conservation evaluation purposes. However, it seems highly desirable to attempt comprehensive surveys for a number of *reference sites* in order to provide a means of assessing the validity of assessments based on sample surveys only. The use of reference sites to allow interpretation of data obtained in sample surveys is discussed below.

12.4   SAMPLE SURVEYS

The purpose of a sample survey of a site must be to produce diversity data that can be used to rank the site in relation to other comparable sites. As such surveys are limited, they will be assessing the diversity of samples and not diversity of the total faunas. They must, however, aim to procure samples whose diversity measures are directly related to those for the total faunas of the sites. Therefore the requirement is for

replicated samples from each site, sampled in order to be able to assess
the significance of any differences between measures of diversity based
on these different sets of samples. It is equally necessary to ensure that
samples from different sites are strictly comparable. Differences in the
diversity estimates for the different sites must reflect differences in their
faunas, not differences in the sampling procedures. The aim is to
discover whether variability between sites is greater than the variability
within a set of samples from the same site.

12.4.1 Sampling problems

Samples may not be comparable for various reasons. If different sites are
sampled on different occasions (different days or different times of day)
then the vagaries of the weather or circadian rhythms of activity will be
likely to give rise to results reflecting differences due to these extraneous
factors rather than differences in species richness. If the different occasions are too far apart then seasonal changes in the available fauna will
influence the results. If the effectiveness of the sampling procedure
chosen varies with some feature of the habitats then differences in
diversity measurements may be largely due to this feature. This consideration, however, tends to be more of a problem for the ecologist
than the conservationist. The former frequently wishes to compare the
faunas of different habitat types, or the different parts of a habitat
exhibiting some pattern of variation in conditions. The latter wishes to
rank a series of sites of approximately equivalent habitat type, or provide an integrated measure of diversity for a habitat with a range of
conditions. Nevertheless small differences between similar sites may
make a significant difference to the comparability of sets of samples from
these different sites.

12.4.2 The terms 'collecting-success' and 'collecting-efficiency'

In discussing the problem of the choice of appropriate trap types for
sampling populations of forest mammals, and also for sampling river-
crabs, two terms were employed (Disney, 1968, 1976a). The term *trap-
efficiency* was used to refer to the extent to which returns for a particular
type of trap truly reflected the size or age range of the population being
sampled. The term *trap-success* was used to refer to the number of
animals caught for a given effort. Thus a particular type of trap might
give a high trap-success but a poor trap-efficiency as only middle-sized
crabs were caught. These basic concepts can be adapted for the purposes
of sampling species from a total fauna, as opposed to sampling indi-

viduals from particular populations of a single species. *Collecting-success* is the return for a given effort in terms of the number of species procured. It can be readily measured for different collecting methods being operated at the same place over the same period. *Collecting-efficiency* is the percentage of species caught of the total number of species available in the habitat during the period of collection. It is far from easy to measure this in practice. It is important to realize, however, that the same collecting method used in two different situations may give results showing differences in collecting-success which are, in reality, largely due to differences in collecting-efficiency. Some over-looked factor affecting the efficiency of the method employed may mislead one into believing one has demonstrated differences in species richness.

The ideal is to select a collecting method that not only produces replicated samples but which has a collecting-success which gives a high return for a small effort. In addition, when the method is employed simultaneously in two or more sites, the collecting-efficiencies must be equivalent. As long as this requirement is fulfilled, an increase in collecting-success will also result in a desirable increase in collecting-efficiency.

There has been a tendency for students of insect faunas of terrestrial habitats to pay undue attention to maximizing collecting-success at the expense of equivalence of collecting-efficiency. Frequently it seems that as long as a particular method collected plenty of specimens there has been no attempt to assess either relative collecting-success or consistency of collecting-efficiency when compared with some alternative method.

Students of freshwater invertebrate faunas have frequently adopted a more critical approach. For example Furse *et al.* (1981) demonstrated that collecting-efficiency for pond-net samples varied with both site and operator. Likewise Elliott and Drake (1981) studied the collecting-efficiencies of seven different grabs for sampling benthic invertebrates and, among other things, showed that collecting-efficiency varied with the mean particle size of the substratum. In studies on sampling *Simulium* larvae in rivers in the West African rain forest it was shown (Disney, 1972) that collecting-efficiency was greatly influenced by the behaviour of the river (whether it was rising or falling in level). As tropical storms are frequently very local this can mean the behaviours of two rivers only a kilometre apart may be very different. Hence, apparently equivalent sampling procedures being operated in two adjacent rivers simultaneously may not be comparable at all. It was partly for this reason that the simultaneous survey of the *Simulium* populations at eleven sites on a river system was designed to sample pupae rather than larvae (Disney,

1976b). The pupal samples represented settlements of larvae over several days, during which time several spates may have come and gone.

For terrestrial habitats it would seem that a standardized series of soil cores taken at approximately the same time on the same day at a number of sites might avoid many of the problems relating to a failure to achieve equivalent collecting-efficiencies at different sites. The selected taxa (for example mites or Collembola) could be extracted and the mean number of species per sample for each site used as a measure of diversity. If, however, the types of soil were dissimilar, it is highly probable that the efficiencies of extraction would vary (Phillipson, 1971). However, this will mainly affect the numbers of individuals extracted. Only if the proportion of the available species extracted varied would the collecting-efficiencies be rendered non-equivalent for the different sites.

12.4.3   Choice of collecting methods

Many authors have reviewed particular methods for collecting various invertebrate taxa. Southwood (1978b) provides a critical review of a range of methods, for a range of insects, in a variety of habitat types. Most reviews are more of a lucky-dip selection of methods (for example Laliberté, 1976). The most detailed reviews cover families of organisms of applied significance (e.g. Service, 1976, 1977). Others restrict their coverage by concentrating on a particular habitat, such as grassland (for example Clements, 1982). The conservationist will, likewise, need to confine attention to selected taxa, but will be wanting to survey several families, a large order, or several orders simultaneously. Thus the conservationist is seeking out methods that will give a high collecting-success over a large range of taxonomic groups. Most of the literature, however, is concerned with the sampling of a limited number of species for the purposes of population and other ecological studies based on the scoring of individual organisms. The conservationist, by contrast, is trying to assess whole faunas by counting species, not individual organisms.

Some of those who have tried to survey large taxonomic groups (that is, groups with large numbers of species, such as Diptera) have reported preliminary attempts to compare their collecting methods. Penny and Arias (1982) surveyed the insects of an area of Amazonian forest and compared various methods in the process. However, their analysis is not at the species level. Bährmann (1976) surveyed Diptera in grassland and takes his comparison of methods further. Von Tschirnhaus (1981) also made useful comparisons of a variety of collecting methods at the species level, for Chloropidae and Agromyzidae (Diptera), and has demonstrated differential selectivity for even closely related species.

Likewise Younan and Hain (1982) compared five trap designs for insects attracted to severed pines and demonstrated that each design was selective for a particular group of insects. A more deliberate attempt to select a simple, but effective, method for sampling Dipteran faunas, for conservation evaluation purposes, has been reported by Disney *et al.* (1982). It was found that a set of white water traps (that is, white plastic bowls part-filled with very dilute detergent in water) produced a good return for a relatively small effort. A malaise trap exhibited a higher collecting-success than a single water trap. However, it was found that not only is the effort of sorting the catch a formidable task, but it has proved difficult to replicate sets of malaise traps that are equivalent in collecting-efficiency. Small differences in design, or even the state of cleanliness of the malaise trap, can produce significant differences in the catch. Likewise small differences in the siting of the traps make large differences in the catch. In other words, with a set of malaise traps the mean for the catches has a large variance. To offset this one can, of course, increase the number of traps. This, however, makes the task of sorting the catches even more formidable. By contrast, a set of white water traps, despite a lower collecting-success, produces a more acceptable return for a smaller effort. For some families, yellow traps were more successful. Kirk (1984) has explored the selectivity of different coloured water traps further. Among other things he contrasts true white and white that appears black in UV light. For other taxonomic groups other methods are better. For a number of arthropod groups (for example Phalangida) pit-fall traps are to be preferred. Dobson (1978) reported that about 70% of the Coleoptera species collected in a survey using pit-fall traps at a classic site had not been recorded by those using the traditional coleopterist's collecting methods over a number of years. Indeed, for certain families of Coleoptera (for example Carabidae) pit-fall traps seem to be very effective (see Butterfield and Coulson, 1983). It is clear that the wider the range of taxa one wishes to survey simultaneously the greater the variety of collecting methods one will need to employ simultaneously.

The complex problem of integrating the returns from a variety of different sorts of trap, each with its own peculiar characteristics in terms of collecting-success and collecting-efficiency, is easily overcome. A sampling unit can be a set of different types of trap, or sets of traps, rather than an individual trap. For example, a sampling unit might comprise an 18 × 18 cm white water trap, plus a 23 cm diameter yellow water trap, plus a set of ten 5 cm diameter glass pit-fall traps. This sampling unit can then be replicated ten times in each of the sites being sampled. The measure of diversity employed then becomes the mean number of species per sampling unit, rather than the mean number of species per trap.

The suggested figure of ten sampling units is derived from experience (for example Henshaw, 1984). It is a compromise between an adequate number of samples required for statistical analysis, an allowance for accidents to some units, and a desire to minimize the effort required to achieve some valid results.

12.4.4   Achieving equivalence in terms of collecting-efficiency and habitat type

The conservationist evaluating sites is primarily concerned with ranking a set of similar ecosystems. For example, the aim might be to rank a series of oak woods rather than to compare the fauna of an oak wood with that of a peat bog.

In the last analysis every parcel of habitat is unique. Recognition of the 'equivalence' of two sites in terms of habitat type is, therefore, a first stage in the grouping of unique entities by ignoring certain differentiating features. The zoologist, however, can readily draw upon the perceptions of the botanist in his recognition of distinct vegetation types. Thus an oak wood should be compared with other oak woods and a spruce plantation with other spruce plantations. When subdividing mixed woodlands some arbitrary, but commonsense, divisions may be required to achieve equivalent habitat units.

An important consideration when seeking to compare equivalent sites is area. For terrestrial habitats in particular the structural complexity of the habitats must be equated. The minimum requirement is recognition of Elton's major habitat categories of open ground, field type, scrub and woodland (Elton, 1966). However two woods may possess very different structural characteristics, perhaps because one is grazed by livestock whilst the other is ungrazed. This latter consideration is important not only in terms of equivalence of habitat type, but also, possibly, in terms of equivalence of collecting-efficiency. For example, in a comparison of a stand of sycamores, whose ground flora was subjected to regular grazing (Wood A), with an adjacent piece of ungrazed, mixed woodland with frequent sycamores (Wood B), the following results were obtained. Diptera species were scored (excluding females of Chironomidae and Sciaridae and males of Calyptrata) in a set of ten white water traps (circular bowls of 31 cm diameter) set in each site on the ground for 48 hours. The results are presented in Table 12.2. Wood A produced a mean number of species per trap of 38.7 species, and Wood B produced a mean of 32.7 species. The difference is not significant ($P > 0.05$). Over precisely the same period ten 18×18 cm white water traps were attached to the trunks of sycamores 1.5 m from the ground, on the south aspect of each tree, in the two sites. The mean for Wood A was 17.0 species whilst that for Wood B was 26.4 species. This difference is

Table 12.2   Data obtained when white water traps were set in two woods in two sets: ten traps on trees and ten traps on the ground in each site. The figures are the number of species of Diptera (excluding female Chironomidae and Sciaridae and male Calyptrata). See text for an interpretation of the results.

Trap number, and calculated statistics	Wood A		Wood B	
	On tree	On ground	On tree	On ground
1	16	32	15	29
2	18	37	33	38
3	10	33	26	57
4	21	42	25	31
5	22	39	27	26
6	11	54	42	30
7	19	34	25	31
8	23	40	21	43
9	13	31	32	24
10	17	45	18	18
mean	17.0	38.7	26.4	32.7
standard error	1.43	2.23	2.48	3.48

significant ($P < 0.02$). The interpretation is that by placing white water traps on the ground there was a failure to achieve equivalent collecting-efficiencies. In the grazed wood traps were more exposed, because of the short ground flora, and therefore achieved a higher collecting-efficiency. By contrast the tall ground vegetation of Wood B not only reduced the degree of exposure of the traps generally but did so differen-tially. This not only reduced the mean number of species procured in the richer site but also increased the variance. The traps set on the tree trunks are thought to be equivalent in terms of collecting-efficiency. They are not only equally exposed but, being all attached to the same species of tree, their shading by the canopy is more uniform. The variance of the catches is smaller than that of the catches from the bowls on the ground. The difference between the means is thought to reflect real differences between the faunas of the two sites.

A cause of increased variance in trap returns for a site may be spatial heterogeneity of microhabitats. One may have a mosaic of two or more sub-units within what is otherwise a recognizable vegetation type. The simplest way of dealing with such a situation is to increase the number of traps employed, but not the number of units analysed. That is to say one may have ten replicated sets of traps rather than ten individual traps in each site. The mean number of species per set of traps can then be used as one's measure of diversity.

If one is having to set traps some distance from a road the labour of carrying even lidded water-traps becomes tiresome. It can be circum-vented by pouring the contents of each trap through a numbered square of cloth in a funnel. Each cloth is carefully folded for transport to the laboratory, where the insects can be washed off into a bowl for sorting and scoring.

The supreme advantage of using traps is that they will yield useful data even during periods of wet weather, when many other collecting methods cannot be employed.

12.4.5 The comparison of data obtained with sets of traps

To calculate the significance (P values) of the differences between the two data sets given in Table 12.2, the Mann–Whitney U-test was employed (see Siegel, 1956; Elliott, 1977). This non-parametric test is appropriate since counts of species are seldom normally distributed. The calculations for the traps on the trees follows.

The counts obtained in each sample are ordered separately into a sequence of ascending values as in the first two lines of Table 12.3. These counts are now assigned ranks starting with the lowest value, and hence the ten species in a trap in Wood A takes the rank 1, and so on up to 42 species in a trap in Wood B taking rank 20 (see Table 12.3). Where two (or more) counts are the same, they are each assigned the mean of the ranks (that is, the two counts of 18 species would have the 7th and 8th ranks, and hence they are given rank 7½). The U-value is calculated from the largest R value, thus

$$\text{if } R_1 > R_2, \text{ then } U = R_{max} - R_1;$$

$$\text{if } R_2 > R_1, \text{ then } U = R_{max} - R_2.$$

R_{max} is the maximum value that R can take. In this example it would be if the series of 10 samples from one wood had the ten top ranks (that is,

Table 12.3 The operation of the Mann–Whitney U-test for the tree data in Table 12.2. The calculation of test statistics is described in the text. Note that $R_1 + R_2 = 69 + 141 = 210$, and that $R_1 + R_2 = 1 + 2 + 3 + \ldots + 20 = 210$. As a further check this sum is given by ½L(L + 1) = 10 × 21 = 210, where L is the largest rank (20 in this example).

Wood	Data										Total
	Raw data ordered from smallest to largest										
A	10	11	13	16	17	18	19	21	22	23	170
B	15	18	21	25	25	26	27	32	33	42	264
	Raw data ranked (1 = smallest, 20 = largest)										
A	1	2	3	5	6	7½	9	10½	12	13	69(R_1)
B	4	7½	10½	14½	14½	16	17	18	19	20	141(R_2)

Table 12.4 The Mann–Whitney U-test. Significance levels (P values) that obtain when U is less
than the tabulated figure. Note that n_1 is the number of counts in sample 1 and n_2 is the number of
counts in sample 2. For fuller tables see Siegel
(1956), Neave (1981) or Lindley and Scott (1984).

n_1/n_2	P values (2-tailed test)		
	0.05	0.02	0.002
10/10	23	19	10
9/10	20	16	8
9/9	17	14	7
8/10	17	13	6
8/9	15	11	5
8/8	13	9	4

ranks 11, 12, . . . , 20), and hence $R_{max} = 155$. Thus $U = 155 - 141 = 14$.

This U-value is compared with the tabulated values in Table 12.4 for
$n_1 = 10$ and $n_2 = 10$. Since 14 lies between 10 and 19, P is between 0.02 and
0.002. Conventionally a difference between means is regarded as significant when P is less than 0.05.

12.4.6 Ranking of sites

Assuming one has chosen collecting methods that procure an adequate collecting-success with equivalent collecting-efficiency, one must
employ them simultaneously in a series of sites one wishes to rank in
terms of conservation value. The rank values will be derived from the
mean number of species per sampling unit as a measure of faunal
diversity. The higher one's collecting-efficiency the more reliance can be
placed on the rankings obtained. However, even with a high collecting-
efficiency one may be sampling only a small proportion of the fauna
available over the whole year. Macan (1957) concluded from his studies
of stream Ephemeroptera that 'a true picture of the fauna of a stream
cannot be obtained unless collections are made twice: in July and August
and again sometime during the first five months of the year'. While his
conclusions relate to the attainment of a comprehensive survey they are
not without relevance in relation to limited surveys. With the latter, the
requirement is that evaluations on a particular occasion should be more
or less consistent with those made on a subsequent occasion later in the
year. There is an obvious need for investigations of a range of habitat
types in which evaluations of a series of sites at different seasons over
more than one year are repeated several times in order to assess consistency of the rankings obtained from such limited surveys. Where a lack
of consistency is found, with certain habitat types, the need is to
determine how many times limited surveys need to be repeated before

the means of the rank values (derived from these means – see below) obtained on each occasion become a reliable method of ranking the sites.

The question as to whether diversity is an adequate measure of wildlife conservation value is really a question concerning the selection of taxa used for evaluation purposes. A good case can be made for using Diptera, as it is a large order with the greatest variety of larval habits of any order of terrestrial invertebrates. An assessment of dipteran diversity reflects the richness of the ecosystem as a whole more fully than an assessment based on an exclusively phytophagous order. The latter assessment might tell you little more than a more rapidly accomplished survey of the higher plant flora. On the other hand, total Diptera may embrace a lot of species which fall into Moran and Southwood's (1982) guild of 'tourists'. Opinion is divided as to whether this is important. Limited data suggest that the outcome of rankings are little affected if the tourist species are left in or if they are excluded. If this is so, then on grounds of economy of effort there could be a good case for excluding certain taxa and only scoring the remaining Diptera. However, before deciding too hastily on particular 'rules of the game', it seems desirable to obtain data for a series of sites scoring the different families separately. The effect on the rankings, when different families are excluded from the analysis, can then be examined.

With less experienced entomologists working under the guidance of the more expert, it has been found that much material can be rapidly sorted to species on their 'jizz' alone. The accuracy of such sorting is very high if females are excluded from certain families of Diptera (for example, the Chironomidae) and Hymenoptera-Parasitica are substituted for the Calyptrata. With Parasitica the exclusion of the males leads to relatively straightforward sorting to species on their 'jizz', even if subsequent attempts to name individual species present formidable problems. The use of Diptera, exclusive of Calyptrata, plus Parasitica, provides a package of taxa that interact with a very broad spectrum of ecological niches and microhabitats. Diversity measures, therefore, are likely to be closely related to conservation value.

Finally, there remains the problem of how to relate the rankings obtained for a particular series of sites on a particular occasion with those obtained for a different series of sites on a different occasion. Differences in weather and seasonal patterns of occurrence will undoubtedly affect collecting-success such that the two exercises cannot be directly compared. A simple solution to this problem has been proposed (Disney, 1982, 1986). For each habitat type a particular *reference site* is selected for the region under consideration. In each survey carried out in the region the reference site is included in the set of sites being compared. In each particular survey the means for each site are divided by the mean obtained for the reference site in that particular exercise. For

Table 12.5 The evaluation of four woods using data from two surveys performed on different occasions. In June, Woods X, Y and C were compared: site C was designated the reference site. In July, site Z was compared with the reference site C. Despite the mean number of species per sampling unit obtained at site C in July being twice that obtained in June, the rank values allow the results for site Z to be directly related to those for sites X and Y (see text for details). P values from U-tests indicate that the June results for Woods X and Y, and for Y and C, significantly differ (P<0.02). Likewise the July results for Woods Z and C significantly differ (P<0.02). Rank values (mean for a site divided by mean for site C on the same sampling date, multiplied by 100) are rounded to the nearest integer.

Survey date	Woods	Mean number of Diptera species	Rank values
June	X	29.8	186
June	Y	24.1	151
June	C	16.0	100
July	Z	43.2	129
July	C	33.3	100

convenience each result could be multiplied by 100, and consequently the reference site would always have a rank value of 100. An example of this method is given in Table 12.5. Three woods were compared on the first occasion using ten pairs of white watertraps in each wood. The mean number of Diptera species per sampling unit is shown in Table 12.5. Site C was the reference site and therefore its mean was divided into the means for the other two sites and the results multiplied by 100 to give the rank values (expressed as a percentage of the reference site). The differences between the data sets were assessed, using the Mann–Whitney U-test, having first arranged the sites in descending sequence of the means. The results for Wood X were significantly more diverse than those for Wood Y, which was also more diverse than Wood C. On the subsequent July survey, Wood Z was compared with the reference site (Wood C), and the results are also presented in Table 12.5. On this occasion site C produced a mean number of species twice that in the first survey, and the mean for site Z is larger than any obtained in June. This difference was mainly due to an improvement in the weather. The rank values, however, suggest site Z possesses a less diverse fauna than either site X or Y, though site Z cannot be compared statistically with Woods X or Y as it can with C. In an attempt to obtain some independent check on the validity of this procedure ten white watertraps were set on tree trunks in each of a set of woods. For two woods, inventory surveys of the snail faunas had been carried out. These indicated that the number of snail species (12) in the more ancient Wood H was precisely twice the number (6) recorded for the younger Wood S. Scoring all Diptera and Hymenoptera-Parasitica species rank values of

153 and 84 were obtained for the two woods. These values give a ratio of 1.8:1 for the difference in species richness of the insect groups scored. This is surprisingly close to the 2:1 ratio for snails, particularly as one would expect the figures for the relatively immobile snails to exhibit a closer relationship to the age of the woods.

Such a procedure for ranking sites on the basis of limited surveys assumes a measure of consistency in rank orders based on the use of the mean number of species per sampling unit as a measure of faunal diversity. The greater the collecting-efficiencies achieved, the more likely it is that this assumption is acceptable. Where this assumption proves to be untenable it is probable that sample surveys of invertebrate faunas are of little value in the evaluation of sites for conservation purposes. Where it does hold limited surveys can be used for ranking sites, and rankings can be re-assessed at intervals in order to monitor changes. As the knowledge of the fauna of a reference site increases, the results of a limited survey of a set of sites (that include the reference site) on any particular occasion can be interpreted better.

12.5 CONCLUSION

While it is evident that there are immense problems encountered, when trying to evaluate invertebrate faunas for conservation purposes, these are not insuperable. Certain groups, such as snails, have considerable potential for use in the assessment of sites on the basis of comprehensive (inventory) surveys.

Limited (or sample) surveys of insects can yield data that allow sites to be ranked in terms of species richness. The use of Diptera plus Hymenoptera-Parasitica would appear to be particularly useful as indicators of conservation value. More work is needed to test the adequacy and consistency of the rank values obtained with the simple procedures proposed.

Finally it needs to be emphasized that an evaluation of the invertebrate faunas of a set of sites will normally constitute only part of an evaluation exercise. At the very least such an exercise would normally include an analysis of the vegetation as well.

12.6 SUMMARY

When the list of criteria on which to base an evaluation is considered, only four – naturalness, rarity, area and diversity – would appear to be particularly useful when working with invertebrates. The chapter concentrates on one of these criteria, diversity, assuming that value for conservation is positively correlated with species richness.

In a comprehensive survey, an attempt is made to list all of the species of one or more taxonomic groups that occur within a site: this is rarely achieved, since to approach this goal, a large number of trapping and searching methods have to be employed over a long period of time. Sample surveys are cheaper in terms of time and resources, but emphasis is placed on understanding *collecting-success* (the number of species collected for a given effort) and *collecting-efficiency* (the proportion of the total number of species present that are actually caught).

The chapter concentrates on methods of comparing the invertebrate faunas of two or more sites. So that reasonably consistent evaluations can be made at different times of the year, or during different weather conditions, it is important to compare all results against those of a reference site. Various statistical procedures for undertaking comparisons, either between candidate sites or between a candidate site and a reference site, are illustrated in the text and tables.

PART FOUR

General principles

CHAPTER 13

Conservation evaluation in practice

CHRISTOPHER R. MARGULES

13.1 Introduction
13.2 Two evaluation exercises in North Yorkshire, England
    13.2.1 Methods
    13.2.2 Results and discussion of the evaluation studies
13.3 General discussion of the use of criteria
    13.3.1 The role of diversity
    13.3.2 The importance of site area
    13.3.3 The concept of rarity
    13.3.4 Criteria reflecting ecological vulnerability and the threat of destruction
    13.3.5 Representativeness
13.4 An operational framework for conservation evaluation
13.5 Conclusions
13.6 Summary

Wildlife Conservation Evaluation. Edited by Michael B. Usher.
Published in 1986 by Chapman and Hall Ltd, 11 New Fetter Lane, London EC4P 4EE
© 1986 Chapman and Hall

13.1 INTRODUCTION

The aim of this chapter is to consider what procedures people follow when they assess conservation value and, in the light of that review, to re-examine some criteria and to suggest a general procedure for assessing conservation value. There are four specific questions:

(1) How do people intuitively combine estimates or measurements of different criteria?
(2) What weights do people actually give different criteria?
(3) How are criteria of conservation value related to actual values people place on sites?
(4) Is there an overall index of conservation value which models the responses of people assessing conservation values?

To answer these questions two studies investigated what a group of people with experience in assessing conservation value actually do when faced with having to assign values to a group of potential conservation sites. Drawing on the results, and generalizing, there is a discussion of topics such as the role of diversity, the importance of site area, different interpretations of the criterion of rarity, and the significance of criteria relating to ecological fragility and the threat of human interference.

Conservation value is taken here to be a derived value rather than a primitive (or unanalysable) value. There are a range of human values which will be enhanced or threatened by decisions on conservation issues and it is these values from which judgements for conservation value are derived. The primitive value, beauty, for example, gives rise to derived values concerning the existence of natural communities of plants and animals, or natural landscapes, ignoring for the time being how such things might be defined. Naturalness is the resultant standard by which conservation value is judged. The value of rare species is reflected in the conservation of *Orchis militaris* in the UK, a value derived in turn from the primitive values of beauty and humanity (seeing in humanity the caretaker attitude to nature explored by Black, 1970). In this case, rarity is the criterion for judging conservation value.

Sometimes, conservation value can be assessed using only one criterion. More often, though, more than one criterion is involved. Indeed, sites meeting many criteria tend to be more highly valued than sites meeting few criteria. The existence of many criteria and the various ways in which they can be combined, pose problems for the explicit assessment of conservation value.

Seen in this way, as a product of human values, the tangible results of nature conservation, for example nature reserves and national parks, the persistence of native species in the landscape, etc., conform to the definition of resources introduced by Zimmerman in 1933. Resources are an expression or reflection of human appraisal. They perform some function for the benefit of those making the appraisal (Hunker, 1964). Thus, conservation resources, just like agricultural resources, forest resources, mineral resources, etc., have been appraised by man for his benefit.

It is important to appreciate that whilst primitive values tend to persist through time and from place to place, derived values may change. Conservation evaluation schemes used in Sweden (for example Statens Naturvårdsverk, 1980) and Australia (for example Austin and Miller, 1978), generally weight naturalness heavily, either implicitly or explicitly, whilst in the Netherlands (for example Vera, 1980) and Great Britain (for example Ratcliffe, 1977) criteria referring to the vulnerability of fragments of previously more widespread ecosystems and their species, such as fragility and rarity, are given prominence. Given present rates of land use change similar criteria no doubt will, in time, attain greater significance in Sweden and Australia.

The two evaluation exercises described below, which form the basis for the ideas discussed in this chapter, were carried out in northern England. All sites involved were relatively small, and all, apart from the two moorland sites, were well defined by abrupt changes in land use or by fences.

13.2 TWO EVALUATION EXERCISES IN NORTH YORKSHIRE, ENGLAND

Two studies, to investigate the extent of agreement on site value and to compare the relative weights of criteria for judging site value, were conducted in North Yorkshire, England. The studies are reported in detail elsewhere (Margules, 1981; Margules, 1984a; Margules and Usher, 1984). Only a brief resumé of methods and results is given here.

13.2.1 **Methods**

The first study used sites representing different habitats in or near the North York Moors National Park. The eight sites comprised three woodlands, two moorlands, two limestone grasslands and an abandoned limestone quarry. The second study used eight sites which were all enclosed grasslands on soils derived largely from limestone, in the Yorkshire Dales National Park (Table 13.1). In both studies a group of people with experience in assessing conservation value, termed an assessors' panel, visited the eight sites. Panel members were asked to give each site a score out of 20, and, subsequently, to indicate the importance of each criterion, from a list supplied, in arriving at that

Table 13.1 Summary description of the sites used in both evaluation studies. In the first study diversity was measured as the number of plant species in five 1 m² quadrats. In the second it was the number of plant species recorded by Bunce et al. (1985).

Site number	Site name	Area (ha)	Diversity	Description
First exercise				
Summer 1980				
1.	Dawsons Wood	33	17	Oak woodland
2.	Saintoft Grange Quarry	4	40	Abandoned limestone quarry
3.	Gundale	2.5	32	Limestone grassland
4.	Haugh Wood	45	27	Mixed woodland
5.	Yatt's Farm	8	30	Limestone grassland
6.	Fylingdales Moor	400	25	Acid moorland
7.	Levisham Moor	688	18	Acid moorland
8.	Mell Bank Wood	31	22	Ash woodland
Second exercise				
Spring 1981				
1.	Grass Wood Pastures	18	69	Grassland with introduced pasture species
2.	Deepdale Pastures	12	61	Grassland with few introduced pasture species
3.	Middleham Moor	25	48	Limestone grassland with acid moorland knoll
4.	Capplebank	3	31	Grassland resulting from old woodland clearing
5.	Worton Scar	1	52	Grassland and limestone cliff
6.	Countersett Pastures	4	47	Grassland with slight acid influence
7.	Widdale	80	46	Grassland with strong acid influence
8.	Sleights Pasture	30	79	Grassland on lower slopes of Ingleborough with limestone pavement

value. Nine people comprised the panel for the first study. They were given a list of 18 criteria (taken from Margules and Usher, 1981) which could take weights from 0 for little or no importance to 3 for great importance. Fourteen people comprised the panel for the second study, and they were given a list of 11 criteria (reduced from 18 after analysing the results of the first study), which could take weights from 0 for no importance to 4 for great importance. Not all panel members from the first exercise participated in the second, though some did.

Thus, two kinds of data were collected: the site scores (scores out of 20 from each panel member for each site), and weights from each panel member for each criterion at each site. The criteria weights were transformed to what were called importance values by summing the weights given to each criterion by all assessors at each site and then expressing the sum for each criterion as a percentage of the total for that site (Margules and Usher, 1984). Table 13.2 lists the criteria supplied in each exercise, gives the number of panel members who used each criterion at least once (i.e. gave it a weight of one or more at, at least, one site) and gives the mean importance values for each criterion.

The data were analysed in three ways. First, the site scores were compared between sites and between assessors by estimating values with the linear model and subtracting them from the observed values to obtain the differences, called discrepancies (Snedecor, 1956). The magnitude of the discrepancies indicate the extent of agreement between panel members at sites and between sites for panel members. Secondly, criteria were compared by relating them pairwise in regression and altogether in ordination, using the importance values as variables. Thirdly, criteria were related to site scores in multiple regression using criteria weights as independent variables and site scores as dependent variables.

Table 13.2 The criteria provided in each exercise, the number of panel members who gave each criterion a weight of 1 or more at least once, and the mean importance values calculated from the criteria weights. (A full explanation of importance values is given in Margules and Usher, 1984.)

Criterion	First study		Second study	
	No. panel members (max. 9)	Mean importance value	No. panel members (max. 14)	Mean importance value
Diversity	9	11.0	14	14.7
Rarity	9	10.0	—	—
Habitat rarity	—	—	14	10.5
Species rarity	—	—	14	10.0
Representativeness	9	9.1	12	10.8
Area	8	9.3	12	10.1
Naturalness	8	7.7	12	7.8
Threat of human interference	8	5.7	11	7.4
Ecological fragility	9	6.4	11	6.6
Scientific value	7	8.5	11	7.1
Wildlife reservoir potential	5	4.8	11	8.6
Position in ecological geographical unit	5	4.6	9	6.3
Uniqueness	9	7.0	—	—
Potential value	6	3.9	—	—
Management factors	5	4.6	—	—
Replaceability	4	3.6	—	—
Amenity value	3	2.6	—	—
Recorded history	3	1.0	—	—
Educational value	2	1.0	—	—
Availability	1	0.3	—	—

13.2.2 Results and discussion of the evaluation studies

There was general agreement on sites of highest and lowest value in both studies but disagreement on sites having intermediate value, especially sites of small size and relatively high diversity (measured as the

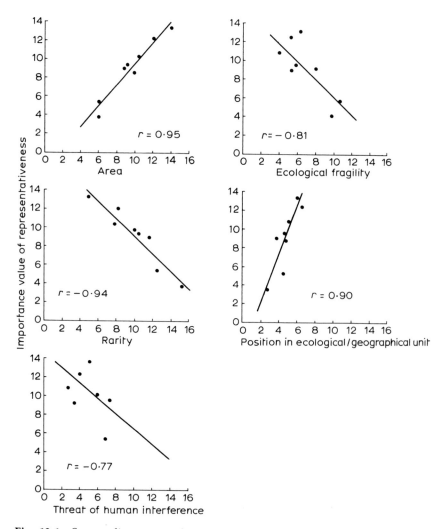

Fig. 13.1 Scatter diagrams with regression lines, of the importance values of representativeness against the importance values of five other criteria (from the first study using different habitats). Area and position in ecological/geographical unit are positively correlated with representativeness; rarity, ecological fragility and threat of human interference are negatively correlated with representativeness.

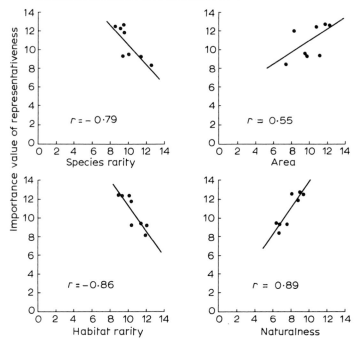

Fig. 13.2 Scatter diagrams, with regression lines, of the importance values of representativeness against the importance values of four other criteria from the study on enclosed grasslands. Unlike the study of sites of different habitats (Fig. 13.1), representativeness and area are not correlated whilst representativeness and naturalness are positively correlated.

number of plant species). In the first study sites 2 and 3 showed the greatest discrepancies from estimated values in both directions, and in the second study site 5 showed the greatest discrepancy. Each of these sites is small and relatively rich in plant species (Table 13.1). Thus, the assessors' panels were capable of agreement on extremes of conservation value but could not agree when features suggesting high value (species richness) occurred with features suggesting low value (small size).

In both studies, two contrasting groups of criteria emerged, one based on concepts of rarity and vulnerability and the other on concepts of size, representativeness and, to a lesser extent, naturalness. When criteria in the first group were given high weights, criteria in the second were not and vice versa. The two groups can be seen in Figs 13.1 and 13.2, and in Table 13.3 where the importance values of some criteria have been related pairwise. Fig. 13.1 displays results from the study of sites of different habitats where a contrast between representativeness, area and position in ecological/geographical unit (Ratcliffe, 1977) on the one hand

Table 13.3 Correlation coefficients (r) derived from relating the importance values of criteria to one another from both conservation evaluation exercises. Because rarity was divided into species rarity and habitat rarity for the second study (enclosed grasslands) some comparisons could not be made and these are shown by a dash. Statistical significance is indicated by the following: n.s., not significant; *,$0.05 \geqslant P > 0.01$; **$0.01 \geqslant P > 0.001$; ***$P \leqslant 0.001$.

Criteria	Different habitats	Enclosed grasslands
Diversity, area	−0.30 n.s.	−0.89**
Species rarity, habitat rarity	—	0.84**
Species rarity, area	—	−0.75*
Habitat rarity, area	—	−0.76*
Rarity, area	−0.91***	—
Species rarity, representativeness	—	−0.79*
Habitat rarity, representativeness	—	−0.86**
Rarity, representativeness	−0.94***	—
Species rarity, position, in ecological/geographical unit	—	−0.79*
Rarity, position in ecological/geographical unit	−0.86**	—
Habitat rarity, naturalness	—	−0.77*
Rarity, naturalness	−0.43 n.s.	—
Representativeness, naturalness	0.28 n.s.	0.89**
Representativeness, area	0.95***	0.55 n.s.
Ecological fragility, representativeness	−0.81*	−0.31 n.s.
Threat to human interference, representativeness	−0.77*	0.12 n.s.
Ecological fragility, threat of human interference	0.81*	0.65 n.s.

and rarity, ecological fragility and threat of human interference on the other, accounted for approximately 50% of the variance in the matrix of importance values (Margules and Usher, 1984). Figure 13.2 displays results from the study on enclosed grasslands where the contrast between representativeness, area, position in ecological/geographical unit and naturalness on the one hand and species rarity and habitat rarity on the other, account for approximately 50% of the variance in the matrix of importance values, and ecological fragility accounted for a further 20% (Margules, 1984a).

There was a number of differences in the results obtained from the two studies. Most of them were small differences in relationships between importance values which can be attributed mainly to the fact that one study used only enclosed grasslands whilst the other used sites from a range of habitats. For example, the criteria of representativeness and area were positively correlated in the study on sites of different habitats but not correlated in the study on enclosed grasslands, possibly due to the fact that the range of sizes of the enclosed grasslands is much less than that of the sites of different habitats. Similarly, the criteria of ecological fragility and threat of human interference were positively correlated for sites of different habitats but not correlated for enclosed grasslands where ecological characteristics are essentially the same and the potential for man-induced changes, such as the application of fertilizer, are essentially the same.

The most telling difference in the results, however, was the relationship between a measure of the actual plant species richness of the sites and the scores out of 20 that the sites received. In the study of sites of different habitats there was no relationship between the number of plant species and the site score. In the second study, however, there was a strong positive correlation between the number of plant species and site score ($r = 0.894$), with sites of essentially the same habitat but with more species considered to be of greater value. Actual diversity (measured as species richness here) only exerts an influence on sites of the same habitat.

However, diversity also features prominently in the similarities between the two studies. On balance diversity was the most important

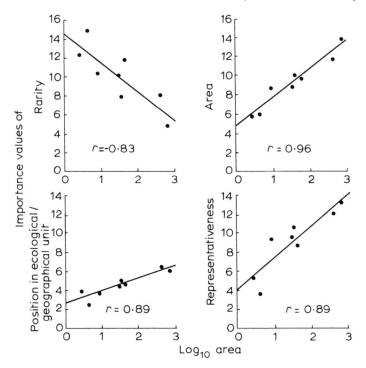

Fig. 13.3 Scatter diagrams, with regression lines, of the importance values of four criteria against the logarithm of the area of each site in the study of sites of different habitats.

criterion in both exercises. It consistently contributed strongly to site evaluations (Table 13.1). Thus, diversity accounts for little variation in the matrix of importance values. It emerges as an extremely important criterion yet not one to facilitate the comparative assessment of a group of potential conservation sites unless they are all of the same habitat type (for example all grasslands).

Area *per se* had no influence on site score. In neither study were larger sites consistently given higher scores. The only influence of site size on sites of different habitats seemed to be to change the criteria by which they were judged. Figure 13.3 shows that there is a strong positive correlation between the criteria of area, representativeness and position in ecological/geographical unit and site area, and that there is a strong negative correlation between rarity and site area. If the two heather moorland sites, which are remote and relatively ecologically stable, are excluded then both ecological fragility and threat of human interference also are negatively correlated with site area. However, no such relationships were found for enclosed grasslands though the fact that there were no very large enclosed grasslands (Table 13.1) may have precluded the possibility of finding such a relationship.

The idea in relating criteria to site scores in multiple regressions was to see if a general model existed which might, given sufficient knowledge of the attributes of a possible conservation site, enable an accurate prediction of conservation value to be made. No such model was discovered because, whilst most regression coefficients were not significantly different, the site scores predicted by those coefficients, were significantly different. This indicates that if the criteria weights were the same, the site scores would be different. This lack of generality really is to be expected in a situation where personal values and beliefs play such a crucial role, and whilst it does not help solve the problem of making valid comparisons of possible conservation sites, it would be somewhat disappointing if everyone thought in exactly the same way! The encouraging result is that there was substantial agreement on a group of criteria which contributed significantly to the evaluation of each site. This, with the other results, enabled the development of the explicit hierarchical evaluation procedure which is proposed and discussed in the following section.

13.3   GENERAL DISCUSSION OF THE USE OF CRITERIA

In the light of the two evaluation exercises, and remembering they were conducted on relatively small sites whose boundaries generally represented an abrupt change in land use, it is possible to make some general comments on the practice of conservation evaluation. In particular, various criteria of conservation value and their roles in the assessment process can be discussed.

13.3.1 The role of diversity

Diversity is the single most important criterion of conservation value. Even so, the diversity of a particular site should not be used alone to judge conservation value. Biological diversity varies from place to place depending on a complex set of environmental variables and interactions between them. Thus, for example, limestone grasslands are inherently more species rich than heather moorlands in the north of England. To judge value on diversity (species richness in this case) alone would mean ignoring whole suites of species, consequently lowering the maximum genetic diversity attainable (Margules, Higgs and Rafe, 1982). In this respect, the idea of representativeness, discussed below, is a more appropriate criterion. Diversity provides no basis for comparison unless the sites being compared are of the same habitat type.

However, even looking at a comparison of sites in what might be considered to be one habitat, enclosed grasslands of the Yorkshire Dales, criteria other than diversity play a significant role. Whilst the study above found a strong positive correlation between scores awarded by an assessors' panel and the number of plant species recorded on limestone grasslands, it is other criteria such as representativeness and rarity which account for most variation in the matrix of importance values. This suggests that evaluations intuitively are made in a staged hierarchical way with diversity being used sometime after the first stage during which a preliminary ordering or classification of sites is made on the basis of other major criteria. This does not mean that diversity is less important, but it does mean that some suitable context within which diversity can be assessed validly must be provided. Bearing this in mind, Smith (1983), for example, has based a conservation evaluation of haymeadows in northern England on an hierarchical procedure in which diversity is used at a third stage 'rather than the construction of a compound index of dubious validity'. Similarly, Mitchell (1983) has based an evaluation of remnant woodlots in the Mount Lofty Ranges near Adelaide, South Australia, on an hierarchical procedure in which diversity is used at a second stage, along with rarity and area, after the woodland communities have been classified and representative woodlots identified at the first stage.

13.3.2 The importance of site area

Area has been linked with diversity through the species–area relationship. This topic is discussed in Chapters 1 and 14, and it is not appropriate to discuss it further here. However, area also has been used as a criterion of conservation value in its own right. In fact, it seems that no panel members in either exercise considered the species–area relationship or the equilibrium theory of island biogeography in their evaluations; at least not explicitly. One assessor in the study on sites of different habitats and two assessors in the study on enclosed grasslands did not consider area to be relevant to a conservation assessment and did not use it at all. However, when examples of small fragmented habitats are under consideration it seems reasonable for larger examples to be valued more highly simply because they are the largest remaining. It is likely that, for the most part, area was used in such a way in the two evaluation studies because there was a strong positive correlation between the actual area of a site and the importance of area as a criterion. It is interesting to note, however, that very small sites were given values just as high as, if not higher than, large sites but different criteria were used to judge site value. This was especially noticeable in the study of sites of different habitats where criteria relating to fragility, rarity and the threat of destruction were used at small sites and criteria relating to representativeness and size were used at larger sites (Fig. 13.3).

13.3.3   The concept of rarity

Rarity is a difficult concept to define ecologically. A variety of factors, both natural and man-induced, can contribute to the rarity or common-ness of a species. The life history strategies of some species ensure that they will never be abundant (for example Rabinowitz, 1978). Because of these difficulties, Adams and Rose (1978) have gone so far as to suggest that rarity should not be a criterion of conservation value. However, although difficult, it is possible to define rarity in ecological terms as studies on plant population dynamics have shown (for example Williams and Roe, 1975; Rabinowitz, 1978). In addition, to deny the importance of rarity is to ignore one of the most fundamental human values associated with conservation: the caretaker attitude to nature explored fully by Black (1970) and implied by Ehrenfeld (1976).

Whilst it is possible to define rarity it is extremely difficult to measure it precisely. To do so it is necessary to define the context in which the term is being used and the scale and boundaries of the study area (see Chapter 1). It is then necessary to know the distributions and abun-dances of all species within the study area. In the absence of such data, the presence of a known rare species may add some conservation value to a site, but cannot be used in a comparative way. However, in the biogeographically well-known parts of the world, of which North York-shire is one, usually it is possible to say with considerable confidence that species A, for example, is less common than species B and that species B in turn is rarer than species C, etc. Also habitat or community rarity is usually easier to determine than species rarity as it requires a general familiarity with the area in question but not necessarily an intimate knowledge. Thus, rarity of species, habitats or communities can be determined with some relative accuracy and used in a qualitative way to compare possible conservation sites where sufficient survey work has been carried out.

13.3.4 Criteria reflecting ecological vulnerability and the threat of destruction

Ecological fragility is a criterion of conservation value that appears in only two of the 16 assessment schemes reviewed by Margules (1981), both of which are from Britain. Threat of human interference is used much more widely and apparently there is some confusion between the two. Indeed, Ratcliffe (1977) describes 'threat' as forming a second main element in his criterion of ecological fragility. However, fragility can be thought of as referring to intrinsic features of a habitat, community or species, whereas threat refers to extrinsic factors such as land use change. Clearly, more fragile sites can be thought of as more vulnerable to land use change and so at greater risk than more resilient sites. Fragility easily applies to sites which represent fragmented, dwindling ecosystems and to seral stages of ecosystems which are likely to require constant management if they are to be maintained.

Looking at the study of sites of different habitats, the panel did distinguish between ecological fragility and threat of human interference to some extent along the lines suggested above. Ecological fragility was most important at the two smallest sites which did represent man-induced seral stages. However, whilst threat of human interference was seen as most important at one of those two sites, its next highest importance value was at a limestone grassland site on farmland where a change in agricultural practice might radically alter the ecology of the site. However, these are minor differences. Ecological fragility and threat of human interference were strongly positively correlated ($r = 0.813$) and were grouped in the ordination of importance values. Also, in the study of enclosed grasslands, there was little change in the importance of ecological fragility or threat of human interference from site to site. The panel tended to consider enclosed grasslands to all be about as fragile as each other and all to be about equally threatened with land use change.

Given that there is some confusion between the two and that they were related closely in the two evaluation exercises, it might be sensible to reduce them to one criterion of likelihood of damage by human interference, recognizing that more fragile sites or communities are more easily affected by land use change. It seems logical that ecologically more fragile sites or communities should not necessarily be favoured intrinsically for conservation over more resilient sites or communities. However, ecological fragility as a component of threat of interference might be a valid help in determining priorities for reserve acquisition after a 'shopping list' of preferred sites for a reserve system has been compiled according to other criteria.

13.3.5 Representativeness

Representativeness was an important criterion in both evaluation exercises. Sites were assessed in terms of how well they represented their habitat type. However, two panel members did not use it to evaluate enclosed limestone grasslands, taking the view perhaps that all sites represented that habitat equally well. This raises the question of the most appropriate level of classification. At one scale, enclosed limestone grasslands can be recognized as a relatively homogeneous habitat in contrast to, say, acid moorlands or oak woodlands, and it might be reasonable to assume in such situations that one enclosed grassland represents that habitat as well as the next. However, at another scale, there are different kinds of enclosed limestone grasslands as indicated in Table 13.1, which should be recognized when working at that scale. In fact, the eight sites used in the study were selected to encompass the range of enclosed grasslands in the Yorkshire Dales, with the exception of haymeadows. Thus, a particular site could represent a kind of enclosed grassland but not enclosed grasslands as a whole unless it was a large site including fields from every kind of enclosed grassland in the Yorkshire Dales.

This apparent confusion points to the need to define the area of concern precisely; to set boundaries to the study area. It also suggests, as for diversity, that having defined the study area, an initial classification, or at the very least some kind of sorting of the sites to be assessed, is necessary. Because it is unlikely that any one habitat fragment will represent the range of that habitat type fully, it may not be appropriate to apply representativeness to particular habitat fragments. As representativeness is based on the idea that a system of reserves for conservation in a given region should encompass the range of natural variation in that region it is a criterion that can be applied more easily to a group of sites. This idea is incorporated in the procedure for assessing conservation value suggested in the next section.

13.4   AN OPERATIONAL FRAMEWORK FOR CONSERVATION EVALUATION

Conservation evaluation essentially is a process of comparison. The most explicit way of making comparisons and, therefore, the most communicable way, is by using numbers. Numbers have the same meaning for everybody: their use minimizes sources of confusion and ambiguity. It is desirable then for criteria of conservation value to be expressed numerically. Diversity can be expressed as species richness, a count of the number of species, or as a more complex statistic incorporating proportional abundance (for example Pielou, 1975). Similarly, rarity can be expressed using proportional abundance or frequency. Area can be measured. Ways of expressing other criteria numerically are more difficult, though some suggestions regarding representativeness have been made in Chapter 2 and the possibilities of quantifying some other criteria have been discussed in Chapter 1.

Whilst criteria should be expressed numerically if possible, there is no obvious way of fixing weights or of integrating criteria into an index of conservation value. Some of the difficulties of doing so have been described above where it was shown that, although criteria are weighted and combined intuitively by people making assessments, into personal indices of conservation value, there is no overall index, even for a relatively small group of people in a relatively well-known part of the world. Also, it was suggested that evaluations probably are made in such a way that some criteria are only considered in what can be thought of as a later stage of an evaluation process, especially the criterion of diversity which mostly cannot be used appropriately until some sorting of the sites in question has been made on other criteria.

In view of the preceding discussion, the operational framework shown in Table 13.4 is proposed for relatively small, isolated patches of habitat.

Stage 1 is a pre-evaluation stage, at which some sort of classification of the sites to be considered should be made. A numerical classification using the kinds of methods outlined in Chapter 2 will produce groups

Table 13.4 Stages in a suggested conservation evaluation procedure with examples of the kinds of criteria appropriate at each stage.

Stage 1	Pre-evaluation classification or sorting stage
Stage 2	Representativeness
Stage 3	Threshold criteria (text examples are naturalness and area)
Stage 4	Ranking criteria (diversity and rarity are suggested. Naturalness and area could also be used here, as could any other ecological or biological criteria)
Stage 5	Pragmatic criteria (factors of practical significance or immediate importance, such as threat of interference, could be used here to set priorities for the acquisition of the sites identified down to and including stage 4 above)

with a set of explicit descriptors to facilitate communication. However, numerical methods are not always available so any explicit classification will suffice. Once all the sites involved have been sorted into groups on the basis of shared attributes the evaluation can proceed as there is now a logical basis for comparison.

Stage 2 uses only the criterion of representativeness. It might be thought of as a sub-stage of stage 1, but is so important that it has been highlighted in this evaluation procedure. Thus, given the classification from the first stage, the sites which fall into each class should be listed. The very act of doing so ensures that all classes are represented in subsequent stages. At stage 3, all other criteria which can have the effect of removing some sites entirely from consideration should be used taking each class from stage 2 separately. Thus, for example, if there is some size below which a site would not be considered, it can be eliminated at this stage. Similarly, if naturalness is used as a criterion then sites not meeting some suitable definition of naturalness (say, for Britain, self-sown native vegetation which has not been fertilized) can be dismissed from further consideration.

Stage 4 completes the 'shopping list' of sites which would constitute the most preferred network of conservation reserves. Ecological or biological criteria such as diversity and rarity can be applied to rank the remaining sites within classes. Stage 3 can be thought of as a kind of 'all or nothing' stage during which a site either qualifies or it does not. Stage 4, on the other hand, is a ranking stage. Priorities for acquisition are set at stage 5 where criteria such as threat of interference may be applied appropriately. Also, criteria referring to geographic location especially in relation to existing or other potential reserves, for example, Ratcliffe's (1977) position in ecological/geographical unit criterion, could be used at this final stage.

At stages 3 and 4 there is no obligation to use the criteria suggested in Table 13.4. For example, naturalness need not be used in the way described above, with a threshold below which sites are not considered to be natural. If naturalness is defined as having degrees, so that one site can be more natural than another, then naturalness could be used at stage 4 rather than at stage 3. It could also be used at both stages with sites not meeting some minimum threshold being eliminated at stage 3 and others ranked according to their degree of naturalness during stage 4. Similarly, if there was no size below which a site would be considered at all, then area too can be used at stage 4 rather than stage 3. Other criteria may be added or substituted at either stage. Smith (1983), for example, uses a criterion of traditionality at a stage comparable to 3 in his assessment of haymeadows in northern England because it is the meadows which have been managed in a traditional way which are of interest.

Table 13.4 is a generalized framework designed to operate in geographical or ecological situations involving relatively small patches of land with conservation interest separated by land subject to other uses. Assessors are encouraged, however, to develop more specific procedures to meet their particular problems which might involve a rearrangement of Table 13.4 or the use of other criteria. Table 13.4 provides a model, a basis for assessments of more specific kinds of sites.

The procedure in Table 13.4 maintains flexibility in that no specific weights are assigned to criteria and no index of conservation value is derived. It must be stressed again that the location of criteria in different stages does not signify any relative importance, rather that it is not appropriate to use them at other stages (though criteria in stages 3 and 4 may be interchangeable). The exception is representativeness whose use prior to the other main criteria implicitly weights representativeness above those other criteria since its use is designed to ensure, as far as possible, that the entire range (or examples of the entire range) of biological variation in a given area is conserved. Particularly diverse sites or sites containing rare species will still be accorded high conservation status, but as highly valued representatives of their class.

13.5   CONCLUSIONS

There is no simple index for calculating or estimating conservation value, not even in one fairly well-defined habitat type, such as the enclosed grasslands on the carboniferous limestones of the Yorkshire Dales. Two main reasons for this can be suggested. In the first place, conservation value is an expression of human appraisal and whilst people generally can agree on the ideals behind the evaluation, inevitably they will differ on the weights given to different criteria for judging conservation value. Since the product of a conservation evaluation exercise is for the benefit of those making the appraisal (or for others on whose behalf they are acting) any proposal of an overall index, involving as it would prior weightings of criteria, would be unlikely to gain widespread acceptance. Secondly, there is no simple way of integrating the criteria into an overall index. Usher (1980) recognized this problem when he pointed out that the correlation between number of species and area might overweigh those two in any index involving a number of other criteria. Also, whilst it is desirable to express criteria numerically if possible, usually it is not appropriate to combine them. As Smith (1983) points out, a difference of one on the criterion of species richness means one species but on the representativeness criterion (which he measured as the distance from the mean of the first two axes of an ordination of sample sites) it is a unit of length on an ordination diagram. Correlations between criteria, such as those illustrated in Figs 13.1 and 13.2, reinforce the conclusion that criteria are not additive.

The advantages of a single index, if one did exist, would be that valid comparisons of possible conservation sites could be made very easily and the results of an evaluation would be clear and unambiguous. Since no such index exists, other explicit ways of making comparisons must be found. Table 13.4 outlines one possible way.

Reviews of published assessment schemes (Margules, 1981; Margules and Usher, 1981), the results of the two evaluation exercises described above, and a general consideration of conservation evaluation all lead to the criteria listed in Table 13.4. Those six criteria – representativeness, diversity, rarity, naturalness, area, threat of interference – are widely accepted and used to judge conservation value and they appear to account for most variation in conservation value. Other criteria, in a general sense, add little. However, in many situations it is envisaged that other criteria will either be added to or be substituted for those in Table 13.4. The substitution of traditionality for naturalness in the assessment of haymeadows was given as an example above. Other criteria might be used depending on the purpose of the assessment and the land use planning options available. If successional ecosystems are to be conserved for example, criteria relating to their ecological potential or their value to scientific research might be added, though their potential contribution to a representation of the biota of a region should see them recognized at stage 1 and included at stage 2. By following a procedure such as that outlined in Table 13.4 the assessment process is made explicit and the communication of results to decision-makers and land use planners is made easier.

13.6 SUMMARY

Two studies on the actual procedures people follow when assessing conservation value provide the basis for a critical examination of some of the most widely used criteria for judging conservation value: diversity, area, rarity, ecological fragility, threat of interference and representativeness.

Drawing on the results of the two studies, a generalized procedure in five distinct stages is proposed for assessing the conservation value of remnant habitat patches. The procedure does not assign weights to criteria (though assessors may do so if they wish) and no index of conservation value is derived. However, the procedure does enable the assessment of conservation value to be made explicitly.

CHAPTER 14

Design of nature reserves

DANIEL SIMBERLOFF

14.1 Introduction
14.2 Minimum viable population sizes
14.3 Applying equilibrium island biogeographical
theory to refuge design
14.3.1 'SLOSS' – single large or several small?
14.3.2 Refuge shape and the peninsula effect
14.4 Discussion
14.5 Summary

Wildlife Conservation Evaluation. Edited by Michael B. Usher.
Published in 1986 by Chapman and Hall Ltd, 11 New Fetter Lane,
London EC4P 4EE
© 1986 Chapman and Hall

14.1 INTRODUCTION

The field of refuge design *per se* did not really exist until 1971. This is not to say that there were no wildlife refuges, national parks or game reserves, or that no thought was given to their situation and management. Even a nation as young and development orientated as the United States established its first national park in 1872 and its National Park System as early as 1916 (Stone, 1965), while in Europe and Asia hunting reserves of the aristocracy served the same goals as national parks (maintenance of wildlife) centuries earlier.

The principles of siting and managing such refuges were deemed self-evident. Every species has some characteristic range of habitats in which it will thrive, and many have other species (food plants, mutualistic symbionts, hosts, etc.) that are necessary to their survival. These habitat requirements and species interactions may be very subtle, and laborious field research may be required to elucidate them. Furthermore, there may be concatenated effects rippling through a chain of several species.

For example, the red-cockaded woodpecker (*Dendrocopus borealis*) requires old and dying trees of longleaf pine (*Pinus palustris*) and loblolly pine (*Pinus taeda*) for nesting. Longleaf pine, however, cannot persist without intermittent fire that eliminates seedlings of other species that would otherwise outcompete it (Schiff, 1962). Such fires are greatly facilitated by wiregrass (*Aristida stricta*), to the extent that it is difficult to maintain longleaf pine in its absence. Conservation of red-cockaded woodpeckers, at least in north Florida, thus requires maintenance of wiregrass, and it entails the preservation of an entire community of organisms adapted to periodic fires.

Not all species are so inextricably intertwined with so many other species, but almost all have at least a few obligatory interactants, so that a refuge must always be designed for at least a few species, not just one. The research required to elucidate the longleaf pine-wiregrass-red-cockaded woodpecker story was arduous (cf. Stoddard, 1936; Thompson, 1971; Christensen, 1981; Wood, 1983), but no shortcut would have allowed the design of a suitable refuge. In principle, however, a sufficiently perceptive field investigation can determine enough about the habitats and interacting species of any target species for intelligent refuge design to begin. This is how refuges were designed until the last decade. In the sections on site selection and refuge management in standard conservation texts from about ten years ago (for example Cox, 1974; Owen, 1975) it is taken for granted that the key to conservation is learning the autecology of species of interest, and then conserving enough of its habitat.

Sometimes more than just setting aside and protecting a site is required. For example, without occasional fires, longleaf pine, wiregrass, and species associated with them are succeeded by less fire-tolerant species that are otherwise competitively superior. In north Florida two centuries ago, such fires were started every few years by lightning, and because the longleaf pine community stretched quite uninterrupted for tens or hundreds of kilometers, any particular part of the forest would certainly burn every few years.

Two anthropogenous developments changed this situation. First, for at least the past 150 years, north Florida has undergone progressive 'insularization' of the habitat as expanding agriculture divided the landscape into a patchwork of fields and forests of various sorts. Undisturbed forest, dominated by longleaf pine, has been progressively restricted to fewer, smaller, and more isolated 'islands' in a sea of fields. Indeed, similar insularization has occurred over much of North America, Europe, New Zealand, and even the tropics (Burgess and Sharp, 1981; Dawson and Hackwell, 1984). In longleaf pine forests, the isolation alone would prevent natural fires starting in one area from spreading to other areas, and thus would decrease the frequency of fire in any one island, perhaps to the point where succession would ensue and a hardwood forest would replace the pines.

Secondly, fire seems traditionally to inspire fear in human beings. To lay (and even to some professional) observers of nature, especially in Europe and North America, fire has usually been anathema. Many 'primitive' peoples use fire on a routine basis to maintain certain habitats, and it has been known for many years that numerous vegetation communities around the world are maintained by occasional fires (for example Garren, 1943; Shantz, 1947). But the spectacle of burning trees and, occasionally, scorched wildlife seems to inspire visions of horrible suffering and economic disaster. This view of fire makes it very difficult, on psychological grounds alone, to use fire in a controlled way to maintain a habitat. In the United States, for example, a national campaign to eliminate natural forest fires peaked in the 1920s with the passage of the Clarke–McNary Act providing federal assistance to states with fire prevention programmes. The US Forest Service virtually forbade controlled burning at this time, despite a hundred years of scientific study showing that longleaf pine requires fire (Schiff, 1962). Managed refuges were not immune to this pressure; for example, the US National Park Service has been hostile to the use of fire for habitat maintenance for much of its history (Hendrickson, 1973), while in Germany there has been a prohibition on burning in nature reserves and landscape controlled areas since about 1924 (Makowski, 1974).

By 1936, Stoddard (1936) and others had shown how important fire was to the maintenance of certain natural communities, especially those with longleaf pine (cf. Garren, 1943; Schantz, 1947; Schiff, 1962), and by the 1940s the US Forest Service no longer demanded complete proscription of fire. But it was not until a report by a committee of the US National Research Council (Leopold, 1963) that the use of fire in wildlife and habitat maintenance began to make headway in the US National Park Service. However, burning in a way that mimics nature is no trivial matter, especially when fire has been excluded and fuel has built up on the ground. Even with the recent great improvement in the technology of controlled burning (Tall Timbers, 1976), the insularization of habitat renders the process expensive. Starting and controlling fires in myriad small refuges is vastly more difficult and expensive than the same task would have been in one large refuge. A Florida refuge manager has suggested that it is impractical to perform controlled burning on plots of less than about 10 ha.

Although the interest is more often in maintaining climax communities than earlier stages or disclimaxes (like the fire disclimax in north Florida), even climax communities usually require maintenance, or at least protection. And there is increasing realization that there are non-climax communities of conservation interest (Stone, 1965; Pickett and Thompson, 1978; Gilbert, L. E., 1980) that can be maintained only if refuges are actively managed.

One factor in determining how much habitat is 'enough' for conservation purposes is thus the difficulty and expense of management.

14.2 MINIMUM VIABLE POPULATION SIZES

Conservation biologists recognized early that various stochastic genetic factors impose a limit on how small a piece of habitat can be and still function as a refuge. The main such genetic factor is inbreeding depression. Most peoples have incest taboos, and it is often said (for example Wilson, 1978) that these evolved, culturally and/or genetically, in response to the observation of harmful biological effects, the frequencies of which increased when near relatives mated. Whatever the actual basis for incest taboos, animal breeders have recognized for several centuries that continued inbreeding within a herd is associated with deterioration (Darwin, 1868; Wright, 1977). Similarly, among plant breeders, deleterious effects of inbreeding have long been recognized (for example Koelreuter, 1766; von Gaertner, 1849; Darwin, 1876). The formal study of the genetics of inbreeding was initiated in 1906 (Wright, 1977), and sufficient research has since been performed that Ralls and Ballou (1983) conclude that inbreeding depression is likely for animals forced to breed within small groups. Inbreeding depression has two aspects. First, there is an increased frequency of diseases caused by recessive alleles, of one or a few genes, that in homozygous condition are severely debilitating. A common congenital diaphragm defect in the endangered Brazilian golden lion tamarin (*Leontopithecus rosalia*) appears to be such a disease

(Bush *et al.*, 1980). There are fewer than 200 individuals of this species in captivity and 30 in nature (Pinder and Barkham, 1978), so inbreeding is increased and has probably increased the frequency of this disease.

Secondly, a more important aspect of inbreeding depression is heterosis: the more genes for which an individual is heterozygous, the more vigorous and fertile, on average, it is likely to be. This is true even if there is no obvious disease, like the diaphragm defect in the tamarin, present among less heterozygous individuals. Since inbreeding decreases average heterozygosity, and decreases it faster the smaller the population size (Hartl, 1981), decrease in vigour and fertility is a common effect of inbreeding. As Ralls and Ballou (1983) point out, since such effects are subtle, this consequence of inbreeding could be confirmed only by detailed studies and meticulously kept pedigree records, such as those now available for some zoo animals.

For plants, there is less evidence for inbreeding depression, but there is good indication that in at least certain species inbred populations with lower heterozygosity suffer from decreased fertility as well as size and other attributes normally associated with fitness (Schaal and Levin, 1976; Clegg and Brown, 1983; Ledig, Guries and Bonefeld, 1983).

In the wildlife management literature, even in the face of evidence of at least the possibility of inbreeding in small populations such as those on refuges, there is occasional scepticism about its importance. For example, Whitehead (1980) states that 'adverse comment about the ill-effects of inbreeding is . . . often exaggerated'. Such scepticism, though overstated, may not be completely unwarranted. Nevertheless, though its exact importance for particular species remains controversial, the dangers of inbreeding depression in small populations were well known to early conservationists. Gross (1928), discussing the failure of a refuge to save the declining population of heath hens (*Tympanuchus cupido attwateri*) on an island off Massachusetts, said:

> The problem of saving the Heath Hen is not the simple one of providing protection against hawks and cats and supplying food when needed, but is much more complex. There are other factors such as the inadaptability of the species, excessive inbreeding, excess male ratio and disease which are most important in the decline of the Heath Hen.

Franklin (1980) suggests an effective population size of about 50 is the minimum required for long-term prevention of severe inbreeding. Effective population size is often much less than the number of individuals in a population, depending on sex ratio, breeding structure of the population, and other factors (Franklin, 1980). Franklin adds, however, that a population this small would be likely to lose enough alleles by genetic drift for there to be insufficient genetic variability to allow it to respond to the environmental change that would surely occur in the course of a few centuries. He suggests an effective population size of about 500 as a minimum for this purpose, though Berry (1971) feels that 50 is a sufficient population size to retain enough genetic variability.

Shaffer (1981) has included inbreeding depression among a number of stochastic factors that determine a minimum viable population size, C, such that populations with fewer than C individuals are likely to go extinct rather quickly. For refuge design, C can be translated to A_c, the minimum area of suitable habitat required to sustain a population of size C. Shaffer calls a second factor 'demographic stochasticity', which includes variation in a number of demographic traits that is proportionally larger the smaller the population size. For example, the probability that all the offspring in some generation will be male is greater in smaller populations. Models by MacArthur and Wilson (1967), Richter-Dyn and Goel (1972) and others suggest that, for certain types of demographic stochasticity, there is not a linear relationship between population size and probability of extinction, but rather that C represents a threshold: populations less than C will go extinct quickly, and populations greater than C will persist for long periods.

Other stochastic forces, like inbreeding depression, that endanger small populations are not likely to produce a critical threshold size, but rather a general tendency for smaller populations to go extinct more quickly. In addition to inbreeding depression and demographic stochasticity, Shaffer (1981) lists two other factors, environmental stochasticity and natural catastrophes. By 'environmental stochasticity' he means the normal range in variation of various physical parameters, like temperature, and interacting populations, like predators. Such variation will affect a target species and, if the population is already small enough, could cause extinction. For example, the Puerto Rican parrot (*Amazona vittata*) was at one time reduced to only thirteen individuals (Snyder, 1978). One could imagine that a year of generally bad weather combined with a few more predators than normal could eliminate all thirteen birds. By 'natural catastrophes', Shaffer refers not to normal year-to-year variation in the physical environment but to the intermittent disaster, such as a major hurricane. Such events are likely to affect many species, but only if the population is already small and restricted would the danger of extinction be great.

Soulé (1983) adds to this list the dysfunction of social behaviour as a fifth phenomenon that causes small populations to be at greater risk of extinction. For example, many animal species forage in large groups or form large breeding aggregations (Wynne-Edwards, 1962), and stylized behaviour associated with these aggregations can be destroyed when there are too few individuals.

Determination of minimum viable population sizes is in its infancy.

For very few species has there been direct study of both ecology and genetics, or observation of an extinction rate, in given sized populations to attempt to establish minimum viable population sizes. If one had such information, determining A_c would require one to know how much habitat a single individual or pair needs, and even this information is unavailable for most species. Perhaps Shaffer's (1978) study of the grizzly bear (*Ursus arctos*) is the most complete endeavour of this sort. Lovejoy *et al.* (1983) have recently embarked on an ambitious experiment to determine minimum areas for tropical forest birds in Brazil. They censused large forest tracts, then set aside 'islands' of different sizes and cut down all surrounding forest. Now they are continually censusing the forest islands to see which species go extinct in islands of different sizes.

So far, almost all attempts to estimate C or A_c for different species have been indirect, and consisted of scanning sets of sites to see what the smallest site is that any species occupies. Smaller sites are assumed incapable of supporting a population of size C, and thus are believed to be less than A_c. This approach consistently overestimates A_c (Simberloff and Gotelli, 1984). Any set of numbers (for example, the areas of occupied sites) has a minimum, so just observing a minimum does not tell us there are deterministic reasons why that minimum occurred. In other words, even if propagules of a species colonized m out of n sites independently of the areas of the sites, there would still be a smallest occupied site, and which one that was would have been a matter of chance, not because smaller sites were 'unsuitable' for the species.

This random model suggests a simulation to see if the smallest occupied site is larger than one would have expected by chance alone. If it is, one might hypothesize that the reason is that smaller sites simply cannot support the species. But if the smallest occupied site is not larger than chance expectation, there is no reason to think, just from the sizes of occupied and unoccupied sites, that any sites are too small for the species. A computer simulation used n buckets, with sizes proportional to those of the n sites available for the species to colonize. The program simulated randomly thrown balls until exactly m buckets were occupied, corresponding to the m sites in nature that the species occupied. On every throw, the probability that the ball fell in any particular bucket was proportional to the area of that bucket. After m buckets were occupied, by using a Kolmogorov-Smirnov test the question: 'are the occupied sites larger than the randomly occupied buckets?' could be answered. If so, it may be because smaller sites are inimical to the species. Alternatively, does the smallest occupied bucket have a smaller size rank than the smallest occupied site?

When Simberloff and Gotelli (1984) applied this technique to plants of prairie and forest patches of different size in the midwestern United

States, by running the simulation repeatedly to generate confidence limits for the expected size of the smallest occupied site, they found no evidence that any of several hundred herb species were precluded from occupying even the smallest censused sites (some of which were just a few square metres). If they had just looked at the smallest site occupied by each species, they would have concluded that for most species there is an A_c much larger than the area of the smallest site.

Several species may be conserved together, either because the aim is to maintain an entire community or because the species are inextricably linked ecologically, as in the longleaf pine example above. It is then appropriate to find the target species with the largest A_c, and establish a refuge at least that large.

14.3 APPLYING EQUILIBRIUM ISLAND BIOGEOGRAPHICAL THEORY TO REFUGE DESIGN

By the early 1960s, there was already a well-developed science of how to design refuges, based primarily on the identification of appropriate habitat for a target species or group of species, elucidation of various obligatory interactions among species that would necessitate their being conserved together, and consideration of how large a population would be needed to forestall inbreeding depression. Because such habitat requirements and species interactions are notoriously idiosyncratic, and inbreeding depression varies between species, there were few general rules for refuge design, even though the outlines of how to proceed in any given instance were well established. In short, there was a science, but it was an idiographic science, not a nomothetic one like physics, with its relatively few all-encompassing laws. This difference no doubt reflects the fact that there are more 'elementary particles' in ecology than in physics (Strong, 1980); each species, and there are over 3 000 000 of them, differs from each other one, and even populations and individuals within species differ from one another more than do the protons that are components of all nuclei.

The dynamic equilibrium theory of island biogeography (MacArthur and Wilson, 1963, 1967) excited ecologists and biogeographers, because it used a familiar statistic (number of species) and simple mathematics to depict 'Nature' as being dynamic (Simberloff, 1974, 1978a). The theory states that the number of species on an island (or in an insular habitat such as a field in the midst of forest) is a dynamic equilibrium between occasional extinction on the island and occasional immigration from the pool of species not on the island. Immigration rate decreases with increasing distance of the island from source areas, while extinction rate decreases with increasing island size (Fig. 14.1). Many elaborations of this scheme quickly appeared. For example, Whitehead and Jones (1969)

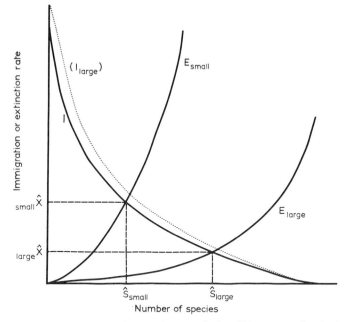

Fig. 14.1 The dynamic equilibrium theory of island biogeography for large and small islands. Immigration rates (I) and extinction rates (E) are monotonic functions of number of species present on an island, and their intersection determines the equilibrium number of species on the island (S) and the 'turnover' rate (X) of continuing extinction. From Simberloff (1976a).

proposed that immigration rate also depends on island size, while Brown and Kodric-Brown (1977) suggested that island isolation could also affect extinction rate. However, the hypothesis of a dynamic equilibrium number of species, with constantly changing species composition, remains at the core of the theory.

The equilibrium model, or a theory very much like it, was an inevitable intersection of two then current areas of ecological research (Simberloff, 1978a). First was the dynamics of single species population growth and pairwise interactions. Second was the distribution of population sizes in samples from natural communities, especially for birds and insects (Williams, 1964). Just as inevitable was the attempt to apply this theory to practical matters. Willis first suggested its application to conservation, when he circulated ideas about refuge design beginning in 1971 (Willis, 1984). His hypotheses are based on the analogy between a network of refuges and an archipelago of islands (and, of course, on the correctness of the theory).

Though Willis (1984) felt that early publications (for example Diamond, 1973) on applications of island biogeographic theory to refuge design do not credit his inspiration, his publication (Wilson and Willis, 1975) was most influential, and two of his suggestions have been repeated frequently (for example Terborgh, 1974, 1975; Diamond, 1975; May, 1975; Diamond and May, 1976; Goeden, 1979; Samson, 1980; Butcher *et al.*, 1981; Whitcomb *et al.*, 1981). First, Wilson and Willis (1975) argue that the theory of island biogeography dictates that single large refuges are preferable to groups of small ones of equal total area. Secondly, they propose that the theory mandates circular refuges rather than long, thin ones. Adherence to both principles, they suggest, will maximize the number of species.

14.3.1 'SLOSS' – single large or several small?

Though Wilson and Willis (1975) base the recommendation of a single large site rather than several small ones on the equilibrium theory, in fact the theory makes no prediction in this matter (Simberloff and Abele, 1976; Simberloff, 1982). For any particular taxon and archipelago, the critical parameters are (i) the gradient of species' colonizing abilities and (ii) the degree of overlap of species lists from different sites. Neither parameter appears directly in the dynamic equilibrium theory, and both must be determined empirically, by studying the species in the pool. Very few such studies exist, particularly for (i). Connor and Simberloff (1978, 1983) and Simberloff (1978b) suggested a way to approximate the gradient in (i) if one knows the species distribution over the entire archipelago, while Higgs and Usher (1980) and Higgs (1981) proposed an index to express the degree of overlap.

Though recommendation of single large refuges is not a consequence of island biogeographic theory, it is a testable proposition. A direct test would require one to take a set of large islands, census them, and then divide each one into a group of smaller islands with minimal area change. The only such direct test is on arboreal arthropods of small mangrove islands in Florida Bay, and the preliminary results show no apparent 'best strategy' for how to divide up a given amount of area to maximize species richness (Simberloff and Abele, 1976).

Several investigators have followed a less direct approach. That is to accept the analogy of refuges to islands and to ask whether, for archipelagoes with species lists for the islands, single large islands do, in fact, have more species than do groups of small islands of equal total area. One is assuming, of course, that there are no other critical differences, especially in habitat, among the islands: only area and the number of islands making up the area change. Abbott (1980) notes that this assumption is often violated, so that such 'experiments' are uncontrolled. Nevertheless, when one chooses refuges from among several potential sites, habitats also change in addition to area and number of sites, so perhaps the island analogy is worth pursuing. Accordingly, the hypothesis will be examined for passerine birds of 16 islands in the Cyclades (Watson, 1964; Simberloff, 1985), a group of rocky islands in the Aegean Sea, south-east of Greece. Results for the Cyclades passerines will then be compared to results of similar studies for other taxa and regions.

To examine the hypothesis, one must first describe the 'species–area relationship', the observation that, for most biota, number of species in a site increases monotonically with the area of the site. Cyclades passerines are no exception; larger islands tend to have more species. Where S is the number of species and A is the area (in km^2), ordinary least squares regression shows that the equation

$$S = 6.47 + 0.058\,A \qquad (14.1)$$

accounts for 71% of the variation in S (Simberloff, 1985). Connor and McCoy (1979) and Connor, McCoy and Cosby (1983) argue that there is no acceptable biological interpretation of the slope and intercept of this equation. Species–area relationships are usually expressed with the variables logarithmically transformed,

$$S = KA^z, \text{ so that } \ln S = \ln k + z \ln A. \qquad (14.2)$$

For the Cyclades passerines, the regression of $\ln S$ on $\ln A$ is not as good as equation (14.1) ($r^2 = 56\%$) and Connor and McCoy (1979) suggest there is no valid reason, other than fit of the equation to the data, to transform either of the variables.

In Connor and McCoy's survey (1979) of 100 species–area relationships, for log-transformed data 50 studies had larger r^2 than the Cyclades passerines do, while for untransformed data, only 28 studies had larger r^2 than equation (14.1). So it is fair to say that area usually explains a significant fraction of the variation in species richness.

One can test hypotheses about 'SLOSS' – 'single large or several small' – by following a procedure of Simberloff and Gotelli (1984). One simulates constructing archipelagoes of 2, 3, 4, and 5 small islands by randomly drawing pairs, trios, quartets, and quintets of islands from the Cyclades, constraining the total area of each archipelago not to exceed the area of the largest island (Naxos, 438 km^2). One thus assembles 16 pairs, 16 trios, etc., of islands to match the 16 single islands. Areas of the archipelagoes are constrained to insure that, for each number of islands, there was a quite uniform distribution of areas between the smallest possible for that number of islands and the largest possible (438 km^2).

For each randomly assembled archipelago, one constructs a species list of all species found anywhere in the archipelago. Figure 14.2 shows the relationship between number of species in an archipelago and total area of the archipelago, for archipelagoes consisting of from one to five

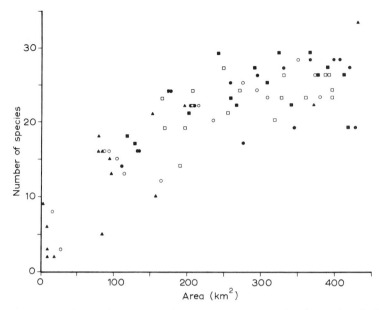

Fig. 14.2 Numbers of passerine bird species on 16 islands in the Cyclades archipelago, and for 64 simulated archipelagoes comprising 2–5 islands each (see text for a discussion of the simulations). The following symbols are used: ▲, 1 island; ○, 2 islands; □, 3 islands; ■, 4 islands; and ●, 5 islands. From Simberloff (1985).

islands. Wilson and Willis (1975) contend that, when area and all other factors are held constant, an archipelago of one island will have more species than will an archipelago with several islands. Figure 14.2 does not show such a pattern. For all 80 archipelagoes consisting of from one to five islands, the regression equation is

$$\ln S = 0.442 + 0.473 \ln A \qquad (14.3)$$

where $r^2 = 70.9\%$ (Simberloff, 1985). Since the 80 archipelagoes are not all independent of one another (some of them contain some of the same islands), the degrees of freedom for the regression are less than 78. However, even with as few as 4 degrees of freedom (surely too few for this situation), the coefficient of determination would be significant with $P < 0.05$. The correlation between number of species and number of islands is much less than that for area and number of species. WN is the number of islands in an archipelago, the correlation between $\ln N$ and $\ln S$ is larger ($r^2 = 28\%$) than that between N and $\ln S$.

Wilson and Willis's contention can be tested by adding the number of islands in an archipelago to the regression of number of species on area. Unexplained variation was not significantly reduced. The regression equation became

$$\ln S = 0.494 + 0.449 \ln A + 0.078 \ln N \qquad (14.4)$$

and R^2 was increased only to 71.3%. The coefficient of $\ln N$ is small but positive (0.078). The hypothesis predicts it will be negative.

This result resembles those in most other analyses (reviewed by Simberloff and Abele (1982) and Simberloff and Gotelli (1984)). For several taxa, including both animals and plants, either there is no consistent difference between single large and several small sites in species richness, or else groups of small sites tend to have more species. Why this should be so is not known. A reasonable hypothesis is that, on average, habitat diversity is greater for groups of separate sites than for single contiguous sites of equal total area, simply because parts of the group are physically more distant than parts of the single site. The influence of habitat diversity on species richness is well known, though not as easily proven as the effects of area because it is more difficult to quantify habitat diversity than it is to measure area. The most frequent explanation for the species–area relationship is that larger sites, on average, have more habitats than do smaller sites, and that different habitats support somewhat different species. In other words, the area effect is viewed as an artifact of a habitat effect (Connor and McCoy, 1979). Other factors may contribute to the species–area relationship (Simberloff, 1976a; Connor and McCoy, 1979), but habitat diversity is probably the most important variable except for very small and homogeneous sites. For Aegean passerines, Watson (1964) interprets the large correlation between S and A as a reflection of greater habitat diversity on larger islands.

There is no published test of the hypothesis that groups of small sites have more species than single large ones, on average, because of greater habitat diversity. Game and Peterken (1984) suggest this hypothesis for results of a study showing that several small woods in England typically contain more herb species than does a single large wood of equal area. For lizards of the Australian wheatbelt, Kitchener *et al.* (1980) argued:

> . . . while scattered small reserves, totalling 1.78×10^4 ha, contain almost all known lizard species in the . . . wheatbelt, a single area . . . in order to contain the same number of species would need to be immensely larger – possibly by a factor of 600. This situation is again believed to reflect the heterogeneity of habitat within the region such that an enormous area is required to encompass all its habitat diversity and consequently to carry representatives of all lizard species in the region.

No clear-cut measure of habitat diversity applies to all ecological analyses. One would need to know what variables are important to the taxon of interest, and to measure them in the field. This endeavour would be more difficult than measuring area on a map, which is probably why the hypothesis has been raised more than once, but not tested.

If the habitat diversity hypothesis correctly explains the observation that groups of small sites often have more species than single large ones, one might expect that the further apart were the sites comprising an archipelago, the greater the habitat diversity encompassed and therefore the more species should be supported. If one were to take one small island in the Cyclades and one in the Hebrides, one would expect a greater number of species than if one examined two islands from the Cyclades alone. At some point, one begins to add together species from biogeographical regions with different evolutionary histories, and the effect on species richness of adding evolutionarily different regions obfuscates the habitat diversity effect (Williams, 1964), but the point is clear: all other things being equal, the more widely separated the sites, the greater the habitat diversity.

For the Cyclades, all distances were measured between pairs of islands. Then for every randomly assembled archipelago, average distance (D) between pairs of islands within the archipelago was calculated (Simberloff, 1985). For an 'archipelago' of only one island, average distance was set to zero. The regression of ln S on ln A, ln N, and D gave

$$\ln S = 0.491 + 0.444 \ln A + 0.021 \ln N + 0.0015 D. \qquad (14.5)$$

There was a very slight increase in variation accounted for ($R^2 = 71.8\%$).

Analogous regressions to those of equations (14.3)–(14.5) with S instead of ln S as dependent variable yielded similar results with slightly less variation explained. Although average distance among islands has little effect on Cyclades passerine species richness, D may still be a good predictor in other systems. Islands of the Cyclades may all be similar physically, no matter where they are. This would rationalize both the slight effect of ln N relative to that in similar studies (for example American prairie plants, Simberloff and Gotelli, 1984) and the slight effect of D. Watson (1964) describes all the Cyclades as rocky and quite barren, and his *ad hoc* index of habitat diversity for all of the central Cyclades, even the very small islands, was the same as that for Naxos, the largest island.

For Cyclades passerines there is no evidence that one large refuge would be better than several small ones or vice versa, if the goal were just to preserve the greatest number of species. The total area, however it is arranged among sites, is the important variable, though area probably acts indirectly, through habitat diversity. This result is limited in several ways. First, analysis of any particular community bears largely on that community, and communities are so idiosyncratic that results cannot automatically be generalized (Simberloff and Abele, 1976, 1982,

1984; Simberloff and Gotelli, 1984). Although this caveat seems apparent to any field biologist, it is nevertheless true that the original application of island biogeographic theory to refuge design (Wilson and Willis, 1975) and most subsequent applications spoke of general laws that dictate design in all systems.

A second caveat is that this approach addresses only how subdivision affects species richness. Conservationists may wish to consider other factors in designing refuges. For example, the main goal of refuge design does not always seem to be maximizing species richness (Simberloff and Abele, 1982; Simberloff, 1982); refuges are often established to conserve species of special interest for one reason or another (for example those of endangered status). In the United States, refuges have been established specifically to protect the California condor (*Gymnogyps californianus*), saguaro cactus (*Carnegia gigantea*), Joshua tree (*Yucca brevifolia*), Kirtland's warbler (*Dendroica kirtlandii*), and other species. The key criterion for such efforts was determining the appropriate habitat. It is ironic that some early opponents of the suggestion that island biogeographic theory has little to offer conservation argued that the suggestion focused only on species richness. For example, Diamond (1976) contended that: 'Conservation strategy should not treat all species as equal but must focus on species and habitats threatened by human activities', and, 'Species must be weighted, not just counted'. Island biogeographical theory, and its application to conservation, for example by Wilson and Willis (1975), Diamond (1975), Diamond and May (1976), do view all species as equal, and just count numbers of species. The key prediction of the theory, that there is a dynamic equilibrium determined by island area and isolation, was only ever stated in terms of numbers of species, not in terms of their identities or traits.

Other considerations are absent from the equilibrium theory treatment of refuge design (Simberloff and Abele, 1982; Simberloff, 1982). For example, cost of acquisition and management may differ systematically between several small and single large refuges. Also, catastrophes like epizootic diseases, fires, and introduced predators would probably be less crucial to a series of small refuges than to a single large one. As an example the Seychelles Islands in the Indian Ocean contained 14 endemic land bird species when Europeans arrived in 1770. Since then, land clearing, a series of fires, and introductions of such predators as rats and cats have devastated these islands (Simberloff, 1978c). However, only the green parakeet (*Psittacula eupatria wardi*) and chestnut-flanked white-eye (*Zosterops mayottensis semiflava*) have been extinguished. This loss is restricted partly because there are several small islands and not just one large one; fires and introduced predators were unable to reach all the islands. The Seychelles magpie robin (*Copsychus sechellarum*), for example, remains only on Frigate Island. The feral cats that destroyed it elsewhere were controlled on Frigate Island, and cannot reinvade because of the water barrier.

A final caveat is that, when the small sites are small enough, extinction must increase greatly and a single large site would be far better (Simberloff and Abele, 1976, 1982; Simberloff and Gotelli, 1984). The species–area relationship is usually viewed as a species–habitat relationship, but the equilibrium theory of island biogeography interprets the species–area relationship differently (Simberloff, 1974, 1976a): smaller sites have higher extinction rates, so that at any instant, more species in the pool are locally and temporarily extinct at a small site than at a large one. Furthermore, for each species, as described above, there is an A_c such that extinction is assured on sites smaller than A_c. Thus, as the component sites in the archipelago get smaller, more species find all sites below A_c, and disappear from the archipelago, but would be perfectly capable of existing on one large site. There is no species pool for which A_c is known for all species. Yet such an estimate would be required if one were to assess the minimum size of refuges in a multi-refuge system, such that, below this size, a single large refuge would be preferable.

14.3.2 Refuge shape and the peninsula effect

Wilson and Willis (1975) also contend that the best refuge shape is circular, and that long, thin refuges should be avoided. Diamond (1975), Diamond and May (1976), and Butcher et al. (1981) also argue that the dynamic equilibrium theory of island biogeography dictates this recommendation. Just as the equilibrium theory is unrelated to SLOSS, the theory also makes no prediction about refuge shape (Simberloff, 1982). The inspiration for the contention (Willis, 1984) was the 'peninsula effect', the observation that peninsulas often have fewer species than do equal-sized tracts of mainland and that species richness often decreases from base to tip of a peninsula. Simpson (1964) first described this effect for mammals of North American peninsulas; similar observations soon followed for North American birds (Cook, 1969) and reptiles (Kiester, 1971). Simpson (1964) very tentatively ascribed the effect to increased extinction at tips of peninsulas (because of decreased area and thus increased extinction) combined with low rates of immigration, so that immigration to balance any extinction takes a long time. MacArthur and Wilson (1967) were not so tentative and argued that the peninsula effect certainly reflects immigration–extinction dynamics. This argument led to Wilson and Willis's (1975) recommendation.

However, there is little direct evidence to support the hypothesis that the peninsula effect, where observed, is caused by increased extinction and decreased immigration. Recent extinctions have rarely been observed. Furthermore, there is an attractive alternative hypothesis for the peninsula effect: there are habitat gradients along many peninsulas, such that on those peninsulas on which a peninsula effect is observed, each site has about as many species as are suited to the site, and it happens that sites at the tip tend to have poorer habitats than do sites at the base. Wamer (1978) for Florida birds, Taylor and Regal (1978a,b) for mammals of Baja California, Seib (1980) for reptiles of Baja California, and Means and Simberloff (in prep.) for the herpetofauna of Florida have all tried to explain observed peninsula effects for specific biotas, while Busack and Hedges (1984) have asked whether the effect even exists for several herpetofaunas.

The results are that some biotas (for example reptiles of Baja California) show no peninsula effect at all, while for biotas that do show a peninsula effect, the habitat gradient hypothesis is reasonable. Only for the mammals of Baja California does it appear that the immigration–extinction hypothesis is viable, and even here it seems less likely than the habitat gradient hypothesis.

Though the equilibrium theory and the originally cited data turn out not to be a foundation for recommending a particular refuge shape, one can still ask if empirical observations suggest that round and not thin refuges would be best. As for SLOSS, passerine birds of the Cyclades (Watson, 1964) can serve as a test system. The species–area relationship is given by equation (14.1).

One then asks if island shape increases the coefficient of determination above that given by area alone. The statistic $B = P/A$ where P is the perimeter of the island, represents island shape. Mandelbrot (1982) shows that measured perimeter increases to infinity as the scale of the map increases, so one must measure all perimeters on maps of the same scale (1:800 000 in the example used here). This procedure should minimize relative measurement error among islands. Because, for figures of the same shape, P increases arithmetically with increase in a linear dimension while A increases with the square of a linear dimension, B remains constant for islands of different sizes but the same shape. Long thin islands, even if curled around, will produce relatively large values of B, while circles have the minimum value of B (3.545). For the 16 measured Cyclades, A and B are uncorrelated ($r = -0.292$, $P > 0.05$).

Shape has little effect on S. For S regressed on $\ln B$, $r^2 = 5.6\%$, and for all other combinations of transformed and untransformed S and B, the coefficients of determination were even smaller. However, B rather than $\ln B$ adds more to the coefficient of determination when added in a stepwise multiple regression after area,

$$S = 5.26 + 0.058\,A + 0.196\,B \tag{14.6}$$

where $R^2 = 71\%$, although the improvement is not significant ($F = 0.037$, 1 and 13 df). The coefficient of B in equation (14.6) is *positive*, opposite to the prediction if round islands had more species.

There have been few other attempts to test Wilson and Willis's hypothesis about the effect of refuge shape on species richness, and those proceeded very similarly to the approach above. Faeth and Kane (1978) used the shape parameter $C = 0.399 \times B$, and found that, after area effects were determined, C added nothing to the prediction of species richness of flies and beetles in nine urban parks near Cincinnati. Blouin and Connor (1985) studied various taxa in 14 archipelagoes of true islands, regressing species richness first on island area, then on one of three shape parameters (including C). Their result was almost identical to that for the Cyclades passerines. None of their three shape variables consistently explained a significant fraction of residual variation in species richness after effects of area were taken into account. There is no suggestion whatever that a long thin shape is disadvantageous for a refuge. If anything, Blouin and Connor's extensive analysis indicates a slight tendency for thin islands to have more species.

If one had predicted in advance what effect island shape would have on species richness, conflicting hypotheses might have resulted depending on what factors were considered most important. Wilson and Willis (1975) based their original recommendation on the notion that there would be decreased immigration to and increased extinction in at least part of long, thin refuges, but there is little evidence on immigration and extinction rates. Willis (1984) notes that he was also influenced by other hypothesized harmful effects of edges, especially the feeling that increased edge would allow increased human access to refuges. He further suggests that costs of fencing and other edge activities would use an inordinate amount of funds in long, thin refuges. The well-known 'edge effect' of ecology, that species suited to an edge or ecotone often differ from those suited to the centre of a habitat, has been advanced by Williamson (1975) and Levenson (1981) as an argument in favour of single large refuges in the debate over SLOSS, and could also be used to argue against long thin refuges. Relative edge is greater for long, thin refuges than for round ones, just as it is greater for groups of small refuges compared to single large ones, when total area is constant. For habitat islands, in which species of the matrix and the islands are not mutually exclusive, this increased edge could work to the disadvantage of the insular species. For oceanic islands, no such effect would be predicted.

Game (1980) observed that, if immigration is indeed important in maintaining species richness in refuges, then a long, thin refuge may be optimal, at least if it were oriented perpendicularly to a likely immigration route. This is because the apparent size of such a 'target' would be greatest to a propagule. Game (1980) also noted that long, thin refuges might contain greater habitat diversity than round ones by virtue of passing through a greater number of habitat types. This is probably true, since habitats must surely usually be clumped. This was the rationale of Johnson and Simberloff (1974) when they included latitudinal range as an independent variable in a regression analysis of determinants of plant species richness in the British Isles. However, in that archipelago latitudinal range correlated so highly with area that it was redundant in multiple regression analysis.

Though studies so far all show shape to have little or no effect, this result depends on scale just as the SLOSS result does. At some small size, a long, thin refuge would be all edge. For some taxa, this size might be very small; Faeth and Kane (1978) used quite small urban parks in their study and still found no effect of park shape on insect richness. But at some point a refuge must be so narrow that only edge species can exist there, or disturbance from the surrounding habitat will cause extinctions. Exactly what that point is can only be determined empirically for each taxon. Levenson (1981), using woody plants in Wisconsin forests, concluded that woodlots smaller than 2.3 ha function essentially as all edge and that certain typical mesic forest species can only persist in the face of disturbance in lots greater than 3.8 ha. Since he seemed to be considering square or round lots, 3.8 ha translates to a minimum distance of about 110 m from a boundary before an interior site can be viewed as suitable for certain species. So refuges for such species should be no narrower than 220 m. Janzen (1983) suggests that tropical animals are likely to disperse seeds from secondary habitats into tree falls in pristine forest up to 5 km away. If such dispersal is common, a tropical forest refuge would have to be at least 10 km wide for its centre to be immune to such effects.

14.4 DISCUSSION

Determinants of Cyclades passerine species richness seem similar to those of other groups on other archipelagoes. The common theme is that variables of island shape and number of islands comprising an archipelago have little or no effect on species richness. Instead, area explains a large fraction of the variation of species richness. Adding variables for island shape and number of islands does not usually significantly improve r^2 (or R^2) above that for the straightforward relationship between S and A. Watson (1964), working with the Aegean islands, and many other authors with different biota, have argued that area acts largely as a surrogate for habitat diversity and/or quality. In this view, except on the most distant oceanic islands, most species absences are caused by inappropriate habitat. On distant islands, an extinction caused by anything – stochastic population fluctuation, unsuitable habitat, catastrophes, etc. – may require a very long time to be redressed since immigration is so rare, but on islands that are not too isolated, such as many 'habitat islands', balancing immigration will not take too long, and the consequent island equilibrium will not be much below that for an equal-sized patch of mainland with the same habitats.

Measuring habitat diversity is not easy. One must know what aspects of the habitat are important for the species of interest, and that is knowledge that is not easily gleaned. Watson's habitat diversity index for Aegean passerines was reasonable but *ad hoc*, and even if his index is valid for these birds, one cannot apply it to another taxon or even, perhaps, to passerines of a different region. The only way that one can tell what habitats are really important to a given species is by field study of the species' autecologies, the most laborious sort of ecological research. Nevertheless, if one really wants to conserve species, such research is the only guarantor of success.

Conservation ecologists face an enormous challenge. Watson (1964) notes that, for the Aegean islands, most of which are small and topographically simple,

> When the detailed distribution of any particular species is considered, availability of suitable habitat is the paramount factor determining its pattern of occurrence . . . This point is of great significance since nearly every Aegean island is unique on the basis of its locales, size, relief, geology, soil, vegetation, and human exploitation. For this reason, adjacent islands may differ strikingly and the species of birds may, therefore, likewise differ.

This view dovetails so completely with the pre-equilibrium theory development of a science of effective refuge design, as outlined in the introduction to this chapter, that one can surely say that, for Aegean birds, at least, we would know exactly how to begin effective conservation: exhaustively study the habitat requirements of the different species, just as the habitat requirements of the red-cockaded woodpecker were studied in the longleaf pine example. Any contribution that might be made to this effort by considerations inspired by island biogeographical theory would be very secondary, at best.

In retrospect, it is not surprising that island biogeographical theory has not been useful in conservation, though it *is* depressing that the main supposed lessons of island biogeographical theory were quickly demonstrated (for example Simberloff and Abele, 1976) not to derive from the theory, and eventually were shown not to hold empirically (for example Simberloff and Abele, 1982), yet are still propagated (for example Iker, 1982). It is also sad that unwarranted focus on the supposed lessons of island biogeographical theory has detracted from the main task of refuge planners, determining what habitats are important and how to maintain them.

When one considers how idiosyncratic species are, and how complicated and varied are their interactions both with one another and with the physical environment, one must view with scepticism any general injunctions about how to preserve them. For application of island biogeographical theory, such scepticism would have been especially healthy. There are seven reasons why this application has been a red herring.

(1) It is increasingly clear that the equilibrium theory of island biogeography itself has not been sufficiently tested (Simberloff, 1976b, 1982; Abbott, 1980; Gilbert, F. S., 1980). The direct evidence is often equivocal, and the most optimistic statement that can now be made is that the theory seems to hold in a few systems and not in others. Its main contention, that there is frequent species extinction at local sites, has not often been demonstrated, and much of the reported 'turnover', published in support of the theory, may be transient movement rather than population extinction (Lynch and Johnson, 1974; Simberloff, 1976b). This is not to say that the theory has not been useful in other ways. Haila and Järvinen (1982) argued that, even though many of the theory's tenets have been falsified, it was important in leading ecologists to erect a conceptual framework for addressing relevant questions that had been asked vaguely, if at all. However, literal acceptance of the theory as 'valid' or 'proven' in some sense is an error, and basing a practical recommendation, like refuge design, on the theory is dangerous.

(2) The species–area relationship that underpins most of the purported conservation applications is not best explained by the dynamic equilibrium theory (Simberloff, 1974). The majority of species–area relationships described in the literature probably arise, not from effects of island size on turnover rates, but from the fact that larger sites tend to have more habitats. Each habitat allows another complement of species to inhabit a site.

(3) The peninsula effect that is claimed to result from equilibrium turnover and to mandate certain refuge shapes is also not necessarily caused by turnover at all, or by decreased immigration to peninsular tips. The peninsula effect occurs only in some taxa on some peninsulas, and such cases are probably consequences of a habitat gradient along the peninsula.

(4) Even if the species–area relationship were caused by the turnover envisioned by the theory, the theory would still make no prediction about SLOSS. One could only approach this matter empirically, and the results from one system need not tell us much about conservation in some other system. Likewise, even if some observed peninsula effect could be shown to be partly due to equilibrium turnover and reduced immigration to the tip, one would have to determine the immigration

and extinction rates for each system to see if these should be taken into account. For each system, one would have to determine the scale at which these effects matter.

(5) Although there is little empirical work on either the SLOSS issue or refuge shape, there is a growing literature in which archipelagoes of oceanic and habitat islands are analogized to refuges, and regression analysis is used to find which variables can, plausibly, be argued to determine species richness. For SLOSS, the result is that either groups of small islands tend to have more species than do single large islands of equal total area, or there is no difference in species richness. The reasons for this pattern are not known, but an apt hypothesis is that habitat diversity tends to be greater for groups of small islands than for single large ones (Simberloff and Abele, 1982). Island shape apparently has little or no effect on species richness (Blouin and Connor, 1985).

(6) Despite these results for insular analyses, one can deduce that, at some small size, which would differ among species, groups of small refuges would not be useful for conservation. Neither would very thin refuges. The sizes at which such effects would arise must be determined empirically, and little such research has been conducted.

(7) Much of the research cited tends to confirm the view of pre-equilibrium theory refuge planners that the major consideration is conserving enough habitat for the target species. The longleaf pine community serves as an example.

Continued propagation of the notion that island biogeographical theory has refuge design implications, and of the two specific recommendations discussed above, does more than just lard the literature and add to individuals' publication lists. Invoking theory that has been shown to be irrelevant can actually be detrimental to conservation. Conservationists may be unwilling or unable to judge the academic literature, and can adopt uncritically a published recommendation, especially if the recommendation is in the guise of a venerated 'theory'. For example, the International Union for Conservation of Nature and Natural Resources (IUCN, 1980) suggested measures that must be taken immediately to prevent a disastrous increase in extinction rate. A key recommendation is that refuge design criteria and management practices must accord with the equilibrium theory of island biogeography. The analysis above shows that such a recommendation is preposterous.

14.5 SUMMARY

Prior to the equilibrium theory of island biogeography, design of wildlife refuges was based on the known or perceived requirements of the species to be conserved. An example of the red-cockaded woodpecker illustrates this approach, and also indicates how conservation of one species may inevitably entail conservation of several others that are all part of an ecological network.

Recent consideration of the merits and verisimilitude of the equilibrium theory forces one to be less enthusiastic about its universal applicability to nature generally and especially to refuge design. In particular, two misconceptions abound in the conservation literature. First, it has been argued that the theory favours a single large reserve rather than several small ones (the SLOSS principle). Second, it has been claimed that the theory suggests that refuges should have as round a shape as possible. In fact, neither of these recommendations derives from equilibrium theory, and data available in the literature support neither assertion.

The number of species in refuges is related to area. Probably the major cause of this species–area relationship for most published data is that the number of habitats increases with area, on average. Equilibrium turnover, as envisioned in the equilibrium theory, probably plays a relatively minor role in most species–area relationships. It is important to dispel dogmatic refuge design principles said to be based on the equilibrium theory.

References

Abbott, I. (1980) Theories dealing with the ecology of landbirds on islands. *Advances in Ecological Research*, **11**, 329–371.

Adams, M.W. and Rose, C.I. (1978) The selection of reserves for nature conservation. *University College, London, Discussion Papers in Conservation*, **20**.

Adamus, P.R. and Clough, G.C. (1978) Evaluating species for protection in natural areas. *Biological Conservation*, **13**, 165–178.

Adriani, M.J. and van der Maarel, E. (1968) *Voorne in de Branding*. Stichting wetenschappelijk duinonderzoek, Oostvoorne.

Alsop, F.J. (1970) A census of a breeding bird population in a virgin spruce-fir forest on Mt Guyot, Great Smoky Mountains National Park. *The Migrant*, **41**, 49–55.

An Foras Forbartha (1981) *Areas of Scientific Interest in Ireland*. An Foras Forbartha, Dublin.

Anderson, M. (1979) The development of plant habitats under exotic forest crops; in *Ecology and Design in Amenity Land Management* (ed. S. E. Wright and G. P. Buckley), Wye College, Kent, pp. 87–108.

Anderson, M.L. (1950) *The Selection of Tree Species*, Oliver and Boyd, London.

Anderson, T.R., Hardy, E.R., Roach, J.T. and Whitmer, R.E. (1976) A land use and land cover classification system for use with remote sensor data. *United States Geological Survey Paper*, **964**.

Anonymous (1971) *De Landinrichting van het Gebied Volthe-De Lutte*, Studiegroep Volthe-De Lutte, Landbouwhogeschool and ICW, Wageningen.

Anonymous (1972) *De Kleuren van Zuidwest-Nederland*, Contactcommissie Natuur- en landschapsbescherming, Amsterdam.

Anonymous (1973) National registry of Natural Landmarks. *Federal Register*, **38**, 23928–23985.

Anonymous (1980) *This is Papua New Guinea*, Papua New Guinea Government Office of Information, Port Moresby.

Anonymous (1981a) *Milieu-effectrapport Waterwinning Zuid-Kennemerland*. Bijlagen deel 2, Ministerie van CRM, Rijswijk.

Anonymous (1981b) The afforestation of the uplands: the botanical interest of areas left unplanted. *University College, London, Discussion Papers in Conservation*, **35**.

Anonymous (1982) *Federal Register*, **47**, 43–48.

Anonymous (1983) *The Conservation and Development Programme for the UK: a Response to the World Conservation Strategy*, Kogan Page, London.

Antonovics, J., Bradshaw, A.D. and Turner, R.G. (1971) Heavy metal tolerance in plants. *Advances in Ecological Research*, **7**, 1–85.

Arnolds, E.J.M. (1975) Een floristisch-oecologische waardebepaling nabij Utrecht ten behoeve van natuurbehoud en planologie. *Gorteria*, **7**, 161–179.

Arrhenius, O. (1921) Species and area. *Journal of Ecology*, **9**, 95–99.

Ash, J.E. and Barkham, J.P. (1976) Changes and variability in the field layer of a coppiced woodland in Norfolk, England. *Journal of Ecology*, **64**, 697–712.

Asherin, D.A., Short, H.L., and Roelle, J.E. (1979) Regional evaluation of wildlife habitat quality using rapid assessment methodologies. *Transactions of the North American Wildlife and Natural Resources Conference*, **44**, 404–424.

Ashton, P.S. (1976) Factors affecting the development and conservation of tree genetic resources in South-east Asia; in *Tropical Trees: Variation, Breeding and Conservation* (eds J. Burley and B. T. Sykes), Academic Press, London, pp. 189–198.

Ashton, P.S. (1981) Techniques for the identification and conservation of threatened species in tropical forests; in *The Biological Aspects of Rare Plant Conservation* (ed. H. Synge), Wiley, Chichester, pp. 155–164.

Atherden, M.A. (1976) The impact of late prehistoric cultures on the vegetation of the North York Moors. *Transactions of the Institute of British Geographers*, **1**, 284–300.

Atkinson-Willes, G.L. (1976) The numerical distribution of ducks, swans and coots as a guide in assessing the importance of wetlands in midwinter; in *Proceedings of the International Conference on the Conservation of Wetlands and Waterfowl, Heiligenhafen, 1974*, International Waterfowl Research Bureau, Slimbridge, pp. 199–254.

Austin, M.P. (1978) Vegetation; in *Land Use on the South Coast of New South Wales, Vol. 2* (eds M. P. Austin and K. D. Cocks), CSIRO, Melbourne, pp. 44–67.

Austin, M.P. (1984) Problems of vegetation analysis for nature conservation; in *Survey Methods for Nature Conservation* (eds K. Myers, C. R. Margules and I. Musto), CSIRO, Canberra, pp. 101–130.

Austin, M.P. and Basinski, J.J. (1978) Biophysical survey techniques; in *Land Use on the South Coast of New South Wales, Vol. 1* (eds M. P. Austin and K. D. Cocks), CSIRO, Melbourne, pp. 24–34.

Austin, M.P. and Cocks, K.D. (1978) *Land Use on the South Coast of New South Wales; A Study in Methods of Acquiring and Using Information to Analyse Regional Land Use Options*, CSIRO, Melbourne, 4 volumes.

Austin, M.P., Cunningham, R.B. and Good, R.B. (1983) Altitudinal distribution of several eucalypt species in relation to other environmental factors in southern New South Wales. *Australian Journal of Ecology*, **8**, 169–180.

Austin, M.P., Cunningham, R.B. and Fleming, P.M. (1984) New approaches to direct gradient analysis using environmental scalars and statistical curve fitting procedures. *Vegetatio*, **55**, 11–27.

Austin, M.P. and Miller, D. J. (1978) Conservation; in *Land Use on the South Coast of New South Wales, Vol. 4* (eds M. P. Austin and K. D. Cocks), CSIRO, Melbourne, pp. 165–195.

Austin, M.P. and Nix, H.A. (1978) Regional classification of climate and its relation to Australian rangeland; in *Studies of Australian Arid Zone. III. Water in Rangelands* (ed. K. M. W. Howse), CSIRO, Melbourne, pp. 9–17.

Austin, M.P. and Yapp, G.A. (1978) Definition of rainfall regions of south-eastern Australia by numerical classification methods. *Archiv für Meteorologie, Geophysik und Bioklimatologie, Series B*, **26**, 121–142.

Bährmann, R. (1976) Verleichende Untersuchungen der Ergebruisse verschiedener Fangverfahren an brachyceren Dipteren aus dem Naturschutzgebiet 'Leutratal' bei Jena (Thur). *Entomologische Abhandlungen*, **41**, 19–27.

Bailey, R.G. (1976) *Ecoregions of the United States (map)*, United States Forest Service, Ogden, Utah.

Bailey, R.G. (1978) *Descriptions of the Ecoregions of the United States*, United States Forest Service, Ogden, Utah.

Ball, D.F. and Stevens, P.A. (1981) The role of 'ancient' woodland in conserving 'undisturbed' soils. *Biological Conservation*, **19**, 163–176.

Barber, D. (1970) *Farming and Wildlife: a Study in Compromise*, Royal Society for the Protection of Birds, Sandy, Bedfordshire.

Barber, D. (1983) The countryside: some aspects of the state of play in mid-1983. The Jeffery Harrison Memorial Lecture, British Association for Shooting and Conservation, Wrexham, Clwyd.

Barrett, B.W. and Chatfield, J.E. (1978) Valley deposits at Juniper Hall, Dorking: a physical and ecological analysis with emphasis on the sub-fossil land snail fauna. *Field Studies*, **4**, 671–692.

Beadle, N.C.W. (1981) *The Vegetation of Australia*, Cambridge University Press, Cambridge.

Berry, R.J. (1971) Conservation aspects of the genetical constitution of populations; in *The Scientific Management of Animal and Plant Communities for Conservation* (eds E. Duffey and A. S. Watt), Blackwell, Oxford, pp. 177–206.

Bezzel, E. (1976) On the evaluation of waterfowl biotopes; in *Proceedings of the International Conference on Wetlands and Waterfowl, Heiligenhafen, 1974*, International Waterfowl Research Bureau, Slimbridge, pp. 294–299.

Bezzel, E. and Reichholf, J. (1974) Die Diversität als Kriterium zur Bewertung der Reichhaltigkeit von Wasservogel-Lebensräumen. (Species diversity as a standard for the richness of waterfowl habitats.)*Journal für Ornithologie*, **115**, 50–61.

Black, J. (1970) *The Dominion of Man: the Search for Ecological Responsibility*, Edinburgh University Press, Edinburgh.

Blana, H. (1980) Rasterkartierung und Bestandsdichteerfassung von Brutvögeln als Grundlage für die Landschaftsplanung – ein Vergleich beider Methoden im selben Untersuchungsgebiet. (Results of atlas work and censuses of breeding birds as a basis for nature conservation and landscape evaluation – a comparison between both methods in the same study area.) *Proceedings of the 6th International Conference on Bird Census Work, Gottingen, 1979* (ed. H. Oelke), pp. 32–54.

Bloomfield, H.E. (1971) *Field and Computer Studies on Succession in Calcicolous Plants*. BA Thesis, University of York.

Blouin, M.S. and Connor, E.F. (1985) Is there a best shape for nature reserves? *Biological Conservation*, **32**, 277–288.

Bolwerkgroep (1979) *Natuurwaarden en Cultuurwaarden in het Landelijk Gebied*, Staatsuitgeverij, Den Haag.

Bonner, W.N. (1984) Conservation and the Antarctic; in *Antarctic Ecology, Vol. 2* (ed. R. M. Laws), Academic Press, London, pp. 821–850.

Booth, T.H. (1978a) Numerical classification techniques applied to forest tree distribution data. I. A comparison of methods. *Australian Journal of Ecology*, **3**, 297–306.

Booth, T.H. (1978b) Numerical classification techniques applied to forest tree distribution data. II. Phytogeography. *Australian Journal of Ecology*, **3**, 307–314.

Bouma, F. and van der Ploeg, S.W.F. (1975) *Functies van der Natuur, een Economisch-oecologische Analyse*, Institute for Environmental Studies, Publication **46**, Free University Press, Amsterdam.

Boyce, S.G. (1977) Management of eastern hardwood forests for multiple benefits (DYNAST-MB). *Southeastern Forest Experiment Station, Asheville, North Carolina, United States Forest Service Research Paper*, SE–168.

Boyce, S.G. (1981) Biological diversity and its use; in *Proceedings of the National Silvicultural Workshop, Hardwood Management* (ed. US Forest Service, Washington, DC), pp. 163–181.

Boycott, A.E. (1934) The habitats of land mollusca in Britain. *Journal of Ecology*, **22**, 1–38.

Braat, L.C., van der Ploeg, S.W.F. and Bouma, F. (1979) *Functions of the Natural Environment*, Institute for Environmental Studies, Publication **79/9**, Free University Press, Amsterdam.

Bradshaw, A.D. (1977) Conservation problems in the future. *Proceedings of the Royal Society of London, series B*, **197**, 77–96.

Bradshaw, A.D. (1983) The reconstruction of ecosystems. *Journal of Applied Ecology*, **20**, 1–17.

Braithwaite, L.W. (1983) Studies on the arboreal marsupial fauna of eucalypt forests being harvested for woodpulp at Eden, N.S.W. I. The species and distribution of animals. *Australian Wildlife Research*, **10**, 219–229.

Braithwaite, L.W., Dudzinski, M.L. and Turner, J. (1983) Studies on the arboreal marsupial fauna of eucalypt forests being harvested for woodpulp at Eden, N.S.W. II. Relationship between the fauna density, richness and diversity and measured variables of the habitat. *Australian Wildlife Research*, **10**, 231–247.

Braithwaite, L.W., Turner, J. and Kelly, J. (1984) Studies of the arboreal marsupial fauna of eucalypt forests being harvested for woodpulp at Eden, N.S.W. III. Relationships between faunal densities, eucalypt occurrence and foliage nutrients, and soil parent materials. *Australian Wildlife Research*, **11**, 41–48.

Bray, J.R. and Curtis, J.T. (1957) An ordination of the upland forest communities of Southern Wisconsin. *Ecological Monographs*, **27**, 325–349.

British Association for Shooting and Conservation (1983) The conservation of wildlife habitat in England by B.A.S.C. members, part 1. Report by the British Association for Shooting and Conservation, Wrexham, Clwyd.

British Lichen Society (1982) Survey and assessment of epiphytic lichen habitats, *Unpublished Report, Nature Conservancy Council*, Peterborough.

Brooker, M. P. (ed.) (1982) *Conservation of Wildlife in River Corridors*, Powys County Council, Welshpool, Powys.

Brown, A.H.F. and Oosterhuis, L. (1981) The role of buried seed in coppice woods. *Biological Conservation*, **21**, 19–38.

Brown, J.H. (1971) Mammals on mountaintops: non-equilibrium insular biogeography. *American Naturalist*, **105**, 467–478.

Brown, J.H. (1978) The theory of insular biogeography and the distribution of boreal birds and mammals. *Intermountain Biogeography: a Symposium. Great Basin Naturalist Memoirs*, pp. 209–227.

Brown, J.H. and Kodric-Brown, A. (1977) Turnover rates in insular biogeography: effect of immigration on extinction. *Ecology*, **58**, 445–449.

Brown, J.M., Goodyear, C.D., Gravatt, G.R. and Villella, R.F. (1983) The endangered species information system: a case history of data base development. *Unpublished paper delivered at the National Workshop on Computer Uses in Fish and Wildlife Programs*. Virginia Polytechnical Institute, Blacksburg, Virginia.

Bryce, D. (1962) Chironomidae (Diptera) from fresh water sediments, with special reference to Malham Tarn (Yorkshire). *Transactions of the Society for British Entomology*, **15**, 41–54.

Buckley, G.P. and Forbes, J.E. (1979) Ecological evaluation using biological habitats: an appraisal. *Landscape Planning*, **5**, 263–280.

Bull, C. (1981) The evaluation of reserves by the public. *Values and Evaluation* (ed. C. I. Rose), *University College, London, Discussion Paper in Conservation*, **36**, pp. 11–21.

Bunce, R.G.H. (1981) The scientific basis of evaluation. *Values and Evaluation* (ed. C. I. Rose), *University College, London, Discussion Papers in Conservation*, **36**, pp. 22–27.

Bunce, R.G.H., Crawley, R., Gibson, R. and Pilling, R. (1985) *The Composition of Enclosed Grasslands in the Yorkshire Dales National Park*, Yorkshire Dales National Park Committee, Bainbridge, North Yorkshire.

Bunce, R.G.H. and Shaw, M.W. (1973) A standardised procedure for ecological survey. *Journal of Environmental Management*, **1**, 239–258.

Bunce, R.G.H. and Smith, R.S. (1978) *An Ecological Survey of Cumbria*, Cumbria County Council and Lake District Special Planning Board, Kendal, Cumbria.

Burgess, R.L. and Sharpe, D.M. (1981) Introduction, in *Forest Island Dynamics in Man-Dominated Landscapes* (eds by R. L. Burgess and D. M. Sharpe), Springer Verlag, New York, pp. 1–5.

Burggraaff, M., van Deijl, L., Meester-Broertjes, H.A. and Stumpel, A.H.P. (1979) *Milieukartering*, PUDOC, Wageningen.

Burley, W.F. (1983) Natural Heritage Programs: a methodology for determining priorities in land-use and conservation, *Unpublished paper presented at the 15th Pacific Science Congress*. Dunedin, New Zealand.

Busack, S.D. and Hedges, B.S. (1984) Is the peninsular effect a red herring? *American Naturalist*, **123**, 266–275.

Buse, A. (1974) Habitats as a recording unit in ecological survey: a field trial in Caernarvonshire, North Wales. *Journal of Applied Ecology*, **11**, 517–528.

Bush, M., Montali, R.J., Kleiman, D.G., Randolph, J., Abramowitz, M.D. and Evans, R.F. (1980) Diagnosis and repair of familial diaphragmatic defects in golden lion tamarins. *Journal of the American Veterinary Medical Association*, **177**, 858–862.

Butcher, G.S., Niering, W.A., Barry, W.J. and Goodwin, R.H. (1981) Equilibrium biogeography and the size of nature preserves: an avian case study. *Oecologia*, **49**, 29–37.

Butterfield, J. and Coulson, J.C. (1983) The Carabid communities on peat and upland grasslands in northern England. *Holarctic Ecology*, **6**, 163–174.

Cahn, R. (1982) *The Fight to Save Wild Alaska*, The Audubon Society, New York.

Cameron, R.A.D. and Down, K. (1980) Historical and environmental influences on hedgerow snail faunas. *Biological Journal of the Linnaean Society*, **13**, 75–87.

Cameron, R.A.D. and Redfern, M. (1972) The terrestrial mollusca of the Malham area. *Field Studies*, **3**, 589–602.

Carnahan, J.A. (1977) *Natural Vegetation. Atlas of Australian Resources, Second Series*, Australian Department of National Resources, Division of National Mapping, Canberra.

Carter, A. (1985) *A Study of the Succession in Wharram Quarry Nature Reserve*, BSc. Thesis, University of York.

Catchpole, C.K. and Tydeman, C.F. (1975) Gravel pits as new wetland habitats for the conservation of breeding bird communities. *Biological Conservation*, **8**, 47–59.

Cheetham, A.H. and Hazel, J.E. (1969) Binary (presence-absence) similarity coefficients. *Journal of Paleontology*, **43**, 1130–1136.

Christensen, N.L. (1981) Fire regimes in southeastern ecosystems; in *Proceedings of the Conference on Fire Regimes and Ecosystem Properties*, US Department of Agriculture, Washington, DC, pp. 112–136.

Christian, C.S. and Stewart, G.A. (1968) Methodology of integrated surveys; in *Proceedings of the Toulouse Conference on Aerial Surveys and Integrated Studies*, UNESCO, Paris.

Clapham, A.R. (ed.) (1969) *Flora of Derbyshire*, Museum and Art Gallery, Derby.

Clapham, A.R., Tutin, T.G. and Warburg, E.F. (1962) *Flora of the British Isles*, Cambridge University Press, Cambridge.

Clapham, A.R., Tutin, T.G. and Warburg, E.F. (1981) *Excursion Flora of the British Isles*, Cambridge University Press, Cambridge.

Clausman, P.H.M.A., van Wijngaarden, W. and den Held, A.J. (1984) *Versprei-ding en Ecologie van Wilde Planten in Zuid-Holland. Vol. 1a: Waarderingspara-meters*, Provinciale Planologische Dienst, Den Haag.

Clegg, M.T. and Brown, A.H.D. (1983) The founding of plant populations; in *Genetics and Conservation* (eds C. M. Schonewald-Cox, S. M. Chambers, B. MacBryde and L. Thomas), Benjamin/Cummings, Menlo Park, California, pp. 216–228.

Clements, R.O. (1982) Sampling and extraction techniques for collecting inverte-brates from grasslands. *Entomologist's Monthly Magazine*, **118**, 133–142.

Clifford, H.T. and Stephenson, W. (1975) *An Introduction to Numerical Classifica-tion*, Academic Press, New York.

Cobham, R.O. (1983) The economics of vegetation management; in *Management of Vegetation* (ed. J. M. Way), British Crop Protection Council, London, pp. 35–66.

Cobham, R.O., Matthews, J.R., McNab, A., Stephenson, E. and Slatter, M. (1984) *Agricultural Landscapes: Demonstration Farms*, Countryside Commis-sion, Cheltenham, Gloucester.

Cobham Resource Consultants (1983) *Country sports: their economic significance.* Standing Conference on Countryside Sports, College of Estate Manage-ment, Reading.

Cocks, K.D. and Austin, M.P. (1978) The land use problem and approaches to its solution; in *Land Use on the South Coast of New South Wales, Vol. 1* (eds M. P. Austin and K.D. Cocks), CSIRO, Melbourne, pp. 1–11.

Cocks, K.D., Baird, I.A. and Anderson, J.R. (1982) Application of the SIRO-PLAN planning method to the Cairns section of the Great Barrier Reef Marine Park, Australia, *CSIRO Division of Water and Land Resources, Division Report*, 83/1.

Coles, C. (1984) Some advantages of small game management in the countryside and the hunter's role on conservation; in *Proceedings of the First World Hunting Congress*, Madrid.

Connell, J.H. and Slatyer, R.O. (1977) Mechanisms of succession in natural communities and their role in community stability and organisation. *Ameri-can Naturalist*, **111**, 1119–1114.

Connor, E.F. and McCoy, E.D. (1979) The statistics and biology of the species–area relationship. *American Naturalist*, **113**, 791–833.

Connor, E.F., McCoy, E.D., and Cosby, B.J. (1983) Model discrimination and expected slope values in species–area studies. *American Naturalist*, **122**, 789–796.

Connor, E.F. and Simberloff, D. (1978) Species number and compositional similarity of the Galapagos flora and avifauna. *Ecological Monographs*, **48**, 219–248.

Connor, E.F. and Simberloff, D. (1983) Interspecific competition and species co-occurrence patterns on islands: null models and the evaluation of evi-dence. *Oikos*, **41**, 455–465.

Cook, B.G. (1978) Computer methods; in *Land Use on the South Coast of New South Wales, Vol. 1*, (eds M. P. Austin and K. D. Cocks), CSIRO, Melbourne, pp. 35–48.

Cook, R.E. (1969) Variation in species density of North American birds. *Systematic Zoology*, **18**, 63–84.

Council on Environmental Quality (1980) *The Global 2000 Report to the President*, United States Government Printing Office, Washington, DC.

Countryside in 1970 (1965) *Report of Study Group No. 2, Training and Qualifications of Professions concerned with Land and Water*, Countryside in 1970, London.

Cox, G.W. (ed.) (1974) *Readings in Conservation Ecology*, 2nd edn, Appleton Century Crofts, New York.

Croft, T.A. (1981) Lake Malawi National Park: a case study in conservation planning. *Parks*, **6(3)**, 7–11.

Crowson, R.A. (1981) *The Biology of Coleoptera*, Academic Press, London.

Cumbria County Council (1978) *An Ecological Survey of Cumbria*, Cumbria County Council and Lake District Special Planning Board, Kendal, Cumbria.

Cushwa, C.T. (1983) Fish and wildlife information management – needs versus availability. *Unpublished paper delivered at the National Workshop on Computer Uses in Fish and Wildlife Programs*, Virginia Polytechnical Institute, Blacksburg, Virginia.

Cushwa, C.T., DuBrock, C.W., Gladwin, D.N., Gravatt, G.R., Plantico, R.C., Rowse, R.N. and Salaski, L.J. (1980) A procedure for describing fish and wildlife: summary evaluation report, *US Fish and Wildlife Service, Office of Biological Services Report*, OBS–79/19–A.

Cyrus, D. and Robson, N. (1980) *Bird Atlas of Natal*, University of Natal Press, Pietermaritzberg.

Czekanowski, J. (1932) "Coefficient on racial likeness" and "durchschnittliche Differenz". *Anthropologischer Anzeiger*, **9**, 227–249.

Dale, M.B., Lance, G.N. and Albrecht, L. (1971) Extensions of information analysis. *Australian Computer Journal*, **3**, 29–34.

Daniel, C. and Lamaire, R. (1974) Evaluating effects of water resource developments on wildlife habitat. *Wildlife Society Bulletin*, **2**, 114–118.

Darwin, C. (1868) *The Variation of Animals and Plants under Domestication*, John Murray, London.

Darwin, C. (1876) *The Effects of Cross-Hand Self-Fertilisation in the Vegetable Kingdom*, John Murray, London.

Dasmann, R.F. (1973) Biotic Provinces of the World. *International Union for the Conservation of Nature and Natural Resources, Occasional Paper*, 9.

Davidson, D.A. (1980) *Soils and Land Use Planning*, Longman, London.

Davis, B.N.K. (1976) Wildlife, urbanisation and industry. *Biological Conservation*, **10**, 249–291.

Davis, B.N.K. (1979) Chalk and limestone quarries as wildlife habitats. *Minerals and the Environment*, **1**, 48–56.

Davis, B.N.K. (ed.) (1982a) *Ecology of Quarries*, Institute of Terrestrial Ecology, Cambridge.

Davis, B.N.K. (1982b) Regional variation in quarries; in *Ecology of Quarries* (ed. B. N. K. Davis), Institute of Terrestrial Ecology, Cambridge, pp. 12–19.

Davis, B.N.K. (1983) Plant succession in chalk and ragstone quarries in south east England. *Proceedings of the Croydon Natural History and Scientific Society*, **17**, 153–172.

Davis, P.H. (1951) Cliff vegetation in the eastern Mediterranean. *Journal of Ecology*, **39**, 63–93.

Dawkins, H.C. and Field, D.R.B. (1978) A long-term surveillance system for British woodland vegetation. *Commonwealth Forestry Institute, Occasional Paper*, 1.

Dawson, D.G. and Hackwell, K.R. (1984) Should nature reserves be large or small? (Unpublished manuscript.)

Day, P. (1978) Derelict land in North Wales, nature conservation interest and importance, *Unpublished report, Nature Conservancy Council, Bangor*.

Day, P. and Deadman, A.J. (1981) The nature conservation interest and importance of derelict land in North Wales, U.K. *Nature in Wales*, **17**, 230–240.

Day, P., Deadman, A.J., Greenwood, B.D. and Greenwood, E.F. (1982) A floristic appraisal of marl pits in parts of north-western England and northern Wales. *Watsonia*, **14**, 153–165.

de Lange, L. and van Zon, J.C.J. (1983) A system for the evaluation of aquatic biotypes based on the composition of the macrophytic vegetation. *Biological Conservation*, **25**, 273–284.

de Soet, F. (ed.) (1976) *De Waarden van de Uiterwaarden*. PUDOC, Wageningen.

Dedon, M.F., Smith, K.A. and Laudenslayer, W.F. (1983) Progress report of California's coordinated wildlife planning and information systems. *Unpublished paper delivered at the National Workshop on Computer Uses in Fish and Wildlife Programs*, Virginia Polytechnical Institute, Blacksburg, Virginia.

Diamond, J.M. (1973) Distributional ecology of New Guinea birds. *Science*, **179**, 759–769.

Diamond, J.M. (1975) The island dilemma: lessons of modern biogeographic studies for the design of natural reserves. *Biological Conservation*, **7**, 129–146.

Diamond, J.M. (1976) Island biogeography and conservation: strategy and limitations. *Science*, **193**, 1027–1029.

Diamond, J.M. and May, R.M. (1976) Island biogeography and the design of natural reserves; in *Theoretical Ecology* (ed. R. M. May), Saunders, Philadelphia, pp. 163–186.

di Castri, F. and Robertson, J. (1982) The biosphere reserve concept: 10 years after. *Parks*, **6(4)**, 1–6.

Disney, R.H.L. (1968) Observations on a zoonosis: leishmaniasis in British Honduras. *Journal of Applied Ecology*, **5**, 1–59.

Disney, R.H.L. (1972) Observations on sampling pre-imaginal populations of blackflies (Dipt., Simuliidae) in West Cameroon. *Bulletin of Entomological Research*, **61**, 485–503.

Disney, R.H.L. (1975) *Environment and Creation*, Chester House Publications, London.

Disney, R.H.L. (1976a) Notes on crab-phoretic Diptera (Chironomidae and Simuliidae) and their hosts in Cameroon. *Entomologist's Monthly Magazine*, **111**, 131–136.

Disney, R.H.L. (1976b) A survey of blackfly populations (Dipt., Simuliidae) in West Cameroon. *Entomologist's Monthly Magazine*, **111**, 211–227.

Disney, R.H.L. (1981) Conservation sites. *Nature*, **290**, 432.

Disney, R.H.L. (1982) Rank Methodists. *Antenna*, **6**, 198.

Disney, R.H.L. (1986) Rapid surveys and the ranking of sites in terms of conservation value; in *Handbook on Biological Surveys of Estuaries and Coasts* (ed. J. Baker), EBSA.

Disney, R.H.L., Coulson, J.C. and Butterfield, J. (1981) A survey of the scuttle flies (Diptera: Phoridae) of upland habitats in Northern England. *The Naturalist*, **106**, 53–66.

Disney, R.H.L., Erzinçlioğlu, Y.Z., Henshaw, D.J.de C., Howse, D., Unwin,

D.M., Withers, P. and Woods, A. (1982) Collecting methods and the adequacy of attempted fauna surveys, with reference to the Diptera. *Field Studies*, **5**, 607–621.

Dobson, R.M. (1978) Beetles in pitfalls at the Black Wood of Rannoch. *Glasgow Naturalist*, **19**, 363–376.

Dony, J.G. (undated) Species–area relationships. *Unpublished report, duplicated by the Natural Environment Research Council, London*.

Dony, J.G. and Denholm, I. (1985) Some quantitative methods of assessing the conservation value of ecologically similar sites. *Journal of Applied Ecology*, **22**, 229–238.

Doody, J.P. (1977) The conservation of the semi-natural vegetation of the Magnesian Limestone. I. The Durham escarpment. *Vasculum*, **62**, 17–32.

Drury, W.H. (1974) Rare species. *Biological Conservation*, **6**, 162–169.

Drury, W.H. and Nisbet, I.C.T. (1973) Succession. *Journal of the Arnold Arboretum*, **54**, 331–368.

DuBrock, C.W. (1983) Pennsylvania fish and wildlife data base: a computerized information system. *Unpublished paper presented at the National Workshop on Computer Uses in Fish and Wildlife Programs*. Virginia Polytechnical Institute, Blacksburg, Virginia.

Duffey, E. (1968) Ecological studies on the large copper butterfly *Lycaena dispar* Haw. *batavus* Obth. at Woodwalton Fen National Nature Reserve, Huntingdonshire. *Journal of Applied Ecology*, **5**, 69–96.

Duffey, E. (1971) The management of Woodwalton Fen: a multidisciplinary approach; in *The Scientific Management of Plant and Animal Communities for Conservation* (eds E. Duffey and A. S. Watt), Blackwell, Oxford, pp. 581–597.

Duffey, E. (1974) *Nature Reserves and Wildlife*, Heinemann, London.

Duffey, E. (1982) *National Parks and Reserves of Western Europe*, MacDonald, London.

Edwards, S.J. (undated) Ayrshire woodland report. *Unpublished Report, Scottish Wildlife Trust*, Edinburgh.

Ehrenfeld, D.W. (1976) The conservation of non-resources. *American Scientist*, **64**, 648–656.

Ehrlich, P.R. and Ehrlich, A.H. (1981) *Extinction: the Causes and Consequences of the Disappearance of Species*, Random House, New York.

Elliott, J.M. (1977) Some methods for the statistical analysis of samples and benthic invertebrates, *Freshwater Biological Association Scientific Publication*, 25.

Elliott, J.M. and Drake, C.M. (1981) A comparative study of seven grabs used for sampling benthic macro-invertebrates in rivers. *Freshwater Biology*, **11**, 99–120.

Elton, C.S. (1966) *The Pattern of Animal Communities*, Methuen, London.

Emberson, R.M. (1985) Comparisons of site conservation value using plant and soil arthropod species. *British Ecological Society Bulletin*, **16**, 16–17.

England Field Unit (1982) A survey of Orlestone Forest. *Nature Conservancy Council, Unpublished Report of the England Field Unit*, **18**.

Evans, J. (1984) Silviculture of broadleaved woodland. *Forestry Commission Bulletin*, **62**.

Evans, J.G. (1975) *The Environment of Early Man in the British Isles*, Elektra, London.

Everts, F.H., de Vries, N.P.J. and Udo de Haes, H.A. (1982) *Een Landelijk Systeem van Ecotooptypen*, CML-mededelingen No. 8, Centrum voor Milieukunde, Rijksuniversiteit, Leiden.

Eyre, F.H. (ed.) (1980) *Forest Cover Types of the United States and Canada*, Society of American Foresters, Washington, DC.

Faaborg, J. (1980) Potential uses and abuses of diversity concepts in wildlife management. *Transactions of the Missouri Academy of Science*, **14**, 41–49.

Faeth, S.H. and Kane, T.C. (1978) Urban biogeography. *Oecologia*, **32**, 127–133.

Federal Committee on Ecological Reserves (1975). Charter of the Federal Committee on Ecological Reserves. *Federal Register*, **40**, 8127–8128.

Federal Committee on Ecological Reserves (1977) *A Directory of Research Natural Areas on Federal Lands of the United States of America*, United States Government Printing Office, Washington, DC.

Federal Committee on Research Natural Areas (1968) *A Directory of Research Natural Areas on Federal Lands of the United States of America*, United States Government Printing Office, Washington, DC.

Fenneman, N.M. (1948) *Physiography of the Eastern United States*, McGraw-Hill, New York.

Finlayson, H.H. (1961) On Central Australian mammals, part 4. The distribution and status of Central Australian species. *Records of the South Australian Museum*, **14**, 141–191.

Fitzpatrick, E. and Nix, H.A. (1970) The climatic factor in Australian grassland ecology; in *Australian Grasslands* (ed. R. M. Moore), Australian National University Press, Canberra, pp. 3–26.

Flint, J.H. (ed.) (1963) The insects of the Malham Tarn area. *Proceedings of the Leeds Philosophical and Literary Society, Science Section*, **9(2)**, 15–91.

Flood, B.S., Sangster, M.E., Sparrowe, R.D. and Baskett, T. S. (1977). A handbook for habitat evaluation procedures. *United States Fish and Wildlife Service Resource Publication*, 132.

Forbes, J.E. (1979) Survey of selected woodland sites in the Weald of Kent and East Sussex. *Unpublished Report, Nature Conservancy Council*, Peterborough.

Frankel, O.H. and Soulé, M.E. (1981) *Conservation and Evolution*, Cambridge University Press, Cambridge.

Franklin, I.R. (1980) Evolutionary change in small populations; in *Conservation Biology – An Evolutionary-Ecological Perspective* (ed. M. E. Soulé and B. A. Wilcox), Sinauer, Sunderland, Massachussetts, pp. 135–149.

Friend, G.R. (1980) Wildlife conservation and softwood forestry in Australia: some considerations. *Australian Forestry*, **43**, 217–225.

Fuller, R.J. (1980) A method for assessing the ornithological interest of sites for conservation. *Biological Conservation*, **17**, 229–239.

Fuller, R.J. (1982) *Bird Habitats in Britain*, Poyser, Calton.

Fuller, R.J. and Youngman, R.E. (1979) The utilization of farmland by golden plovers wintering in southern England. *Bird Study*, **26**, 37–46.

Furse, M.T., Wright, J.F., Armitage, P.D. and Moss, D. (1981) An appraisal of pond-net samples for biological monitoring of lotic macro-invertebrates. *Water Research*, **15**, 679–689.

Galbraith, H., Furness, R.W. and Fuller, R.J. (1984) Habitats and distribution of waders breeding on Scottish agricultural land. *Scottish Birds*, **13**, 98–107.

Game, M. (1980) Best shape for nature reserves. *Nature*, **287**, 630–632.

Game, M. and Peterken, G. F. (1984) Nature reserve selection strategies in the woodlands of central Lincolnshire, England. *Biological Conservation*, **29**, 157–81.

Gandawijaja, D. and Arditti, J. (1983). The orchids of Krakatau: evidence for a mode of transport. *Annals of Botany*, **52**, 127–130.

Garren, K.H. (1943) Effects of fire on vegetation of the southeastern United States. *Botanical Review*, **9**, 617–654.

Gauch, H.G. (1982) *Multivariate Analysis in Community Ecology*, Cambridge University Press, Cambridge.

Gehlbach, F.R. (1975) Investigation, evaluation and priority ranking of natural areas. *Biological Conservation*, **8**, 79–88.

Geier, A.R. and Best, L.B. (1980) Habitat selection by small mammals of riparian communities: evaluating the effects of habitat alterations. *Journal of Wildlife Management*, **44**, 16–24.

Gemmell, R.P. (1982) The origin and botanical importance of industrial habitats; in *Urban Ecology* (ed. R. Bornkamm, J. A. Lee and M. R. D. Seaward), Blackwell, Oxford, pp. 33–39.

Gilbert, F.S. (1980) The equilibrium theory of island biogeography: fact or fiction? *Journal of Biogeography*, **7**, 209–235.

Gilbert, L.E. (1980) Food web organisation and conservation of neotropical diversity; in *Conservation Biology – An Evolutionary-Ecological Perspective* (eds M. E. Soulé and B. A. Wilcox), Sinauer, Sunderland, Massachusssetts, pp. 11–33.

Gillison, A.N. (1983) Tropical savannas of Australia and the south-west Pacific; in *Tropical Savannas* (ed. F. Boulière), Elsevier, Amsterdam, pp. 183–243.

Gleason, H.A. (1922) On the relation between species and area. *Ecology*, **3**, 158–162.

Godwin, H. (1975) *The History of the British Flora. A Factual Basis for Phytogeography*, Cambridge University Press, Cambridge.

Goeden, G.B. (1979) Biogeographic theory as a management tool. *Environmental Conservation*, **6**, 27–32.

Goldsmith, F.B. (1975) The evaluation of ecological resources in the countryside for conservation purposes. *Biological Conservation*, **8**, 89–96.

Goldsmith, F.B. (1983). Evaluating nature; in *Conservation in Perspective* (eds A. Warren and F. B. Goldsmith), Wiley, Chichester, pp. 233–246.

Goode, D.A. (1981) Values in conservation; in *Values and Evaluation* (ed. C. I. Rose), *University College, London, Discussion Papers in Conservation*, 36, pp. 28–37.

Goodfellow, S. and Peterken, G.F. (1981) A method for survey and assessment of woodlands for nature conservation using maps and species lists: the example of Norfolk woodlands. *Biological Conservation*, **21**, 177–196.

Goodman, D. (1975) The theory of diversity-stability relationships in ecology. *Quarterly Review on Biology*, **50**, 237–265.

Gordon, A.D. (1980) *Classification*, Chapman and Hall, London.

Graber, J.W. and Graber, R.R. (1976) Environmental evaluations using birds and their habitats. *Illinois Natural History Survey, Biological Notes*, **97**.

Gray, H. (1982) Plant dispersal and colonisation. *Ecology of Quarries* (ed. B. N. K. Davis), Institute of Terrestrial Ecology, Cambridge, pp. 27–31.

Green, B.H. (1981) *Countryside Conservation*, George Allen & Unwin, London.

Green, R.E. (1984) Prospects for integrated field studies of bird migration. *Ringing & Migration*, **5**, 23–31.

Greenwood, E.F. and Gemmell, R.P. (1978) Derelict industrial land as a habitat for rare plants in South Lancs (V.C. 59) and West Lancs (V.C. 60). *Watsonia*, **12**, 33–40.

GRIM (1974) *Landschapsecologische Basisstudie voor het Streekplangebied Midden-Gelderland*, Werkgroep GRIM; Katholieke Universiteit, Nijmegen & PPD Gelderland, Arnhem.

Grime, J.P. (1979) *Plant Strategies and Vegetation Processes*, Wiley, Chichester.

Grospietsch, T. (1965) *Wechseltierchen* (Rhizopoden), Kosmos-Verlag, Stuttgart.

Gross, A.O. (1928) The Heath Hen. *Memoirs of the Boston Society for Natural History*, **6**, 491–588.

Grubb, P.J. (1977) The maintenance of species richness in plant communities: the importance of the regeneration niche. *Biological Reviews*, **52**, 107–145.

Haila, Y. and Järvinen, O. (1982) The role of theoretical concepts in understanding the ecological theatre: a case study on island biogeography; in *Conceptual Issues in Ecology* (ed. E. Saarinen), D. Reidel, Dordrecht, pp. 261–278.

Halcomb, C.M., Boner, R.R. and Sites, J.W. (1976) The Tennessee Heritage Program: an applied data management system for endangered and threatened species information. *The Association of Southeastern Biologists Bulletin*, **23**, 155–158.

Hall, C.A.S. and Day, J.W. (1977) Systems and models: terms and basic principles; in *Ecosystem Modelling in Theory and Practice* (eds C. A. S. Hall and J. W. Day), Wiley Interscience, New York, pp. 6–36.

Hall, N., Johnston, G.M. and Chippendale, G.M. (1970) *Forest Trees of Australia*, Australian Government Publishing Service, Canberra.

Hamor, W.H. (1970) *Guide for Evaluating the Impact of Water and Related Land Resource Development Projects on Fish and Wildlife Habitat*, US Department of Agriculture, Soil Conservation Service, Lincoln, Nebraska.

Harding, P.T. (1981) The conservation of pasture woodlands; in *Forest and Woodland Ecology* (ed. F. T. Last and A. S. Granger), Institute of Terrestrial Ecology, Cambridge, pp. 45–48.

Harms, W.B. and Kalkhoven, J.T.R. (1979) *Landschapsecologie en Natuurbehoud in Midden-Brabant*, Rapport 'De Dorschkamp' No. 208, Wageningen and Rapport RIN No. 79/15, Leersum.

Harrison, J., Miller, K. and McNeely, J. (1982) The world coverage of protected areas: development goals and environmental needs. *Ambio*, **9**, 238–245.

Hartl, D.L. (1981) *A Primer of Population Genetics*, Sinauer, Sunderland, Massachusetts.

Hartshorn, G. (1982) *Costa Rica: Country Environment Profile: a Field Study*, Tropical Science Center, San José, Costa Rica.

Heath, J. and Harding, P.T. (1981) Biological Recording. *A Handbook for Naturalists* (ed. M. R. D. Seaward), Constable, London, pp. 80–86.

Heath, J. and Scott, D. (1977) *Biological Records Centre, Instructions for Recorders*, Natural Environment Research Council, London.

Heinselman, M.L. (1971) Preserving nature in forested wilderness areas and national parks; in *Proceedings of the Forest Recreation Symposium*. United States Department of Agriculture Forest Service, Syracuse, New York.

Helliwell, D.R. (1972) Some effects of afforestation on the flora and fauna of upland areas. *Quarterly Journal of Forestry*, **66**, 235–238.

Helliwell, D.R. (1973) Priorities and values in nature conservation. *Journal of Environmental Management*, **1**, 85–127.

Helliwell, D.R. (1978) Survey and evaluation of wildlife on farmland in Britain: an 'indicator species' approach. *Biological Conservation*, **13**, 63–73.

Helliwell, D.R. (1982) Assessment of conservation values of large and small organisms. *Journal of Environmental Management*, **15**, 273–277.

Helliwell, D.R. (1983) The conservation value of areas of different size: Worcestershire ponds. *Journal of Environmental Management*, **17**, 179–184.

Hendrickson, W.H. (1973) Fire in the National Parks. *Proceedings of the Annual Tall Timbers Fire Ecology Conference*, **12**, 339–343.

Henshaw, D.J.de C. (1984) Insects as indicators of environmental quality. *Sanctuary* (Conservation Bulletin of the Ministry of Defence), **13**, 23–26.

Hepburn, I. (1942) The vegetation of the Barnack stone quarries. A study of the vegetation of the Northamptonshire Jurassic limestone. *Journal of Ecology*, **30**, 57–64.

Heslop-Harrison, J.W. and Richardson, J.A. (1953) The Magnesian Limestone area of Durham and its vegetation. *Transactions of the Northern Naturalists Union*, **2**, 1–28.

Heyligers, P.C. (1978) Biological and ecological aspects related to the forestry operations of the export woodchip industry; in *Ecological and Environmental Effects of the Export Woodchip Industry. 2, Attachments.* Australian Government Publishing Service, Canberra.

Higgs, A.J. (1981) Island biogeography theory and nature reserve design. *Journal of Biogeography*, **8**, 117–124.

Higgs, A.J. and Usher, M.B. (1980) Should nature reserves be large or small? *Nature*, **285**, 568–569.

Hill, M.O. (1979) The development of a flora in even-aged plantations; in *The Ecology of Even-Aged Plantations* (eds E. D. Ford, D. C. Malcolm and J. Atterson), Institute of Terrestrial Ecology, Cambridge, pp. 175–192.

Hill, M.O. and Jones, E.W. (1978) Vegetation changes resulting from afforestation of rough grazings in Caeo Forest, South Wales. *Journal of Ecology*, **66**, 433–456.

Hinton, R. (1978) A survey of ancient woodland in Hertfordshire. *Unpublished Report, Hertfordshire and Middlesex Trust for Nature Conservation, Hitchin.*

Hirsch, A., Krohn, W.B., Schweitzer, D.L. and Thomas, C.H. (1979) Trends and needs in federal inventories of wildlife habitat. *Transactions of the North American Wildlife and Natural Resources Conference*, **44**, 340–359.

Hodgson, J.G. (1982) The botanical interest and value of quarries; in *Ecology of Quarries* (ed. B. N. K. Davis), Institute of Terrestrial Ecology, Cambridge, pp. 3–11.

Holliday, R.J. and Johnson, M.S. (1979) The contribution of derelict mineral and industrial sites to the conservation of rare plants in the United Kingdom. *Minerals and the Environment*, **1**, 1–17.

Hooper, M. D. (1981) Collecting nature reserves; in *Values and Evaluation* (ed. C. I. Rose), *University College, London, Discussion Paper in Conservation*, **36**, pp. 38–44.

Hopkins, P.J. and Webb, N.R. (1984) The composition of the beetle and spider faunas on fragmented heathlands. *Journal of Applied Ecology*, **21**, 935–946.

Horn, H.S. (1974) The ecology of secondary succession. *Annual Review of Ecology and Systematics*, **5**, 25–37.

Hoskins, W.G. (1970) *The Making of the English Landscape*, Pelican, London.

Howe, C.W. (1971) *Benefit-Cost Analysis for Water System Planning*, Water Resources Monograph No. 2, American Geophysical Union, Washington, DC.

Hubalek, Z. (1982) Coefficients of association and similarity based on binary (presence-absence) data: an evaluation. *Biological Reviews*, **57**, 669–689.

Humphries, R.N. (1980) The development of wildlife interest in limestone quarries. *Reclamation Review*, **3**, 197–207.

Hunker, H.L. (ed.) (1964) *Erich W. Zimmermann's Introduction to World Resources*, Harper and Row, New York.

Hunter, F.A. (1977) Ecology of pinewood beetles; in *Native Pinewoods of Scotland* (eds R.G.H. Bunce and J.N.R. Jeffers), Institute of Terrestrial Ecology, Cambridge, pp. 42–55.

Huxley Committee (1947) *Conservation of Nature in England and Wales*. Report of the Wild Life Conservation Special Committee, Cmd 7122. HMSO, London.

Iker, S. (1982) Islands of life in a forest sea. *Mosaic*, **13(5)**, 24–30.

Inhaber, H. (1976) *Environmental Indices*, Wiley, New York.

International Waterfowl Research Bureau (1980) *Conference on the Conservation of Wetlands of International Importance especially as Waterfowl Habitat, Cagliari, Italy, 24–29 November 1980*, IWRB, Slimbridge.

International Union for Conservation of Nature and Natural Resources (IUCN) (1975) *Red Data Book* (various volumes, dealing with fish, amphibians and reptiles, birds, mammals), updated and revised periodically, International Union for the Conservation of Nature and Natural Resources, Gland, Switzerland.

International Union for Conservation of Nature and Natural Resources (IUCN) (1978) *Categories, Objectives and Criteria for Protected Areas*. Final Report by the Committee on Criteria and Nomenclature, Commission on National Parks and Protected Areas, International Union for the Conservation of Nature and Natural Resources, Gland, Switzerland.

International Union for Conservation of Nature and Natural Resources (IUCN) (1980) *World Conservation Strategy*, International Union for the Conservation of Nature and Natural Resources, Gland, Switzerland.

James, F.C. and Rathbun, S. (1982) Rarefaction, relative abundance and diversity of avian communities. *Auk*, **98**, 785–800.

Janzen, D.H. (1983) No park is an island: increase in interference from outside as park size decreases. *Oikos*, **41**, 402–410.

Järvinen, O. and Väisänen, R.A. (1978). Habitat distribution and conservation of land bird populations in northern Norway. *Holarctic Ecology*, **1**, 351–361.

Jefferson, R.G. (1984a) Quarries and wildlife conservation in the Yorkshire Wolds, England. *Biological Conservation*, **29**, 363–380.

Jefferson, R.G. (1984b) *Ecological Studies of Disused Chalk Quarries with particular reference to Nature Conservation*, D.Phil. Thesis, University of York.

Jenkins, R. (1976) Maintenance of natural diversity: approach and recommendations. *Transactions of the North American Wildlife and Natural Resources Conference*, **41**, 441–451.

Jenkins, R. (1977) Classification and inventory for the perpetuation of ecological diversity. *Classification, Inventory, and Analysis of Fish and Wildlife Habitat* (ed. A. Marmelstein), Office of Biological Services, United States Fish and Wildlife Service, Washington, DC, pp. 41–52.

Jenkins, R. (1982) Planning and developing Natural Heritage Programs. *Unpublished paper presented at the Indo-U.S. Workshop on Biosphere Reserves and Conservation of Biological Diversity*, Bangalore, Karnataka State, India.

Jenkins, R.E. and Bedford, W.B. (1973) The use of natural areas to establish environmental baselines. *Biological Conservation*, **5**, 168–174.

Jenkyn, T.R. (1968) Note on butterflies in a coniferous forest. *Journal of the Devon Trust for Nature Conservation*, **19**, 818–819.

Jermyn, S. (1974) *Flora of Essex*, Essex Naturalists Trust, Fingringhoe, Essex.

Johnson, M.P. and Simberloff, D. (1974) Environmental determinants of island species numbers in the British Isles. *Journal of Biogeography*, **1**, 149–154.

Johnson, M.S. (1978) Land reclamation and the botanical significance of some former mining and manufacturing sites in Britain. *Environmental Conservation*, **5**, 223–228.

Jones, C.A. (1973) *The Conservation of Chalk Downland in Dorset*, Dorset County Planning Department, Dorchester.

Jones, E.W. (1945) The structure and reproduction of the virgin forest of the north temperate zone. *New Phytologist*, **44**, 130–148.

Jordan, C.F. (ed.) (1981) Editor's comments; in *Tropical Ecology*, Hutchinson Ross, Stroudsburg, Pennsylvania, pp. 2–3, 31–34, 90–91, 160–161, 225–228, 243–245, 321–323.

Jowsey, W.H. (ed.) (1978) *Botanical Atlas of the Harrogate District*, Harrogate and District Naturalists' Society, Harrogate, North Yorkshire.

Kalkhoven, J.T.R., Stumpel, A.H.P. and Stumpel-Rienks, S.E. (1976) *Landelijke Milieukartering*, Staatsuitgeverij, Den Haag.

Kelcey, J.G. (1975) Industrial development and wildlife conservation. *Environmental Conservation*, **2**, 99–108.

Kelcey, J.G. (1984) Industrial development and the conservation of vascular plants, with special reference to Britain. *Environmental Conservation*, **11**, 235–245.

Kempton, R.A. (1979) The structure of species abundance and measurement of diversity. *Biometrics*, **35**, 307–321.

Kessell, S.R., Good, R.B. and Potter, M.W. (1982) Computer modelling in natural area management. *Australian National Parks and Wildlife Service, Canberra, Special Publication*, **9**.

Keymer, R.J. (1980) Survey of woods in Central Region, Scotland. *Unpublished Report, Nature Conservancy Council*, Edinburgh.

Kiester, A.R. (1971) Species diversity of North American amphibians and reptiles. *Systematic Zoology*, **20**, 127–137.

King, W.B. (1981) *Endangered Birds of the World: the ICBP Bird Red Data Book*, Smithsonian Institution Press, Washington, DC.

Kirby, K.J. (1982) The broadleaved woodlands of the Duddon Valley (Cumbria). *Quarterly Journal of Forestry*, **74**, 83–91.

Kirby, K.J., Bines, T., Burn, A., MacKintosh, J., Pitkin, P. and Smith, I. (1986) Seasonal and observer differences in vascular plant records from British woodlands. *Journal of Ecology*, **74**.

Kirby, K.J., Peterken, G.F., Spencer, J.W. and Walker, G.J. (1984) Inventories of ancient semi-natural woodland. *Focus on Conservation*, **6**, 1–67.

Kirk, W.D.J. (1984) Ecologically selective colour traps. *Ecological Entomology*, **9**, 35–41.

Kirkpatrick, J.B. (1983) An iterative method for establishing priorities for the selection of nature reserves: an example from Tasmania. *Biological Conservation*, **25**, 127–134.

Kitchener, D.J., Chapman, A., Dell, J., Muir, B.G. and Palmer, M. (1980) Lizard assemblage and reserve size and structure in the Western Australian wheat-belt – some implications for conservation. *Biological Conservation*, **17**, 25–62.

Koelreuter, J.G. (1766) *Vorläufigen Nachricht von einigen das Geschlecht der Pflanzen betreffenden Versuchen und Beobachtungen*, Leipzig.

Kromme Rijn Projekt (1974) *Het Kromme-Rijnlandschap, een Ekologische Visie*, Stichting Natuur en Milieu, Amsterdam.

Krutilla, J.V. and Fisher, A.C. (1975) *The Economics of Natural Environments: Studies in the Valuation of Commodity and Amenity Resources*, John Hopkins University Press, Baltimore, Maryland.

Küchler, A.W. (1964) The potential natural vegetation of the conterminous United States. *American Geographical Society, New York, Special Research Publication*, **36**.

Kulczynski, S. (1927) Die Pflanzen-associationen der Pienenen. *Bulletin International de l'Academie Polonaise des Sciences et des Lettres, Classe des Sciences Mathematiques et Naturelles, Serie B, Supplement*, **2**, 57–203.

LaBastille, A. (1974) Ecology and management of the Atitlán Grebe, Lake Atitlán, Guatemala. *Wildlife Monographs*, **37**, 1–66.

Lack, D. and Lack, E. (1951) Further changes in the Breckland avifauna caused by afforestation. *Journal of Animal Ecology*, **20**, 173–179.

Laird, J. (1984) The first international biosphere reserve congress. *Uniterra*, (Special issue, May 1984), 25–28.

Laliberté, F. (1976) Le piegeage des insectes. *Fabreries*, **11**, 56–60.

Lance, G.N. and Williams, W.T. (1966) A generalised sorting strategy for computer classifications. *Nature*, **212**, 218.

Lance, G.N. and Williams, W.T. (1967) A general theory of classificatory sorting strategies, I. Hierarchical systems. *Computer Journal*, **9**, 373–380.

Lance, G.N. and Williams, W.T. (1968) Note on a new information statistic classificatory program. *Computer Journal*, **11**, 195.

Lancia, R.A., Miller, S.D., Adams, D.A. and Hazel, D.W. (1982) Validating habitat quality assessment: an example. *Transactions of the North American Wildlife and Natural Resources Conference*, **47**, 96–110.

Laursen, K. and Frikke, J. (1984) The Danish Wadden Sea. *Coastal Waders and Wildlife in Winter* (eds P. R. Evans, J. D. Goss-Custard and W. G. Hale), Cambridge University Press, Cambridge, pp. 214–223.

Laut, P. (1984) Lessons from the survey 'Environments of South Australia'. *Survey Methods for Nature Conservation, vol. 2*, (eds K. Myers, C. R. Margules and I. Musto), CSIRO, Canberra, pp. 275–289.

Laut, P., Heyligers, P.C., Keig, G., Löffler, E., Margules, C., Scott, R.M. and Sullivan, M.E. (1977) *Environments of South Australia* (8 Province reports and handbook), CSIRO, Melbourne.

Laut, P., Margules, C. and Nix, H.A. (1975) *Australian Biophysical Regions*, Australian Government Printing Service, Canberra.

Laut, P. and Paine, T.A. (1982) A step towards an objective procedure for land classification and mapping. *Applied Geography*, **2**, 109–126.

Ledig, F.T., Guries, R.P. and Bonefeld, B.A. (1983) The relation of growth to heterozygosity in pitch pine. *Evolution*, **37**, 1227–1238.

Lee, A. and Greenwood, B. (1976) The colonisation by plants of calcareous wastes from the salt and alkali industry in Cheshire, England. *Biological Conservation*, **10**, 131–149.

Leopold, A.S. (ed.) (1963) *A Report by the Advisory Committee to the National Park Service on Research*. National Academy of Sciences, National Research Council, Washington, DC.

Levenson, J.B. (1981) Woodlots as biogeographic islands in southeastern Wisconsin; in *Forest Island Dynamics in Man-Dominated Landscapes* (eds R. L. Burgess and D. M. Sharpe), Springer-Verlag, New York, pp. 13–39.

Lindley, D.V. and Scott, W.F. (1984) *New Cambridge Elementary Statistical Tables*, Cambridge University Press, Cambridge.

Lloyd, C. (1984) A method for assessing the relative importance of seabird breeding colonies. *Biological Conservation*, **28**, 155–172.

Lousley, J.E. (1969) *Wild Flowers of Chalk and Limestone*, Collins, London.

Lovejoy, T.E. (1980) Discontinuous wilderness: minimum areas for conservation. *Parks*, **5(2)**, 13–15.

Lovejoy, T.E. (1982) Hope for a beleaguered paradise. *Garden*, **6**, 32–36.

Lovejoy, T.E., Bierregaard, R.O., Rankin, J. and Schubart, H.O.R. (1983) Ecological dynamics of tropical forest fragments; in *Tropical Rain Forest: Ecology and Management* (eds S. L. Sutton, T. C. Whitmore and A. C. Chadwick), Blackwell, Oxford, pp. 377–384.

Lucas, G. and Synge, H. (1978) *The IUCN Plant Red Data Book*, International Union for the Conservation of Nature and Natural Resources, Gland, Switzerland.

Lynch, J.F. and Johnson, N.K. (1974) Turnover and equilibria in insular avifaunas, with special reference to the California Channel Islands. *Condor*, **76**, 370–384.

Mabey, R. (1980) *The Common Ground: a Place for Nature in Britain's Future?*, Hutchinson, London.

Macan, T.T. (1957) The Ephemeroptera of a stony stream. *Journal of Animal Ecology*, **26**, 317–342.

MacArthur, R.H. (1965) Patterns of species diversity. *Biological Reviews*, **40**, 510–533.

MacArthur, R.H. and Wilson, E.O. (1963) An equilibrium theory of insular zoogeography. *Evolution*, **17**, 373–387.

MacArthur, R.H. and Wilson, E.O. (1967) *The Theory of Island Biogeography*, Princeton University Press, Princeton.

McClure, J.P., Cost, N.D. and Knight, H.A. (1979) Multi-resource inventories – a new concept for forest survey. *United States Forest Service, Southeastern Forest Experiment Station, Asheville, North Carolina, Research Paper*, SE-191.

MacDonald, A.A. (1981) A classification and evaluation of the broadleaved woodlands in the Northumberland National Park. *Unpublished Report, Northumberland National Park*, Hexham.

McHarg, I.L. (1969) *Design with Nature*, Natural History Press, New York.

Mahunka, S. (ed.) (1981) *The Fauna of the Hortobágy National Park, Vol. 1*, Académiai Kiadó, Budapest.

Makowski, H. (1974) Problems of using fire in nature reserves, *Proceedings of the Annual Tall Timbers Fire Ecology Conference*, **13**, 15–17.

Mandelbrot, B.B. (1982) *The Practical Geometry of Nature*, Freeman, San Francisco.

Marchant, J. (1983) *Common Bird Census Instructions*, British Trust for Ornithology, Tring, Hertfordshire.

Marcot, B.G., Raphael, M.G. and Berry, K.H. (1983) Monitoring wildlife habitat and validation of wildlife-habitat relationships models. *Transactions of the North American Wildlife and Natural Resources Conference*, **48**, 315–329.

Margalef, R. (1968) *Perspectives in Ecological Theory*, University of Chicago Press, Chicago.

Margules, C.R. (1981) *Assessment of Wildlife Conservation Values*. D.Phil. Thesis, University of York.

Margules, C.R. (1984a) Conservation evaluation in practice: II. Enclosed grasslands in the Yorkshire Dales, Great Britain. *Journal of Environmental Management*, **18**, 169–183.

Margules, C.R. (1984b) Vegetation inventory and conservation evaluation: a case study from the North York Moors, England. *Applied Geography*, **4**, 293–307.

Margules, C.R., Higgs, A.J. and Rafe, R.W. (1982) Modern biogeographic theory: are there any lessons for nature reserve design? *Biological Conservation*, **24**, 115–128.

Margules, C.R. and Usher, M.B. (1981) Criteria used in assessing wildlife conservation potential: a review. *Biological Conservation*, **21**, 79–109.

Margules, C.R. and Usher, M.B. (1984) Conservation evaluation in practice: I. Sites of different habitats in north-east Yorkshire, Great Britain. *Journal of Environmental Management*, **18**, 153–168.

Massey, M.E. (1974) The effect of woodland structure on the breeding bird community in sample woods in south-central Wales. *Nature in Wales*, **14**, 95–105.

Massey, M.E., Peterken, G.F. and Woods, R.G. (1977) Comparative assessments of woodlands for nature conservation, *Nature Conservancy Council, Peterborough, Chief Scientists Team Note*, **4**.

Matthews, J.R. (1955) *Origin and Distribution of the British Flora*, Hutchinson, London.

May, R.M. (1973) *Stability and Complexity in Model Ecosystems*, Princeton University Press, Princeton, New Jersey.

May, R.M. (1975) Island biogeography and the design of wildlife preserves, *Nature*, **254**, 177–178.

May, R.M. (1976) Patterns of species abundance and diversity; in *Ecology and Evolution of Communities* (eds M. L. Cody and J. M. Diamond), pp. 81–120. Belknap Press, Cambridge, Massachusetts.

Mayo, K. (1972) Program DIVINFRE: a divisive classification on binary data, *CSIRO Division of Land Use Research, Technical Memorandum*, 72/4.

Mayr, E. (1963) *Animal Species and Evolution*, Harvard University Press, Cambridge, Massachusetts.

Meganck, R.A. and Ramdial, B.S. (1984) Trinidad and Tobago cultural parks: an idea whose time has come. *Parks*, **9(1)**, 1–5.

Meijers, E., ter Keurs, W.J. and Meelis, E. (1982) Biologische meetnetten voor het beleid. *WLO-Mededelingen*, **9**, 51–58.

Mellanby, K. (1981) *Farming and Wildlife*, Collins, London.

Mennema, J. (1973) Een vegetatiewaardering van het stroomdallandschap van het Merkske (N.-Br.), gebaseerd op een floristische inventarisatie. *Gorteria*, **6**, 157–179.

Messenger, G. (1971) *The Flora of Rutland*, Leicester Museums, Leicester.

Miller, R.I. and Harris, L.D. (1977) Isolation and extirpations in wildlife reserves. *Biological Conservation*, **12**, 311–315.

Milne, B.S. (1974) Ecological succession and bird life at a newly excavated gravel pit. *Bird Study*, **21**, 263–278.

Mitchell, L.H. (1983) *Conservation Assessment of Remnant Vegetation in the Mount Lofty Ranges, South Australia*, M.Appl.Sc. Thesis, Canberra College of Advanced Education.

Moir, W.H. (1972) Natural areas, *Science*, **177**, 396–400.

Moore, N.W. (1977) *Nature Conservation and Agriculture*, Nature Conservancy Council, London.

Moore, N.W. (1982) What parts of Britain's countryside must be conserved? *New Scientist*, **1289**, 147–149.

Moran, V.C. and Southwood, T.R.E. (1982) The guild composition of arthropod communities in trees. *Journal of Animal Ecology*, **51**, 289–306.

Morris, M.D. (1979) *Measuring the Condition of the World's Poor*, Pergamon Press, New York.

Morris, M.G. and Lakhani, K.H. (1979) Responses of grassland invertebrates to management by cutting. I. Species diversity of Hemiptera. *Journal of Applied Ecology*, **16**, 77–98.

Mörzer Bruyns, M.F. (1967) Value and significance of nature conservation. *Nature and Man*, 1967, 37–47.

Moser, M. and Carrier, M. (1983) Patterns of population turnover in ringed plovers and turnstones during their spring passage through the Solway Firth in 1983. *Wader Study Group Bulletin*, **39**, 37–41.

Moss, D. (1979) Even-aged plantations as a habitat for birds. *The Ecology of Even-Aged Plantations* (eds E. D. Ford, D. C. Malcolm and J. Atterson), Institute of Terrestrial Ecology, Cambridge, pp. 413–427.

Mount, D. (1981) *Nature Conservation in the Forests of Upland Britain*. M.Sc. Thesis, University College, London.

Myers, N. (1979) *The Sinking Ark*, Pergamon Press, Oxford.

Myers, N. (1981) Conservation needs and opportunities in tropical moist forests; in *The Biological Aspects of Rare Plant Conservation* (ed. H. Synge), Wiley, Chichester, pp. 141–154.

National Research Council (1983a) *Butterfly Farming in Papua New Guinea*, Managing Tropical Animal Resources Series, National Academy Press, Washington, DC.

National Research Council (1983b) *Crocodiles as a Resource for the Tropics*, Managing Tropical Animal Resources Series, National Academy Press, Washington, DC.

Nature Conservancy (1977a) *Preserving our Natural Heritage Vol. I. Federal Activities*, US Government Printing Office, Washington, DC.

Nature Conservancy (1977b) *Preserving our Natural Heritage Vol. II. State Activities*, US Government Printing Office, Washington, DC.

Nature Conservancy (1982) *Preserving our Natural Heritage Vol. III. Private, Academic, and Local Government Activities*, US Department of the Interior, Washington, DC.

Nature Conservancy (1983) Element ranking; in *Heritage Operations Manual*. Duplicated, Nature Conservancy, Washington, DC.

Nature Conservancy Council (1977) *Nature Conservation and Agriculture*. Nature Conservancy Council, London.

Nature Conservancy Council (1984) *Nature Conservation in Great Britain*, Nature Conservancy Council, London.

Neave, H.R. (1981) *Elementary Statistical Tables*, George Allen and Unwin, London.

Nelson, R.N. and Salwasser, H. (1982) The Forest Service wildlife and fish habitat relationships program. *Transactions of the North American Wildlife and Natural Resources Conference*, **47**, 174–183.

Newby, H. (1980) *A Green and Pleasant Land*? Penguin, London.

Nichol, J.E. (1982) Parameters for conservation evaluation. *Journal of Environmental Management*, **14**, 181–194.

Nilsson, S.G. and Nilsson, I.N. (1976) Hur skall naturområden värderas? Exempel från fågellivet i sydsvenska sjöar. (Valuation of South Swedish wetlands for conservation with the proposal of a new method for valuation of wetlands as breeding habitats for birds.) *Fauna och Flora*, **71**, 136–44.

Nix, H.A. and Austin, M.P. (1973) Mulga: a bioclimatic analysis. *Tropical Grasslands*, **7**, 9–21.

North, P.M. (1978) How many bird territories are there on a farm? A statistical approach to an ornithological problem. *Mathematical Spectrum*, **10**, 44–48.

O'Connor, R.J. and Fuller, R.J. (1985) Bird population responses to habitat; in *Bird Census and Atlas Studies* (eds K. Taylor, R. J. Fuller and P. C. Lack), British Trust for Ornithology, Tring, Hertfordshire, pp. 197–212.

Odum, E.P. (1971) *Fundamentals of Ecology*, 3rd edn, Wiley, New York.

Opdam, P. and Retel Helmrich, V. (1984) Vogelgemeenschappen van heide en hoogveen: een typologische beschrijving. (Bird communities of heath and peat moors: a typological description.) *Limosa*, **57**, 47–63.

Ovington, J.D. (1955) Studies of the development of woodland conditions under different trees. III. The ground flora. *Journal of Ecology*, **43**, 1–21.

Owen, O.S. (1975) *Natural Resource Conservation, an Ecological Approach*, 2nd edn, Macmillan, New York.

Pádua, M.T.J. and Quintão, A.T.B. (1982) Parks and biological reserves in the Brazilian Amazon. *Ambio*, **9**, 309–314.

Parr, T.W. (1980) *The Structure of Soil Microarthropod Communities with particular reference to Ecological Succession*. D.Phil. Thesis, University of York.

Parrinder, E.R. (1964) Little ringed plovers in Britain during 1960–62. *British Birds*, **57**, 191–198.

Peachey, C.A. (1980) The conservation of butterflies in Bernwood Forest. *Unpublished Report, Nature Conservancy Council*, Newbury.

Pears, N. (1977) *Basic Biogeography*, Longman, London.

Peet, R.K. (1974) The measurement of species diversity. *Annual Review of Ecology and Systematics*, **5**, 285–307.

Pennington, W. (1969) *The History of British Vegetation*, English Universities Press, London.

Penny, N.D. and Arias, J.R. (1982) *Insects of an Amazon Forest*, Columbia University Press, New York.

Perring, F.H. (1970) The last seventy years; in *The Flora of a Changing Britain* (ed. F. H. Perring), Classey, Middlesex, pp. 128–135.

Perring, F.H. and Farrell, L. (1977) *British Red Data Book, I. Vascular Plants*, Society for the Promotion of Nature Conservation, Nettleham, Lincoln.

Perring, F.H. and Farrell, L. (1983) *British Red Data Books: I. Vascular Plants*, 2nd edn, Royal Society for Nature Conservation, Lincoln.

Perring, F.H. and Mellanby, K. (eds) (1977) *Ecological Effects of Pesticides*, Academic Press, London.

Perring, F.H., Sell, P.D., Walters, S.M. and Whitehouse, H.L.K. (1964) *A Flora of Cambridgeshire*, Cambridge University Press, Cambridge.

Perring, F.H. and Walters, S.M. (1962) *Atlas of the British Flora*, Thomas Nelson, London.

Peterken, G.F. (1974) A method for assessing woodland flora for conservation purposes using indicator species. *Biological Conservation*, **6**, 239–245.

Peterken, G.F. (1977) Habitat conservation priorities in British and European woodlands. *Biological Conservation*, **11**, 223–236.

Peterken, G.F. (1981) *Woodland Conservation and Management*, Chapman and Hall, London.

Peterken, G.F. and Harding, P.T. (1975) Woodland conservation in eastern England: comparing the effects of changes in three study areas since 1946. *Biological Conservation*, **8**, 279–298.

Peterken, G.F. and Tubbs, C.R. (1965) Woodland regeneration in the New Forest, Hampshire, since 1650. *Journal of Applied Ecology*, **2**, 159–170.

Phillipson, J. (ed.) (1971) *Methods of Study in Quantitative Soil Ecology: Population, Production and Energy Flow*, Blackwell, Oxford.

Pianka, E.R. (1966) Latitudinal gradients in species diversity: a review of concepts. *American Naturalist*, **100**, 33–46.

Pianka, E.R. (1974) *Evolutionary Ecology*, Harper and Row, New York.

Pickett, S.T.A. and Thompson, J.N. (1978) Patch dynamics and the design of nature reserves. *Biological Conservation*, **13**, 27–37.

Pielou, E.C. (1975) *Ecological Diversity*, Wiley, New York.

Pienkowski, M.W. (1983) Identification of relative importance of sites by studies of movement and population turnover; in *Shorebirds and Large Waterbirds Conservation* (eds P. R. Evans, H. Hafner and P. L. L'Hermite), Proceedings of the Durham EEC Conference, pp. 52–67.

Pienkowski, M.W. and Evans, P.R. (1982) Breeding behaviour, productivity and survival of colonial and non-colonial shelducks *Tadorna tadorna*. *Ornis Scandinavica*, **13**, 101–116.

Pigott, C.D. (1969) The status of *Tilia cordata* and *T. platyphyllos* on the Derbyshire limestone. *Journal of Ecology*, **57**, 491–504.

Pigott, C.D. (1983) Regeneration of oak-birch woodland following exclusion of sheep. *Journal of Ecology*, **71**, 629–646.

Pilling, R., Gibson, R. and Crawley, R. (1978) A detailed survey and ecological evaluation of some of the broadleaved woods of the Yorkshire Dales National Park. *Unpublished Report, Yorkshire Dales National Park, Bainbridge, Yorkshire*.

Pinder, N.J. and Barkham, J.P. (1978) An assessment of the contribution of captive breeding to the conservation of rare mammals. *Biological Conservation*, **13**, 187–245.

Pires, J.M., Dobzhansky, T. and Black, G.A. (1953) An estimate of the number of species of trees in an Amazonian forest community. *Botanical Gazette*, **114**, 467–477.

Pollard, E. (1982) Monitoring butterfly abundance in relation to the management of a nature reserve. *Biological Conservation*, **24**, 317–328.

Pollard, E., Elias, D.O., Skelton, M.J. and Thomas, J.A. (1975) A method of assessing the abundance of butterflies in Monks Wood National Nature Reserve in 1973. *Entomologists' Gazette*, **26**, 79–88.

Prance, G.T. (1977) The phytogeographic subdivisions of Amazonia and their influence on the selection of biological reserves; in *Extinction is Forever: Threatened and Endangered Species of Plants in the Americas and their Significance in Ecosystems Today and in the Future* (eds G. T. Prance and T. S. Elias), New York Botanical Garden, Bronx, New York, pp. 195–213.

Preston, F.W. (1962) The canonical distribution of commonness and rarity. *Ecology*, **43**, 185–215, 410–432.

Proctor, M.C.F., Spooner, G.M. and Spooner, M.F. (1970) Changes in Wistman's Wood, Dartmoor: photographic and other evidence. *Transactions of the Devon Association for the Advancement of Science*, **112**, 43–79.

Rabinowitz, D. (1978) Abundance and diaspore weight in rare and common prairie grasses. *Oecologia*, **37**, 213–219.

Rackham, O. (1976) *Trees and Woodlands in the British Landscape*, Dent, London.

Rackham, O. (1980) *Ancient Woodland*, Edward Arnold, London.

Rafe, R.W. (1983) *Species–area Relationships in Conservation*. D.Phil. Thesis, University of York.

Rafe, R.W. and Jefferson, R.G. (1983) The status of *Melanargia galathea* (Lepidoptera: Satyridae) on the Yorkshire Wolds. *Naturalist*, **108**, 3–7.

Rafe, R.W., Usher, M.B. and Jefferson, R.G. (1985) Birds on reserves: the influence of area and habitat on species richness. *Journal of Applied Ecology*, **22**, 327–335.

Ralls, K. and Ballou, J. (1983) Extinction: lessons from zoos; in *Genetics and Conservation* (eds C. M. Schonewald-Cox, S. M. Chambers, B. MacBryde and L. Thomas), Benjamin/Cummings, Menlo Park, California, pp. 164–184.

Ralph, C.J. and Scott, J.M. (1981) Estimating numbers of terrestrial birds. *Studies in Avian Biology*, **6**. Allen Press, Lawrence, Kansas.

Randall, M., Coulson, J.C. and Butterfield, J. (1981) The distribution and biology of Sepsidae (Diptera) in upland regions of northern England. *Ecological Entomology*, **6**, 183–190.

Ranson, C.E. and Doody, J.P. (1982) Quarries and nature conservation – objectives and management; in *Ecology of Quarries* (ed. B. N. K. Davis), Institute of Terrestrial Ecology, Cambridge, pp. 20–26.

Ratcliffe, D.A. (1968) An ecological account of Atlantic bryophytes in the British Isles. *New Phytologist*, **57**, 365–439.

Ratcliffe, D.A. (1971) Criteria for the selection of nature reserves. *Advancement of Science*, London, **27**, 294–296.

Ratcliffe, D.A. (1974) Ecological effects of mineral exploitation in the United Kingdom and their significance to nature conservation. *Proceedings of the Royal Society of London, Series B*, **339**, 355–372.

Ratcliffe, D.A. (1976) Thoughts towards a philosophy of nature conservation. *Biological Conservation*, **9**, 45–53.

Ratcliffe, D.A. (ed.) (1977) *A Nature Conservation Review, Vols. 1 and 2.* Cambridge University Press, Cambridge.

Ratcliffe, D.A. (1981) *The Peregrine Falcon.* Poyser, Calton.

Reed, T.M. (1983) The role of species–area relationships in reserve choice: a British example. *Biological Conservation*, **25**, 263–271.

Reed, T.M., Langslow, D.R. and Symonds, F.L. (1983) Breeding waders of the Caithness flows. *Scottish Birds*, **12**, 180–186.

Reichholf, J. (1981) The structure of the water bird community and wetland classification. *Proceedings of the Symposium of the Mapping of Waterfowl Distributions, Migrations and Habitats, Moscow*, pp. 172–175.

Rice, J., Ohmart, R. D. and Anderson, B. W. (1983) Turnovers in species composition of avian communities in contiguous riparian habitats. *Ecology*, **64**, 1444–1455.

Richter-Dyn, N. and Goel, N.S. (1972) On the extinction of a colonizing species. *Theoretical Population Biology*, **3**, 406–433.

Rishbeth, J. (1948) The flora of Cambridge walls. *Journal of Ecology*, **36**, 136–148.

Robinson, A.H. and Bari, A. (1982) Komodo National Park: progress and problems. *Parks*, **7(2)**, 10–12.

Robinson, M.H. (1978) Is tropical biology real? *Tropical Ecology*, **19**, 30–50.

Roche, L. (1979) Forestry and the conservation of plants and animals in the tropics. *Forest Ecology and Management*, **2**, 103–122.

Rose, F. (1976) Lichenological indicators of age and environmental continuity in woodlands; in *Lichenology, Progress and Problems* (eds D. H. Braun, D. L. Hawksworth and R. H. Bailey), Academic Press, London, pp. 279–307.

Rowell, T.A., Walters, S.M. and Harvey, H.J. (1982) The rediscovery of the fen violet, *Viola persicifolia* Schreber, at Wicken Fen, Cambridgeshire. *Watsonia*, **14**, 183–184.

Royal Agricultural College (1979) *Eysey Farm: Farming and Wildlife Study*, Royal Agricultural College, Cirencester, Gloucestershire.

Russell, J.S. and Moore, A.W. (1970) Detection of homoclimates by numerical analysis with reference to the Brigalow region (eastern Australia). *Agricultural Meteorology*, **7**, 455–479.

Saeijs, H.L.F. and Baptist, H.J.M. (1977) Wetland criteria and birds in a changing delta. *Biological Conservation*, **11**, 251–266.

Salwasser, H., Hamilton, C.K., Krohn, W.B., Lipscomb, J.F. and Thomas, C.H. (1983) Monitoring wildlife and fish: mandates and their implications. *Transactions of the North American Wildlife and Natural Resources Conference*, **48**, 297–307.

Salwasser, H. and Laudenslayer, W.F. (1981) *California Wildlife and Fish Habitat Relationships (WFHR) System: Products and Standards for Wildlife.* California Interagency Wildlife Task Group, Sacramento, California.

Salwasser, H., Mealey, S.P. and Johnson, K. (1984) Wildlife population viability: a question of risk. *Transactions of the North American Wildlife and Natural Resources Conference*, **49**, 421–439.

Salwasser, H. and Tappeiner, J.C. (1981) An ecosystem approach to integrated timber management and wildlife habitat management. *Transactions of the North American Wildlife and Natural Resources Conference*, **46**, 473–487.

Salwasser, H., Thomas, J.W. and Samson, F.B. (1982) Applying the diversity concept to national forest management; in *Natural Diversity in Forest Ecosystems* (ed. J. L. Cooley and J. H. Cooley), Institute of Ecology, University of Georgia, Athens, Georgia, pp. 59–69.

Samson, F.B. (1980) Island biogeography and the conservation of nongame birds. *Transactions of the North American Wildlife and Natural Resources Conference*, **45**, 245–251.

Samson, F.B. and Knopf, F.L. (1982) In search of a diversity ethic for wildlife management. *Transactions of the North American Wildlife and Natural Resources Conference*, **47**, 421–431.

Sanders, H.L. (1968) Marine benthic diversity: a comparative study. *American Naturalist*, **102**, 243–282.

Schaal, B.A. and Levin, D.A. (1976) The demographic genetics of *Liatris cylindracea* Michx. (Compositae). *American Naturalist*, **110**, 191–206.

Schamberger, M. and Krohn, W.B. (1982) Status of the habitat evaluation procedures. *Transactions of the North American Wildlife and Natural Resources Conference*, **47**, 154–164.

Schantz, H.L. (1947) *Fire as a Tool in the Management of the Brush Ranges of California*, California Division of Forestry, Sacramento.

Schiff, A.L. (1962) *Fire and Water – Scientific Heresy in the Forest Service*, Harvard University Press, Cambridge, Massachusetts.

Schroeder, R.L. (1982) Habitat suitability index models: yellow warbler. *USDI Fish and Wildlife Service, Western Energy and Land Use Team, Fort Collins, Colorado, Report*, FWS/OBS-82/10.27.

Schweitzer, D.L. and Cushwa, C.T. (1978) A national assessment of wildlife and fish. *Wildlife Society Bulletin*, **6**, 149–152.

Schweitzer, D.L., Cushwa, C.T. and Hoekstra, T.W. (1978) The 1979 assessment of wildlife and fish, a progress report. *Transactions of the North American Wildlife and Natural Resources Conference*, **43**, 266–273.

Schweizerische Vogelwarte Sempach (1980) *Verbreitungs Atlas der Brutvogel der Schweiz*. Schweizerische Vogelwarte Sempach, Zurich.

Seaber, P.R., Kapinos, F.P., Shanton, J.A. and Moss, S.V. (1974) *National Project to Depict Hydrological Units*, WRD Bulletin, United States Geological Survey, Reston, Virginia.

Seib, R.L. (1980) Baja California: a peninsula for rodents but not for reptiles. *American Naturalist*, **115**, 620–631.

Seitz, W.K., Kling, C.L. and Farmer, A.H. (1982) Habitat evaluation: a comparison of three approaches on the Northern Great Plains. *Transactions of the North American Wildlife and Natural Resources Conference*, **47**, 82–95.

Selman, P.H. (1981) Planners and conservationists, bridging the communications gap; in *Values and Evaluation* (ed. C. I. Rose), *University College, London, Discussion Papers in Conservation*, **36**, pp. 45–51.

Service, M.W. (1976) *Mosquito Ecology Field Sampling Methods*, Applied Science Publishers, London.

Service, M.W. (1977) Methods for sampling adult Simuliidae, with special reference to the *Simulium damnosum* complex. *Centre for Overseas Pest Research, Tropical Pest Bulletin*, **5**, 1–48.

Shaffer, M.L. (1978) *Determining Minimum Viable Population Sizes: a Case Study of the Grizzly Bear (Ursus arctos L.)*. PhD Thesis, Duke University.

Shaffer, M.L. (1981) Minimum population sizes for species conservation. *Bio-Science*, **31**, 131–134.

Shannon, C.E. and Weaver, W. (1963) *The Mathematical Theory of Communication*. University of Illinois Press, Urbana, Illinois.

Sharrock, J.T.R. (1976) *The Atlas of Breeding Birds in Britain and Ireland*, British Trust for Ornithology and Irish Wildbird Conservancy, Tring, Hertfordshire.

Sheail, J. (1976) *Nature in Trust*, Blackie, Glasgow and London.

Sheppard, J.L., Wills, D.L. and Simonson, J.L. (1982) Project applications of the Forest Service Rocky Mountain Region Wildlife and Fish Habitat Relationships System. *Transactions of the North American Wildlife and Natural Resources Conference*, **47**, 128–141.

Shoard, M. (1980) *The Theft of the Countryside*, Temple Smith, London.

Short, H.L. (1982) Development and use of a habitat gradient model to evaluate wildlife habitat. *Transactions of the North American Wildlife and Natural Resources Conference*, **47**, 57–72.

Short, H.L. and Burnham, K.P. (1982) Technique for structuring wildlife guilds to evaluate impacts on wildlife communities. *USDI Fish and Wildlife Service, Fort Collins, Colorado, Special Scientific Report*, 244.

Short, H.L. and Williamson, S.C. (in press) Evaluating the structure of habitat for wildlife; in *Wildlife 2000: Modelling Habitat Relationships of Terrestrial Vertebrates* (ed. E. Stienberg), University of Wisconsin Press, Madison, Wisconsin.

Siderits, K. and Radtke, R.E. (1977) Enhancing forest wildlife habitat through diversity. *Transactions of the North American Wildlife and Natural Resources Conference*, **42**, 425–434.

Siebert, S.F. (1984) Conserving tropical rainforests: the case of Leyte Mountains National Park, Philippines. *Mountain Research and Development*, **4**, 272–276.

Siegel, S. (1956) *Nonparametric Statistics for the Behavioural Sciences*, McGraw-Hill/Kogakusha, London.

Simberloff, D. (1972) Properties of the rarefaction diversity measurements. *American Naturalist*, **106**, 414–418.

Simberloff, D. (1974) Equilibrium theory of island biogeography and ecology. *Annual Review of Ecology and Systematics*, **5**, 161–182.

Simberloff, D. (1976a) Experimental zoogeography of islands: effects of island size. *Ecology*, **57**, 629–648.

Simberloff, D. (1976b) Species turnover and equilibrium island biogeography. *Science*, **194**, 572–578.

Simberloff, D. (1978a) Colonisation of islands by insects: immigration, extinction and diversity; in *Diversity of Insect Faunas* (eds L. A. Mound and N. Waloff), Blackwell, Oxford, pp. 139–153.

Simberloff, D. (1978b) Using island biogeographic distributions to determine if colonization is stochastic. *American Naturalist*, **112**, 713–726.

Simberloff, D. (1978c) Islands and their species. *Nature Conservancy News*, **28(4)**, 4–10.

Simberloff, D. (1982) Island biogeographic theory and the design of wildlife refuges. *Ekologiya*, **4**, 3–13.

Simberloff, D. (1985) Design of wildlife refuges – what can island biogeographic principles tell us? *Biologia Gallo-Hellenica*, (in press).

Simberloff, D. and Abele, L.G. (1976) Island biogeography theory and conservation practice. *Science*, **191**, 285–286.

Simberloff, D. and Abele, L.G. (1982) Refuge design and island biogeographic theory: effects of fragmentation. *American Naturalist*, **120**, 41–50.

Simberloff, D. and Abele, L.G. (1984) Conservation and obfuscation: subdivision of reserves. *Oikos*, **42**, 399–401.

Simberloff, D. and Gotelli, N. (1984) Effects of insularisation on plant species richness in the prairie-forest ecotone. *Biological Conservation*, **29**, 27–46.

Simpson, G.G. (1964) Species diversity of North American recent mammals. *Systematic Zoology*, **13**, 57–73.

Slobodkin, L.B. and Sanders, H.L. (1969) On the contribution of environmental predictability to species diversity. *Brookhaven Symposia in Biology*, **22**, 82–95.

Smart, M. (ed.) (1976) Recommendations for criteria to be used in identifying wetlands of international importance; in *Proceedings of the International Conference on the Conservation of Wetlands and Waterfowl, Heiligenhafen 1974*, IWRB (International Waterfowl Research Bureau), Slimbridge, pp. 470–471.

Smith, K.W. (1983) The status and distribution of waders breeding on wet lowland grasslands in England and Wales. *Bird Study*, **30**, 177–192.

Smith, L., Bridges, E., Durham, D., Eagar, D.C., Lovell, A., Pearsall, S., Smith, T. and Somers, P. (1983) *The Highland Rim in Tennessee: Development of a Community Classification and Identification of Potential Natural Areas*, Tennessee Department of Conservation, Nashville, Tennessee.

Smith, M.E. (1981) Broadleaved woodlands in East Gwynedd, *Unpublished Report, Nature Conservancy Council*, Bangor.

Smith, R.S. (1983) Northern haymeadows. *Report to the Yorkshire Dales National Park Committee.*

Sneath, P.H. and Sokal, R.R. (1973) *Numerical Taxonomy*, Freeman, San Francisco.

Snedecor, G.W. (1956) *Statistical Methods, 5th edn*, Iowa State University Press, Ames, Iowa.

Snyder, N.F.R. (1978) Puerto Rican parrots and nest-site scarcity; in *Endangered Birds – Management Techniques for Preserving Threatened Species* (ed. S. A. Temple), University of Wisconsin Press, Madison, pp. 47–60.

Soulé, M.E. (1983) What do we really know about extinction?; in *Genetics and Conservation* (eds C. M. Schonewald-Cox, S. M. Chambers, B. MacBryde and L. Thomas), Benjamin/Cummings, Menlo Park, California, pp. 111–124.

Soulé, M.E. and Wilcox, B.A. (eds) (1980) *Conservation Biology: an Evolutionary– Ecological Perspective*, Sinauer Associates, Sunderland, Massachusetts.

Southwood, T.R.E. (1978a) The components of diversity; in *Diversity in Insect Faunas* (eds L. A. Mound and N. Waloff), Blackwell Scientific Publishers, Oxford, pp. 19–40.

Southwood, T.R.E. (1978b) *Ecological Methods*, Chapman and Hall, London.

Sousa, P.J. (1982) Habitat suitability index models: Veery. *USDI Fish and Wildlife Service, Western Energy and Land Use Team, Fort Collins, Colorado, Report*, FWS/ OBS-82/10.22.

Sparrowe, R.D. and Sparrowe, B.F. (1978) Use of critical parameters for evaluating wildlife habitat; in *Classification, Inventory, and Analysis of Fish and Wildlife Habitats* (ed. United States Fish and Wildlife Service), Washington, DC, pp. 385–405.

Specht, R.L. (1981) Foliage projective cover and standing biomass; in *Vegetation Classification in Australia* (eds A. N. Gillison and D. J. Anderson), Australian National University Press, Canberra, pp. 10–21.

Specht, R.L., Roe, E.M. and Boughton, V.H. (1974) Conservation of major plant communities in Australia and Papua New Guinea. *Australian Journal of Botany, Supplement 7.*

Spellerberg, I.F. (1981) *Ecological Evaluation for Conservation*, Arnold, London.

Stace, C.A. (1984) Chromosome numbers of British Plants, 7. *Watsonia*, **15**, 38–39.

Standing Conference on Countryside Sports (1983) *Countryside sports and their economic significance: summary report.* College of Estate Management, Reading.

Statens Naturvårdsverk (National Environment Protection Board) (1980) Rangordning ock analys av värderingskriterier inom urskogsinventeringen. *Statens Naturvårdsverk, Stockholm, Working Paper*, 1980-08-27.

Stephens, G.R. and Waggoner, P.E. (1970) The forests anticipated from 40 years of natural transitions in mixed hardwoods. *Bulletin of the Connecticut Agricultural Experimental Station*, No. 707.

Steven, H.M. and Carlisle, A. (1959) *The Native Pinewoods of Scotland*, Oliver and Boyd, Edinburgh.

Stoddard, H.L. (1936) Relation of burning to timber and wildlife. *Proceedings of the First North American Wildlife Conference*, **1**, 1–4, American Wildlife Institute, Washington, DC.

Stone, E.C. (1965) Preserving vegetation in parks and wilderness. *Science*, **150**, 1261–1267.

Strong, D.R. (1980) Null hypotheses in ecology. *Synthese*, **43**, 271–286.

Stubbs, A.E. (1972) Wildlife conservation and dead wood. *Quarterly Journal of the Devon Trust for Nature Conservation*, **4**, 169–182.

Stumpel-Rienks, S.E. (1974) De botanische waardering van ecotopen als bijdrage tot een globale waardering van het natuurlijk milieu. *Gorteria*, **7**, 91–98.

Summers, R.W. and Buxton, N.E. (1983) Winter wader populations on the open shores of northern Scotland. *Scottish Birds*, **12**, 206–211.

Svensson, S. (1977) Land use planning and bird census work with particular reference to the application of the point sampling method. *Polish Ecological Studies*, **3**, 99–117.

Sykes, J.M. (1981) Monitoring in woodlands; in *Forest and Woodland Ecology* (eds F. T. Last and A. S. Gardiner), Institute of Terrestrial Ecology, Cambridge, pp. 32–40.

Szijj, J. (1972) Some suggested criteria for determining the international importance of wetlands in the western Palaearctic. *Proceedings of the International Conference on Conservation of Wetlands and Waterfowl, Ramsar, 1971*, International Waterfowl Research Bureau, pp. 111–119.

Tall Timbers (1976) *Proceedings of the Annual Tall Timbers Fire Ecology Conference*, **15**, Tall Timbers Research Station, Tallahassee, Florida.

Tans, W. (1974) Priority ranking of biotic natural areas. *The Michigan Botanist*, **13**, 31–39.

Tansley, A.G. (1939) *The British Islands and their Vegetation*, Cambridge University Press, Cambridge.

Taylor, B.W. (1957) Plant succession on recent volcanoes in Papua. *Journal of Ecology*, **45**, 233–242.

Taylor, L.R., Kempton, R.A. and Woiwood, I.P. (1976) Diversity statistics and the log-series model. *Journal of Animal Ecology*, **45**, 255–272.

Taylor, R.J. and Regal, P.I. (1978a) The peninsula effect on species diversity and the biogeography of Baja California. *American Naturalist*, **112**, 583–593.

Taylor, R.J. and Regal, P.I. (1978b) Erratum. *American Naturalist*, **112**, 658.

Teixeira, R.M. (1979) *Atlas van de Nederlandse Broedvogels*. Natuurmonumenten, 's Graveland and SOVON, Deventer.

Terborgh, J. (1974) Preservation of natural diversity: the problems of extinction-prone species. *BioScience*, **24**, 715–722.

Terborgh, J. (1975) Faunal equilibria and the design of wildlife preserves. *Tropical Ecological Systems: Trends in Terrestrial and Aquatic Research* (eds F. Golley and E. Medina), Springer-Verlag, New York, pp. 369–380.

Terborgh, J. and Winter, B. (1983) A method for siting parks and reserves with special reference to Colombia and Ecuador. *Biological Conservation*, **27**, 45–58.

Thalen, D.C.P. (1979) On photographic techniques in permanent plot studies. *Vegetatio*, **39**, 185–190.

Thibodeau, F.R. (1983) National programs to protect genetic diversity – the US example. *The Environmentalist*, **3**, 39–44.

Thomas, J.W. (ed.) (1979) *Wildlife Habitats in Managed Forests – The Blue Mountains of Oregon and Washington*. United States Department of Agriculture Handbook No. 553.

Thomas, J.W. (1982) Needs for and approaches to wildlife habitat assessment. *Transactions of the North American Wildlife and Natural Resources Conference*, **47**, 35–46.

Thomas, J.W., Miller, R.J., Black, H., Rodiek, J.E. and Maser, C. (1975) Guidelines for maintaining and enhancing wildlife habitat in forest management in the Blue Mountains of Oregon and Washington. *Transactions of the North American Wildlife and Natural Resources Conference*, **41**, 452–476.

Thompson, R.L. (ed.) (1971) *The Ecology and Management of the Red-cockaded Woodpecker*, US Dept of the Interior, Bureau of Sport Fisheries and Wildlife, Washington, DC.

Tittensor, R. (1980) Ecological history of yew (*Taxus baccata* L) in Southern England. *Biological Conservation*, **17**, 243–266.

Tittensor, R. (1981) A sideways look at nature conservation in Britain. *University College, London, Discussion Papers in Conservation*, **29**.

Tjallingii, S.P. and de Veer, A.A. (eds.) (1981) *Perspectives in Landscape Ecology*, PUDOC, Wageningen.

Tomiałojć, L., Walankiewicz, W. and Wesołowski, T. (1977) Methods and preliminary results of the bird census work in primeval forest of Białowieza National Park. *Polish Ecological Studies*, **3**, 215–223.

Tubbs, C.R. and Blackwood, J.W. (1971) Ecological evaluation for planning purposes. *Biological Conservation*, **3**, 169–172.

Tucker, J.J. and Fitter, A.H. (1981) Ecological studies of Askham Bog nature reserve – 2. The tree population of Far Wood. *Naturalist*, **106**, 3–14.

Udvardy, M.D.F. (1975) A classification of the biogeographical provinces of the world. *IUCN Occasional Paper*, **18**.

UNESCO (1974) Task force on criteria and guidelines for the choice and establishment of biosphere reserves. *Man and the Biosphere Report*, 22.

United States Army Corps of Engineers (1980) *HES: a Habitat Evaluation System for Water Resources Planning*, Environmental Analysis Branch, Planning Division, Lower Mississippi Valley Division, Vicksburg, Mississippi.

United States Department of the Interior (USDI) Fish and Wildlife Service (1976) *Habitat Evaluation Procedures*, Division of Ecological Services, Washington, DC.

United States Department of the Interior (USDI) Fish and Wildlife Service (1980a) Habitat as a basis for environmental assessment. *Division of Ecological Services, Washington, DC., Ecological Services Manual*, 101.

United States Department of the Interior (USDI) Fish and Wildlife Service (1980b) Habitat evaluation procedure. *Division of Ecological Services, Washington, DC, Ecological Services Manual*, 102.

United States Department of the Interior (USDI) Fish and Wildlife Service (1981) Standards for the development of habitat suitability index models. *Division of Ecological Services, Washington, DC, Ecological Services Manual*, 103.

United States Department of State (USDS) (1984) *United States Activities Related to in situ Conservation of Genetic Resources.* Bureau of Oceans and International Environmental and Scientific Affairs, Washington, DC.

Usher, M.B. (1973) *Biological Management and Conservation*, Chapman and Hall, London.

Usher, M.B. (1976) Natural communities of plants and animals in disused quarries. *Papers of the Land Reclamation Conference held at the Civic Hall, Grays, Essex, 5th, 6th & 7th October, 1976* (ed. J. Blunden), Thurrock Borough Council, Grays, Essex, pp. 401–420.

Usher, M.B. (1979a) Natural communities of plants and animals in disused quarries. *Journal of Environmental Management*, **8**, 223–236.

Usher, M.B. (1979b) Changes in the species–area relations of higher plants on nature reserves. *Journal of Applied Ecology*, **16**, 213–215.

Usher, M.B. (1980) An assessment of conservation values within a large Site of Special Scientific Interest in North Yorkshire. *Field Studies*, **5**, 323–348.

Usher, M.B. (1981) Modelling ecological succession, with particular reference to Markovian models. *Vegetatio*, **46**, 11–18.

Usher, M.B. (1983) Species diversity: a comment on a paper by W. B. Yapp. *Field Studies*, **5**, 825–832.

Usher, M.B. (1985a) Population and community dynamics in the soil ecosystem; in *Ecological Interactions in Soil: Plants, Microbes and Animals* (eds A. H. Fitter, D. Atkinson, D.J. Read and M.B. Usher), British Ecological Society Special Publication No. 4, pp. 243–265.

Usher, M.B. (1985b) Implications of species–area relationships for wildlife conservation. *Journal of Environmental Management*, **21**, 181–191.

Usher, M. B. and Williamson, M.H. (eds) (1974) *Ecological Stability*, Chapman and Hall, London.

Väisänen, R.A. and Järvinen, O. (1977) Dynamics of protected bird communities in a Finnish archipelago. *Journal of Animal Ecology*, **46**, 891–908.

van Dijk, G. (1983) De populatie-omvang (broedparen) van enkele weidevogelsoorten in Nederland en de omringende landen. *Het Vogeljaar*, **31**, 117–133.

van Horne, B. (1983) Density as a misleading indicator of habitat quality. *Journal of Wildlife Management*, **47**, 893–901.

van Leeuwen, C.G. (1966) A relation theoretical approach to pattern and process in vegetation. *Wentia*, **15**, 25–46.

van der Maarel, E. (1970) De Ooijpolder, een biologische waardering van natuur en landschap. *Natuur en Landschap*, **24**, 201–233.

van der Maarel, E. (1971) Florastatistieken als bijdrage tot de evaluatie van natuurgebieden. *Gorteria*, **5**, 176–188.

van der Maarel, E. (1978) Ecological principles for physical planning; in *The Breakdown and Restoration of Ecosystems* (eds M. W. Holdgate and M. J. Woodman), Plenum Press, London, pp. 413–450.

van der Maarel, E. and Dauvellier, P.L. (1978) *Naar een Globaal Ecologisch Model voor de Ruimtelijke Ontwikkeling van Nederland*. Staatsuitgeverij, Den Haag.

van der Maarel, E. and Vellema, K. (1975) Towards an ecological model for physical planning in the Netherlands; in *Ecological Aspects of Economic Development Planning*. Report on a Seminar of the Economic Commission for Europe, Geneva, pp. 128–143.

van der Meijden, R., Arnolds, E.J.M., Adema, F., Weeden, E.J. and Plate, C.L. (1983) *Standaardlijst van de Nederlandse Flora 1983*, Rijksherbarium, Leiden.

van der Ploeg, S.W.F. and Vlijm, L. (1978) Ecological evaluation, nature conservation and land use planning with particular reference to methods used in The Netherlands. *Biological Conservation*, **14**, 197–221.

van der Weijden, W.J. and van der Zande, A.N. (1980) Ecological evaluation – controversy among Dutch scientists. *International Journal for Environmental Studies*, **15**, 62–65.

van der Zande, A.N., Saris, F.J.A., Tips, W., Deneef, R. and van der Brent, P. (1981) Evaluation of ecological data for planning: insights from controversies in the Netherlands and Belgium. *WLO-Mededelingen*, **8**, 16–23.

VAWN (1981) *Randstad en Broedvogels*. Vogelwerkgroep Avifauna West-Nederland. Gianotten, Tilburg.

Vera, F.W.M. (1980) The Oostvaardersplassen: possible ways of preserving and further developing the ecosystem. *State Forest Service in the Netherlands, Report*, 1980-1.

von Gaertner, C.F. (1849) *Versuche und Beobachfungen über die Bastarderzeugung im Pflanzenreich*, Stuttgart.

von Tschirnhaus, M. (1981) Die Halm- und Minierfliegen im Grenzbereich Land-Meer der Nordsee. Ein ökologische Studie mit Beschreibung von zwei neuen Arten und neuen Fangund Konservierungsmethoden (Diptera: Chloropidae und Agromyzidae). *Spixania, Supplement*, **6**, 1–405.

Walker, J. and Gillison, A.N. (1982) Australian savannas; in *Ecology of Tropical Savannas* (eds B. J. Huntley and B. H. Walker), Springer-Verlag, Berlin, pp. 5–24.

Wamer, N.O. (1978) *Avian Diversity and Habitat in Florida: an Analysis of a Peninsula Diversity Gradient*. M.S. Thesis, Florida State University.

Ward, L.K. and Lakhani, K.H. (1977) The conservation of junipers: the fauna of food plant islands in southern England. *Journal of Applied Ecology*, **14**, 121–135.

Ward, S.D. and Evans, D.F. (1976) Conservation assessment of British limestone pavements based on floristic criteria. *Biological Conservation*, **9**, 217–233.

Watson, G. (1964) *Ecology and Evolution of Passerine Birds on the Islands of the Aegean Sea*. PhD Thesis, Yale University.

Webb, D.A. (1985) What are the criteria for presuming native status? *Watsonia*, **15**, 231–236.

Webb, N.R. (1982) The diversity of invertebrates on fragmented heathland in Dorset; in *Institute of Terrestrial Ecology, Annual Report, for 1981*, 11–13.

Webb, N.R., Clarke, R.T. and Nichols, J.T. (1984) Invertebrate diversity of fragmented *Calluna*-heathland: effects of surrounding vegetation. *Journal of Biogeography*, **11**, 41–46.

Webb, N.R. and Hopkins, P.J. (1984) Invertebrate diversity of fragmented *Calluna* Heathland. *Journal of Applied Ecology*, **21**, 921–933.

Wellman, F.L. (1962) A few introductory features of tropical plant pathology. *Phytopathology*, **52**, 928–930.

Werkgroep GRAN (1973) *Biologische kartering en evaluatie van de groene ruimte in het gebied van de stadsgewesten Arnhem en Nijmegen*, Rapport afdeling Geobotanie, Katholieke Universiteit, Nijmegen.

Werkgroep Methodologie (1983) *Landinrichtingsstudie Midden-Brabant*. PUDOC, Wageningen.

Westhoff, V. (1968) Die "ausgeraumte" Landschaft; in *Handbuch für Landschaftspflege und Naturschutz, Vol. 2* (eds K. Buchwald and W. Engelhardt), Bayerischer Landwirtschaftsverlag, München, pp. 1–10.

Westhoff, V. (1970) New criteria for nature reserves. *New Scientist*, **46**, 108–113.

Westhoff, V. and den Held, A.J. (1969) *Plantengemeenschappen in Nederland*, Thieme, Zutphen.

Westmacott, R. and Worthington, T. (1974) *New Agricultural Landscapes: Report of a Study Undertaken on Behalf of the Countryside Commission*, Countryside Commission, Cheltenham, Gloucestershire.

Westmacott, R. and Worthington, T. (1984) *Agricultural Landscapes: A Second Look*, Countryside Commission, Cheltenham, Gloucestershire.

Wetterberg, G.B., Prance, G.T. and Lovejoy, T.E. (1981) Conservation progress in Amazonia: a structural review. *Parks*, **6(2)**, 5–10.

Whitcomb, R.F., Robbins, C.S., Lynch, J.F., Whitcomb, B.L., Klimkiewicz, M.K. and Bystrak, D. (1981) Effects of forest fragmentation of avifauna of the eastern deciduous forest; in *Forest Island Dynamics in Man-Dominated Landscapes* (eds R. L. Burgess and D. M. Sharpe), Springer-Verlag, New York, pp. 125–205.

Whitehead, D. (1982) Ecological aspects of natural and plantation forests. *Forestry Abstracts*, **43**, 615–624.

Whitehead, D.R. and Jones, C.E. (1969) Small islands and the equilibrium theory of insular biogeography. *Evolution*, **23**, 171–179.

Whitehead, G.K. (1980) Captive breeding as a practical aid to preventing extinction and providing animals for reintroduction. *Deer*, **5**, 7–13.

Whittaker, R.H. (1967) Gradient analysis of vegetation. *Biological Reviews*, **42**, 207–264.

Whittaker, R.H. (1972) Evolution and measurement of species diversity. *Taxon*, **21**, 213–251.

Whittaker, R.H. (ed.) (1978a) *Ordination of Plant Communities*, 2nd edn, Junk, The Hague.

Whittaker, R.H. (ed.) (1978b) *Classification of Plant Communities*, 2nd edn, Junk, The Hague.

Wibberley, G. (1982) Public pressures on farming: the conflict between agricultural and conservation policies. *Farm Management*, **4**, 373–379.

Wildlife Link (1983) *Habitat Report, No. 2*, Wildlife Link, London.

Willard, B.E. and Marr, J.W. (1970) Recovery of alpine tundra under protection after damage by human activities in the Rocky Mountains of Colorado. *Biological Conservation*, **3**, 181–190.

Williams, C.B. (1964) *Patterns in the Balance of Nature*, Academic, London.

Williams, G. (1980) An index for the ranking of wildfowl habitats, as applied to eleven sites in West Surrey, England. *Biological Conservation*, **18**, 93–99.

Williams, G.L., Russell, K.R. and Seitz, W.K. (1977) Pattern recognition as a tool in the ecological analysis of habitat; in *Classification, Inventory, and Analysis of Fish and Wildlife Habitat – the Proceedings of a National Symposium* (ed. A. Marmelstein), US Fish and Wildlife Service, Office of Biological Services, Washington, DC, pp. 521–531.

Williams, O.B. and Roe, R. (1975) Management of arid grasslands for sheep: plant demography of six grasses in relation to climate and grazing. *Proceedings of the Ecological Society of Australia*, **9**, 142–156.

Williams, W.T. (1971) Principles of clustering. *Annual Review of Ecology and Systematics*, **2**, 303–326.

Williams, W.T. (ed.) (1976) *Pattern Analysis in Agricultural Science*, CSIRO, Melbourne, and Elsevier, Amsterdam.

Williams, W.T., Lambert, J. and Lance, G.N. (1966) Multivariate methods in plant ecology. V. Similarity analysis and information analysis. *Journal of Ecology*, **54**, 427–445.

Williamson, M. (1975) The design of wildlife preserves. *Nature*, **256**, 519.

Williamson, M.H. (1973) Species diversity in ecological communities; in *The Mathematical Theory of the Dynamics of Biological Populations* (eds M. S. Bartlett and R. W. Hiorns), Academic Press, London, pp. 325–335.

Willis, E.O. (1984) Conservation, subdivision of reserves, and the anti-dismemberment hypothesis. *Oikos*, **42**, 396–398.

Wilson, E.O. (1978) *On Human Nature*, Harvard University Press, Cambridge, Massachusetts.

Wilson, E.O. and Willis, E.O. (1975) Applied biogeography in *Ecology and Evolution of Communities* (eds M. L. Cody and J. M. Diamond), Harvard University Press, Cambridge, Massachusetts, pp. 522–534.

Wishart, D. (1978) *CLUSTAN User Manual*, Programme Library Unit, University of Edinburgh, Edinburgh.

Wittig, R. and Schreiber, K.F. (1983) A quick method for assessing the importance of open spaces in towns for urban nature conservation. *Biological Conservation*, **26**, 57–64.

Wood, D.A. (ed.) (1983) *Proceedings of the Red-cockaded Woodpecker Symposium, II*. Florida Game and Freshwater Fish Commission, Tallahassee.

Woolhouse, M.E.J. (1981) *A Description of a Species–area Relationship: a Theoretical Analysis, and a Discussion of its Broader Implications*. MSc. Thesis, University of York.

World Bank (1982) *World Bank Atlas*, Washington, DC.

Wright, S. (1977) *Evolution and the Genetics of Populations, Vol. 3. Experimental Results and Evolutionary Deductions*, University of Chicago Press, Chicago.

Wright, S.J. and Hubbell, S.P. (1983) Stochastic extinction and reserve size: a focal species approach. *Oikos*, **41**, 466–476.

Wynne-Edwards, V.C. (1962) *Animal Dispersion in Relation to Social Behaviour*, Hafner, New York.

Yapp, W.B. (1979) Specific diversity in woodland birds. *Field Studies*, **5**, 45–58.

Younan, E.G. and Hain, F.P. (1982) Evaluation of five trap designs for sampling insects associated with severed pines. *Canadian Entomologist*, **114**, 789–796.

Zonneveld, I.S., Tjallingii, S.P. and Meester-Broertjes, H.A. (1975) *Landschapstaal*, WLO-Mededelingen, Delft.

Author index

Abbott, I. 324, 335
Abele, L.G. 324, 327, 328, 329, 330, 334, 336
Abramowitz, M.D. 319
Adams, D.A. 130, 131
Adams, M.W. 8, 185, 251, 308
Adamus, P.R. 114, 208, 257, 259
Adema, F. 168
Adriani, M.J. 164, 172, 174
Albrecht, L. 51, 56
Alsop, F.J. 125
Anderson, B.W. 131, 132
Anderson, J.R. 61
Anderson, M. 212
Anderson, M.L. 4
Anderson, T.R. 127
An Foras Forbatha 13, 90
Anonymous 5, 104, 113, 166, 169, 170, 172, 173, 178, 210
Antonovics, J. 71
Arditti, J. 84
Arias, J.R. 284
Armitage, P.D. 283
Arnolds, E.J.M. 167, 168
Arrhenius, O. 20
Ash, J.E. 204
Asherin, D.A. 129
Ashton, P.S. 208, 209
Atherden, M.A. 26
Atkinson-Willes, G.L. 263
Austin, M.P. 45, 49, 54, 58, 59, 60, 61, 62, 63, 298

Bährmann, R. 284
Bailey, R.G. 123
Ball, D.F. 27, 202
Baird, I.A. 61

Ballou, J. 318, 319
Baptist, H.J.M. 263, 264
Barber, D. 234, 243
Bari, A. 100
Barkham, J.P. 204, 319
Barrett, B.W. 274
Barry, W.J. 324, 330
Basinski, J.J. 49
Baskett, T.S. 121
Beadle, N.C.W. 47, 48
Bedford, W.B. 25
Berry, K.H. 119, 130, 131, 132
Berry, R.J. 320
Best, L.B. 207
Bezzel, E. 264
Bierregaard, R.O. 209, 321
Bines, T. 216, 217, 218
Black, G.A. 97
Black, H. 119, 125, 126
Black, J. 4, 5, 298, 308
Blackwood, J.W. 203, 209
Blana, H. 248, 257, 260
Bloomfield, H.E. 87
Blouin, M.S. 332, 336
Bolwerkgroep 166
Bonefeld, B.A. 319
Boner, R.R. 115, 116
Bonner, W.N. 10, 31
Booth, T.H. 50, 56
Boughton, V.H. 48, 61
Bouma, F. 163, 164, 171
Boyce, S.G. 127, 129
Boycott, A.E. 207
Braat, L.C. 163, 164
Bradshaw, A.D. 70, 71, 73, 91
Braithwaite, L.W. 62, 63, 64, 65
Bray, J.R. 51, 57

Bridges, E.   129
British Association of Shooting and
    Conservation   243
Brooker, M.P.   281
Brown, A.H.D.   319
Brown, A.H.F.   204
Brown, J.H.   21, 323
Brown, J.M.   127
Bryce, D.   276
Buckley, G.P.   209
Bull, C.   202
Bunce, R.G.H.   182, 184, 209, 217, 300
Burgess, R.L.   317
Burggraaff, M.   164, 165, 166, 178
Burley, W.F.   116
Burn, A.   216, 217, 218
Burnham, K.P.   121, 125
Busack, S.D.   331
Buse, A.   209
Bush, M.   319
Butcher, G.S.   324, 330
Butterfield, J.   279, 285
Buxton, N.E.   266
Bystrak, D.   324

Cahn, R.   114
Cameron, R.A.D.   207, 279
Carlisle, A.   202
Carnahan, J.A.   26, 47, 48
Carrier, M.   254
Carter, A.   90
Catchpole, C.K.   71, 256
Chapman, A.   327
Chatfield, J.E.   274
Cheetham, A.H.   50
Chippendale, G.M.   56
Christensen, N.L.   316
Christian, C.S.   52
Clapham, A.R.   4
Clarke, R.T.   278
Clausman, P.H.M.A.   174, 175, 176
Clegg, M.T.   319
Clements, R.O.   284
Clifford, H.T.   50, 51
Clough, G.C.   114, 208, 257, 258, 259
Cobham, R.O.   223, 226, 231, 243, 244
Cobham Research Consultants   243
Cocks, K.D.   59, 61

Coles, C.   243
Connell, J.H.   70, 84
Connor, E.F.   20, 79, 324, 325, 327,
    332, 336
Cook, B.G.   59, 330
Cosby, B.J.   325
Cost, N.D.   127
Coulson, J.C.   279, 285
Council on Environmental
    Quality   101
Countryside in 1970   4
Cox, G.N.   316
Crawley, R.   215, 300
Croft, T.A.   101
Crowson, R.A.   276
Cumbria County Council   52
Cunningham, R.B.   63
Curtis, J.T.   51, 57
Cushwa, C.T.   119, 127, 131
Cyrus, D.   255
Czekanowski, J.   63

Dale, M.B.   50, 56
Daniel, C.   121
Darwin, C.   318
Dasmann, R.F.   46
Dauvellier, P.L.   164, 168
Davidson, D.A.   212
Davis, B.N.K.   71, 73, 78, 82, 90
Davis, P.H.   85
Dawkins, H.C.   212
Dawson, D.G.   317
Day, J.W.   119
Day, P.   81
Deadman, A.J.   81
de Boer, R.J.   169
Dedon, M.F.   127
de Lange, L.   24, 31
Dell, J.   327
Deneef, R.   163, 165
den Held, A.J.   169, 174, 175, 176
Denholm, I.   24, 25
de Soet, F.   164, 170, 173
de Veer, A.A.   163
de Vries, N.P.J.   166, 169, 170, 171,
    173
Diamond, J.M.   323, 324, 329, 330
di Castri, F.   103

Disney, R.H.L.   271, 272, 273, 279, 282, 283, 285, 290
Dobson, R.M.   285
Dobzhansky, T.   97
Dony, J.G.   20, 24, 25
Doody, J.P.   78, 90
Down, K.   207
Drake, C.M.   283
Drury, W.H.   23, 70, 255
DuBrock, C.W.   127, 130
Dudzinski, M.L.   63
Duffey, E.   23, 162, 224, 234
Durham, D.   111, 129

Eagar, D.C.   111, 129
Edwards, S.J.   210
Ehrenfeld, D.W.   308
Ehrlich, A.H.   97
Ehrlich, P.R.   97
Elias, D.O.   235
Elliott, J.M.   283, 288
Elton, C.S.   207, 281, 286
Emberson, R.M.   11, 12
England Field Unit   212
Erzinçlioğlu, Y.Z.   279
Evans, D.F.   32, 35, 81
Evans, J.   209
Evans, J.G.   224
Evans, P.R.   254
Evans, R.F.   319
Everts, F.H.   166, 169, 170, 171, 173, 174
Eyre, F.H.   113

Faaborg, J.   130
Faeth, S.H.   332, 333
Farmer, A.H.   126
Farrell, L.   23, 27, 153
Federal Committee on Ecological Reserves   112
Federal Committee on Research Natural Areas   112
Fenneman, N.M.   114, 129
Field, D.R.B.   212
Finlayson, H.H.   24
Fisher, A.C.   124
Fitter, A.H.   90
Fitzpatrick, E.   54

Fleming, P.M.   63
Flint, J.H.   280
Flood, B.S.   121
Forbes, J.E.   209, 210
Frankel, O.H.   97, 102, 109, 130, 132, 154, 255
Franklin, I.R.   319
Friend, G.R.   212
Frikke, J.   253
Fuller, R.J.   247, 252, 253, 254, 257, 259, 261, 267, 268
Furness, R.W.   268
Furse, M.T.   283
FWAG   233

Galbraith, H.   268
Game, M.   204, 207, 327, 332
Gandawijaja, D.   84
Garren, K.H.   317
Gauch, H.G.   51
Gehlbach, F.R.   114
Geier, A.R.   207
Gemmell, R.P.   71
Gibson, R.   215, 300
Gilbert, F.S.   335
Gilbert, L.E.   318
Gillison, A.N.   54
Gladwin, D.N.   127, 131
Gleason, H.A.   20
Godwin, H.   26
Goeden, G.B.   324
Goel, N.S.   320
Goldsmith, F.B.   183, 184, 188, 189, 209
Good, R.B.   62, 63
Goode, D.A.   182, 184, 217
Goodfellow, S.   203, 215, 218, 219
Goodman, D.   14
Goodwin, R.H.   324, 343
Goodyear, C.D.   127
Gordon, A.D.   50
Gotelli, N.   321, 325, 327, 328, 329, 330
Graber, J.W.   248
Graber, R.R.   248
Gravatt, G.R.   127, 131
Gray, H.   90
Green, B.H.   182

Green, R.E.   254
Greenwood, B.   71
Greenwood, B.D.   81
Greenwood, E.F.   71
GRIM   176
Grime, J.P.   78, 85
Grospietsch, T.   276
Gross, A.O.   319
Grubb, P.J.   18, 204
Guries, R.P.   319

Hackwell, K.R.   317
Haes, E.C.M.   275, 277
Haila, Y.   335
Hain, F.P.   285
Halcomb, C.M.   115, 116
Hall, C.A.S.   119
Hall, N.   56
Hamilton, C.K.   119, 130, 131, 132
Hamor, W.H.   121
Harding, P.T.   207, 234, 276
Hardy, E.R.   127
Harms, W.B.   165, 171
Harris, L.D.   21
Harrison, J.   103
Hartl, D.L.   319
Hartshorn, G.   106
Harvey, H.J.   85
Hazel, D.W.   130, 131
Hazel, J.E.   50
Heath, J.   276
Hedges, B.S.   331
Heinselman, M.L.   25
Helliwell, D.R.   21, 24, 153, 208, 209, 267
Hendrickson, W.H.   317
Henshaw, D.J. de C.   279, 286
Hepburn, I,   78
Heyligers, P.C.   18, 55
Heymann, M.   231
Higgs, A.J.   105, 109, 204, 307, 324
Hill, M.O.   204, 212
Hinton, R.   215
Hirsch, A.   119, 125, 127, 131
Hoekstra, T.W.   127
Holliday, R.J.   71
Hooper, M.D.   184, 204
Hopkins, P.J.   184, 185

Horn, H.S.   70
Hoskins, W.G.   224
Howe, C.W.   124
Howse, D.   79
Hubalek, Z.   50
Hubbell, S.P.   20
Humphries, R.N.   73
Hunter, F.A.   207

Idle, E.T.   181
International Union for Conservation
    and Natural Resources
    (IUCN)   5, 46, 101, 244, 336

James, F.C.   252, 265
Janzen, D.H.   333
Järvinen, O.   252, 257, 259, 260, 335
Jefferson, R.G.   19, 69, 72, 73, 78, 79, 83, 84, 87
Jenkins, R.   116
Jenkins, R.E.   25
Jenkyn, T.R.   211
Johnson, K.   132
Johnson, M.P.   333
Johnson, M.S.   71
Johnson, N.K.   335
Johnston, G.M.   56
Jones, C.A.   226
Jones, C.E.   322
Jones, E.W.   205, 212
Jordan, C.F.   96
Jowsey, W.H.   239

Kalkhoven, J.T.R.   164, 165, 168, 171, 172, 174
Kane, T.C.   332, 333
Kapinos, F.P.   124
Keig, G.   55
Kelcey, J.G.   71
Kelly, J.   63, 65
Kempton, R.A.   16
Kerney, M.P.   275
Kessell, S.R.   62
Keymer, R.J.   210
Kiester, A.R.   330
King, W.B.   255
Kirby, K.J.   201, 209, 216, 217, 218
Kirk, W.D.J.   285

Kirkpatrick, J.B.   204
Kitchener, D.J.   327
Kleiman, D.G.   319
Klimkiewicz, M.K.   324
Kling, C.L.   126
Knight, H.A.   127
Knopf, F.L.   130
Kodric-Brown, A.   323
Koelreuter, J.G.   318
Krohn, W.B.   119, 120, 124, 125, 127,
    130, 131, 132
Kromme Rijn Projekt   174
Krutilla, J.V.   124
Küchler, A.W.   130
Kulczynski, S.   50, 56

LaBastille, A.   101
Lack, D.   209
Lack, E.   209
Laird, J.   103
Lakhani, K.H.   17, 21
Laliberté, F.   284
Lamaire, R.   121
Lambert, J.   51
Lance, G.N.   50, 51, 56
Lancia, R.A.   130, 131
Langslow, D.R.   247, 267
Laudenslayer, W.F.   127, 132
Laursen, K.   253
Laut, P.   52, 54, 55, 56, 57
Ledig, F.T.   319
Lee, A.   71
Leopold, A.S.   318
Levenson, J.B.   332, 333
Levin, D.A.   319
Lipscomb, J.F.   119, 130, 131, 132
Lloyd, C. 253, 258
Löffler, E.   55
Lovejoy, T.E.   102, 103, 107, 108, 209,
    321
Lovell, A.   129
Lucas, G.   101
Lynch, J.F.   324, 335

MacArthur, R.H.   85, 207, 320, 322,
    330
McClure, J.P.   127
McCoy, E.D.   20, 79, 325, 327

MacDonald, A.A.   210
McHarg, I.L.   272
MacIntosh, J.   216, 217, 218
McNab, A.   231, 243, 244
McNeely, J.   103
McNeil, R.J.   95
Mabey, R.   185
Mahunka, S.   281
Makowski, H.   317
Mandelbrot, B.B.   331
Marchant, J.   235
Marcot, B.G.   119, 130, 131
Margalef, R.   4
Margules, C.R.   8, 9, 13, 20, 22, 23,
    25, 26, 27, 28, 43, 45, 52, 54, 55, 56,
    57, 81, 82, 105, 108, 109, 152, 183,
    184, 185, 186, 187, 212, 251, 255,
    274, 297, 299, 300, 301, 304, 309,
    314
Marr, J.W.   171
Maser, C.   119, 125, 126
Massey, M.E.   207, 215, 219
Matthews, J.R.   151, 231, 243, 244
May, R.M.   30, 252, 324, 329, 330
Mayo, K.   55, 59
Mayr, E.   23
Mealey, S.P.   132
Means, J.E.   331
Meelis, E.   166
Meester-Broertjes, H.A.   164, 165,
    166, 174, 178
Meganck, R.A.   104
Meijers, E.   166
Mellanby, K.   273
Mennema, J.   167, 169, 174
Miller, D.J.   59, 60, 299
Miller, K.   103
Miller, R.I.   21
Miller, R.J.   119, 125, 126
Miller, S.D.   130, 131
Milne, B.S.   103, 256
Mitchell, L.H.   307
Moir, W.H.   25
Montali, R.J.   318
Moore, A.W.   54
Moore, N.W.   149, 153, 154
Moran, V.C.   278, 290
Morris, M.D.   98

Morris, M.G.   17
Mörzer Bruyns, M.F.   163
Moser, M.   254
Moss, D.   209, 210, 283
Moss, S.V.   124
Mount, D.   210
Muir, B.G.   327
Myers, N.   96, 209

National Research Council   104
Nature Conservancy   113, 114, 115, 116, 117, 118
Nature Conservancy Council   226, 229, 230
Nelson, R.N.   126, 127, 130
Newby, H.   244
Nicho, J.E.   209
Nicholls, J.T.   278
Niering, W.A.   324, 330
Nilsson, I.N.   257, 259
Nilsson, S.G.   257, 259
Nisbet, I.C.T.   70
Nix, H.A.   52, 54, 56, 57
North, P.M.   239

O'Connor, R.J.   254
Odum, E.P.   171
Ohmart, R.D.   131, 132
Oosterhuis, L.   204
Opdam, P.   257
Ovington, J.D.   212
Owen, O.S.   316

Pádua, M.T.J.   102
Paine, T.A.   52
Palmer, M.   327
Parr, T.W.   73, 88, 89
Parrinder, E.R.   228
Peachey, C.A.   212
Pears, N.   234
Pearsall, S.   111, 129
Peet, R.K.   16
Pennington, W.   26
Penny, N.D.   284
Perring, F.H.   23, 27, 153, 186, 228, 273
Peterken, G.F.   185, 190, 191, 203, 204, 205, 207, 212, 215, 217, 218, 219, 234, 327

Phillipson, J.   284
Pianka, E.R.   97, 119
Pickett, S.T.A.   204, 318
Pielou, E.C.   15, 311
Pienkowski, M.W.   254
Pigott, C.D.   207, 212
Pilling, R.   215, 300
Pinder, N.J.   319
Pires, J.M.   97
Pitkin, P.   216, 217, 218
Plantico, R.C.   127, 131
Plate, C.L.   168
Pollard, E.   212, 235
Potter, M.W.   62
Prance, G.T.   103
Preston, F.W.   20, 22, 174
Proctor, M.C.F.   212

Quintão, A.T.B.   102

Rabinowitz, D.   308
Rackham, O.   207, 208, 215
Radtke, R.E.   129
Rafe, R.W.   19, 20, 21, 78, 87, 105, 109, 207
Ralph, C.J.   206, 218
Ralls, K.   318, 319
Ramdial, B.S.   104
Randall, M.   279
Randolph, J.   319
Rankin, J.   209, 321
Ranson, C.E.   78, 90
Ranwell, D.S.   143, 156
Raphael, M.G.   119, 130, 131, 132
Ratcliffe, D.A.   8, 9, 26, 29, 30, 32, 70, 71, 73, 76, 78, 135, 137, 138, 141, 145, 183, 184, 185, 188, 189, 204, 207, 219, 226, 248, 251, 256, 272, 273, 274, 299, 303, 309, 312
Rathbun, S.   252, 265
Redfern, M   279
Reed, T.M.   42, 267
Regal, P.I.   331
Reichholf, J.   264
Retel, Helmrich V.   257
Rice, J.   131, 132
Richter-Dyn, N.   320
Rishbeth, J.   85
Roach, J.T.   127

Robbins, C.S. 324
Robertson, J. 103
Robinson, A.H. 100
Robinson, M.H. 97, 98
Robson, N. 255
Roche, L. 212
Rodiek, J.E. 119, 125, 126
Roe, E.M. 47, 61
Roe, R. 308
Roelle, J.E. 129
Rose, C.I. 8, 185, 251, 308
Rose, F. 207, 234
Rowe, J. 223
Rowell, T.A. 85
Rowse, R.N. 127, 131
Royal Agricultural College 231, 236
Russell, J.S. 54
Russell, K.R. 126

Saeijs, H.L.F. 263, 264
Salaski, L.J. 127, 131
Salwasser, H. 119, 126, 127, 129, 130, 131, 132
Samson, F.B. 129, 130, 324
Sanders, H.L. 16, 97
Sangster, M.E. 121
Saris, F.J.A. 163, 165
Schaal, B.A. 319
Schantz, H.L. 317
Schamberger, M. 120, 124
Schiff, A.L. 316, 317
Schreiber, K.F. 31
Schroeder, R.L. 122, 123
Schubart, H.O.R. 209, 321
Schweitzer, D.L. 119, 125, 127, 131
Schweizerische Vogelwarte Sempach 255
Scott, D. 276
Scott, J.M. 206, 218
Scott, R.M. 55
Seaber, P.R. 124
Seib, R.L. 333
Seitz, W.K. 126
Selman, P.H. 202
Service, M.W. 284
Shaffer, M.L. 320, 321
Shannon, C.E. 129
Shanton, J.A. 124
Sharpe, D.M. 317

Sharrock, J.T.R. 255
Shaw, M.W. 217
Sheail, J. 136
Sheppard, J.L. 126
Shoard, M. 244
Short, H.L. 120, 121, 125, 129
Siderits, K. 129
Siebert, S.F. 106
Siegel, S. 288
Simberloff, D. 16, 315, 321, 323, 324, 325, 326, 327, 328, 329, 330, 331, 333, 334, 335, 336
Simonson, J.L. 126
Simpson, G.G. 330
Sites, J.W. 115, 116
Skelton, M.J. 235
Slatter, M. 231, 243, 244
Slatyer, R.O. 70, 84
Slobodkin, L.B. 97
Smart, M. 263
Smith, I. 216, 217, 218
Smith, K.A. 127
Smith, K.W. 268, 312, 313
Smith, L. 129
Smith, M.E. 210, 215
Smith, R. 215
Smith, R.S. 209, 307
Smith, T. 129
Sneath, P.H. 50, 51
Snedecor, G.W. 301
Snyder, N.F.R. 320
Sokal, R.R. 50, 51
Somers, P. 129
Soulé, M.E. 97, 102, 109, 130, 132, 154, 255, 320
Sousa, P.J. 125
Southwood, T.R.E. 15, 129, 252, 278, 284, 290
Sparrowe, B.F. 210
Sparrowe, R.D. 121, 210
Specht, R.L. 47, 48, 61
Spellerberg, I.F. 152, 184, 185
Spencer, J.W. 209, 210
Spooner, G.M. 212
Spooner, M.F. 212
Standing Conference on Countryside Sports 243
Statens Naturvårdsverk 299
Stephens, G.R. 90

Stephenson, E. 231, 243, 244
Stephenson, W. 50, 51
Steven, H.M. 202
Stevens, P.A. 27, 202
Stewart, G.A. 52
Stoddard, H.L. 316, 317
Stone, E.C. 316, 318
Strong, D.R. 322
Stubbs, A.E. 207
Stumpel, A.H.P. 164, 165, 166, 168, 171, 172, 174, 178
Stumpel-Rienks, S.E. 164, 168, 171, 172, 174
Sullivan, M.E. 55
Summers, R.W. 266
Svensson, S. 248
Sykes, J.M. 212
Symonds, F.L. 267
Synge, H. 101
Szijj, J. 262, 263

Tall Timbers 318
Tans, W. 114
Tansley, A.G. 139, 141
Tappeiner, J.C. 119, 130
Taylor, B.W. 84
Taylor, L.R. 16
Taylor, R.J. 331
Teixeira, R.M. 168, 251
Terborgh, J. 31, 102, 324
ter Keurs, W.J. 166
Thalen, D.C.P. 212
Thibodeau, F.R. 113, 114
Thomas, C.H. 119, 125, 127, 130, 131, 132
Thomas, J.A. 235
Thomas, J.W. 118, 119, 120, 125, 126, 129, 130, 132
Thompson, J.N. 204, 318
Thompson, R.L. 365
Tips, W. 163, 165
Tittensor, R. 204, 205, 207, 224, 226, 231, 232
Tjallingii, S.P. 163, 166, 174
Tomialojć, L. 256
Tubba, C.R. 203, 204, 209
Tucker, J.J. 90
Turner, J. 63, 65
Turner, R.G. 71

Tutin, T.G. 4
Tydeman, C.F. 71, 256

Udo de Haes, H.A. 166, 169, 170, 171, 173, 174
Udvardy, M.D.F. 46, 48, 49, 103
UNESCO 46, 256
Unwin, D.M. 79
US Army Corps of Engineers 119, 120, 128
US Department of State (USDS) 113, 114
USDI Fish and Wildlife Service 119, 120, 121, 122, 126
Usher, M.B. 3, 4, 5, 7, 8, 9, 17, 19, 20, 21, 22, 23, 25, 26, 28, 29, 30, 32, 35, 40, 41, 42, 43, 69, 81, 82, 84, 89, 90, 108, 109, 132, 183, 184, 185, 186, 187, 204, 212, 251, 255, 274, 299, 300, 301, 304, 313, 314, 324

Väisänen, R.A. 252, 257, 259, 260
van Deijl, L. 164, 165, 166, 178
van Dijk, G. 251
van Horne, B. 119, 120, 129, 130, 131, 132
van Leewen, C.G. 163
van Wijngaarden, W. 176
van Zon, J.C.J. 24, 31
van der Brent, P. 163, 165
van der Maarel, E. 163, 164, 167, 168, 171, 172, 174
van der Meijden, R. 168
van der Ploeg, S.W.F. 161, 163, 164, 165, 166, 169, 171, 172, 173, 176, 177, 248, 251
van der Weijden, W.J. 163, 165
van der Zande, A.N. 163, 165
VAWN 168
Vellema, K. 164
Vera, F.W.M. 299
Villella, R.F. 127
Vlijm, L. 163, 164, 165, 166, 169, 172, 173, 176, 177, 248, 251
von Gaertner, C.F. 318
von Tschirnhaus, M. 284

Waggoner, P.E. 90
Walankiewicz, W. 256

Walker, G.J.   209
Walker, J.   54
Walters, S.M.   85, 153, 186
Wamer, N.O.   331
Warburg, E.F.   4
Ward, L.K.   21
Ward, S.D.   32, 35, 81
Watson, G.   325, 327, 328, 331, 333, 334
Weaver, W.   129
Webb, D.A.   4
Webb, N.R.   184, 185, 278
Weeden, E.J.   168
Wellman, F.L.   96
Werkgroep GRAN   164, 172
Werkgroep Methodologie   166, 178
Wesolowski, T.   256
Westhoff, V.   163, 164, 169
Westmacott, R.   226
Wetterberg, G.B.   103
Whitcomb, B.L.   324
Whitcomb, R.F.   324
Whitehead, D.   205
Whitehead, D.R.   322
Whitehead, G.K.   319
Whitmer, R.E.   127
Whittaker, R.H.   10, 50, 51, 62, 130
Wibberley, G.   244
Wilcox, B.A.   97, 109
Wildlife Link   226
Willard, B.E.   171
Williams, C.B.   323, 328
Williams, G.   257, 258

Williams, G.L.   126
Williams, O.B.   308
Williams, W.T.   50, 51
Williamson, M.   332
Williamson, M.H.   16, 30
Williamson, S.C.   121
Willis, E.O.   323, 324, 326, 329, 330, 332
Wills, D.L.   126
Wilson, E.O.   85, 318, 320, 322, 324, 326, 329, 330, 332
Winter, B.   31
Wishart, D.   51
Withers, P.   279
Wittig, R.   31
Woiwood, I.P.   16
Wood, D.A.   316
Woods, A.   279
Woods, R.G.   215, 219
Woolhouse, M.E.J.   21
World Bank   98
Worthington, T.   226
Wright, J.F.   283
Wright, S.   318
Wright, S.J.   20
Wynne-Edwards, V.C.   320

Yapp, G.A.   54
Yapp, W.B.   7, 17
Younan, E.G.   285
Youngman, R.E.   267

Zonneveld, I.S.   166, 174

Subject index

Italicized page numbers refer to illustrations.

Acacia aneura (mulga)   54
Acacia silvestris   58
Aceras anthropophorum
   (man orchid)   77
Acid soil   219
Acmena smithii   58
Acrobates pygmaens (feathertail
   glider)   64
Acrocephalus palustris (marsh
   warbler)   250
Actaea spicata   32 (table), 33, 35
Adiantum capillus-veneris   74 (table)
Agricultural Development Advisory
   Service (Britain)   232
Agricultural environments   223–46
   agricultural dominance   224–6
   amphibians   235
   birds   235, 236 (table)
   demonstration farms *see*
      Demonstration farms;
      Hopewell House Farm
   documentary material   232
   economic/political environment
      244–5
   grassland   233 (table)
   heaths   233
   hedges   233 (table)
   impact of change   243
   invertebrates   235, 236 (table)
   mammals   235, 236 (table)
   moorlands   233
   plants   6, 7 (table), 235, 236 (table)
   reptiles   235
   survey
      detailed   234–7
      preliminary   232–4

wetland   230 (table), 233 (table);
   see also Wetland
wildlife/husbandry
   relationship   242–3
woodland   233 (table); *see also*
   Woodland
Agrostemma githago (corncockle)   23–4
'Ahmed' (elephant in Kenya)   104
Ajuga chamaepitys   74 (table)
Alcedo atthis (kingfisher)   266
Alkali waste   71
Amazona vittata (Puerto Rican
   parrot)   320
Amboseli Reserve (Kenya)   104
Amenity value   9 (table), 13 (table)
Antarctica
   earliest conservation activities
      9–10
   Specially Protected Areas   32
Aquila chrysaetos (golden eagle)   252,
   255, 267
Archaelogical interest (as a
   criterion)   13 (table)
Area   9 (table), 13 (table), 17–22, 141,
   184, 195–7
   functional unit   *60*
   species–area relationship   17, 20–2,
      79, 90, 325–7
Aristidia stricta (wiregrass)   316
Armeria maritima   71
Arrhenius equation   20
Artificial human disruption   25
Arum maculatum (lords-and-
   ladies)   238
Ash (*Fraxinus excelsior*)   151
Askham Bog Nature Reserve   90

Atitlán (giant pied-billed grebe)   101
Atlases of bird distribution   255
Atlas of the British Flora (Perring and
    Walters, 1962)   153, 186
Attributes   6, 249, 251
Australia
    agriculture   255 (table)
    biogeographical provinces   48, *49*
    geographical regions   47–8
    natural ecosystem   26
    natural vegetation maps   46, 49
    nature conservation evaluation
        46–7
    representativeness assessment
        46–9
    vegetation differentiation   47
Availability (as a criterion)   9 (table),
    13 (table)

Barbados Natural Trust   106
Baseline diversity goal   130
Bedfordshire (England)
    woodlands   24
Bellis perennis (daisy)   174
Berne Convention   152
Bettongia lesueur (Lesueur's rat
    kangaroo)   24
Betula pubescens   71
Biogeographical provinces, Udvardy's
    classification   103
Biological accomodation   97
Biosphere reserves   46, 103
Birds
    conservation evaluation   12, 247–69
    in nature reserves, species–area
        relationship   21
    in small woodlands, species–area
        relationship   21
    on farmland   235, 236 (table), 239
Black hairstreak butterfly (*Strymonidia
    pruni*)   207
Black-tailed godwit (*Limosa
    limosa*)   251
Blackthorn (*Prunus spinosa*)   207
'Blossom' (elephant seal in New
    Zealand)   104
Bluebell (*Endymion non-scriptus*)   151,
    238

Blue Mountains (Oregon), vertebrates
    in national forests   125
Bolivia, protected areas   103
Bray-Curtis association measure   57
Breckland grassland   90
Brazil
    Amazonian tree species   97
    biological reserves   102, 104
    conservation for
        representativeness   103
    national parks   102
    Pleistocene refugia   102
Britain
    agriculture   225 (table), 226–8
    recent changes   226–8
    birds   22, 151, *250*
    butterflies   22
    demonstration farms *see*
        Demonstration farms
    disappearance of
        meadow/hedgerows   226
    evaluation in   10, 135–59, 181–98
    farmland birds   267–8
    grassland   26, 230 (table), 233
        (table)
    international importance of
        sites   150–2
    land snails   22
    lowland regions   226
    lowland waders   268
    moorland birds   266–7
    mountain birds   266–7
    'nature resource'   *149*
    non-breeding wildfowl   257, 258
    ornithological sites
        classification   261
    river systems   266
    semi-natural (near-natural)
        communities   26
    woodland birds   264–5
British Association of Shooting and
    Conservation   243
British Red Data Book (Perring and
    Farrell)   27, 153
British Trust for Conservation
    Volunteers   245
Broad-leaved marsh orchid (*Orchis
    majalis*)   174

Brockham Chalk Quarry (Surrey) 71
Bunium bulbocastanum 74 (table)
Bushtail possum *Trichosurus*
 vulpecula) 64
Butterfly transect recording
 system 212

Cagliari criteria 263
Calcareous dune slacks
 (Lancashire) 71
Calcareous soils over limestones 71
Caledonian Scots pine (*Pinus*
 sylvestris) 264–5
Calidris alpina (dunlin) 267
California condor (*Gymnogyps*
 californianus) 329
Calluna vulgaris (heather) 26–7, 151
Cambridge (England), vascular plants
 colonizing walls 85
Canada, agriculture 255 (table)
Cardamine impatiens 32 (table)
Carex diandra 188
Carex ericetorum 74 (table)
Carex flacca 35, 38
Cargenia gigantea (saguaro cactus) 329
Carum verticillatum 186
Centaurea scabiosa 77 (table)
Cepaea nemoralis (common snail) 275
Cerastium pumilum 74 (table)
Cercatetus nanus (pygmy possum) 64
Chalk grassland species 79
Chamerion angustifolium 87
Charadrius dubius (little ringed
 plover) 228
Chestnut-flanked white-eye (*Zosterops*
 mayottensis semiflava) 329
Chironomid midges, larval 276
Chorthippus vagans 275
Chough (*Pyrrhocorax pyrrhocorax*) 90
Cichlid family 101
Cinclus cinclus (dipper) 266
Circus cyaneus (hen harrier) 267
Cirl bunting (*Emberiza cirlus*) 250
Cirsium helenoides 34, 35
Classification, plant communities
 based 190
Climax communities 70
Climax ecosystem 10

Cloud forest *see* Tropical rain forests
Coastal birds 137
Coastal habitats 230
Coastlands 191
Cochlearia officinalis agg. 71
Collecting efficiency 282–4
Collecting success 282–4
Collembola (spring-tails) 88, *89*, 241
Colliery waste heaps 71, 90
Colombia, endemism areas 31
Common Bird Census technique 235
Common snail (*Cepaea nemoralis*) 275
Community structure, aquatic
 communities 31
Comparative Biological Value
 Index 156–9
Complexity as indicator of wildlife
 values 178
Coniferous plantations 209, 212
 rides in 212
Conservation, definition 4–5
Conservation index 80–2
Copsychus sechellarum (magpie
 robin) 255, 329
Corallorhiza trifida (coralroot) 74
 (table), *187*
Coral reefs 97, 103
Corncockle (*Argostemma githago*) 23–4
Cornfield weeds 23
Costa Rica, national park for sea
 turtles 101
Council on Environment Quality 112
Country Landowners' Association
 (Britain) 245
Countryside Commission
 (Britain) 245
County Conservation Trust nature
 reserves (Britain) 73
County Naturalists' Trusts
 (Britain) 245
Craven Pennines (North
 Yorkshire) 26
Crepis mollis 74 (table)
Crested dog's tail (*Cynosurus*
 cristatus) 26
Crested tit (*Parus cristatus*) 265
Criterion (a) 8, 9 (table), 13 (table),
 183–4, 249

1% 253, 262–3, 264
area 184; *see also* Area
Cagliari 263
conservation 184
ecological *vs* conservation 183
diversity *see* Diversity *and subsequent entries*
endemicity (endemism) 31, 32
evaluation of sites 71–2
fragility 189
Heiligenhafen 263
intrinsic appeal 32, 189
naturalness 25–7, 187–8
Nature Conservation Review 183
period of development 31
popularity poll 9, 13–14
position in geographical unit 189
potential value 189
quantification 8
rarity *see* Rarity
recorded history 189
representativeness 10, 45–67, 143
species richness 7, 143
type locality 31–2
typicalness 29, 142, 143, 188–9; *see also* representativeness
Cryptostigmata 88, *89*
Cushwa's 'Procedures' 127
Cynosurus cristatus (crested dog's-tail) 26

Daisy (*Bellis perennis*) 74
Damage by human use of land and water 136
Danish Wadden Sea, waders in *253*
Daphne mezereum 74 (table)
Dartford warbler (*Sylvia undata*) 250
Darwin's finches 104
Data
  collection 190–2
  dictionary, standard 127
  integration 218–19
  interpretation 218–19
Demographic stochasticity 320
Demonstration Farms (Britain) 6, 226, 227 (table), 229, 231 (table)
  overall approach *237*
  see also Hopewell House Farm

Dendrocopus borealis (red-cockaded woodpecker) 316
Dendroica kirtlandii (Kirtland's warbler) 329
Dendroica petechia (yellow warbler) 122
Desmazeria rigida 87, 88
Dipper (*Cinclus cinclus*) 266
Diptera 12, 280–1, 290–2
  collecting methods 284–5
Disused quarries and chalk pits 72, 73–91
  colonization by
    herbaceous plants 90
    woody vegetation 90
  conservation index 80–2
  ecological island effect 84
  rare plant species 74–5 (table)
  spatially close 84
  surroundings 83
  Yorkshire Wolds 78–82
Diversity 7, 9 (table), 13 (table), 14–17, 141, 185
Diversity index 7–8, 15–16
Diversity levels (Whittaker) 130
DIVINFRE 55, 59
Documentary material 232
Dorset heathlands 185
Doryphora sassafras 58
Dryopteris villarii 32 (table)
Dryas octopetala (mountain avens) 23
Ducks Unlimited 107
Dunlin (*Calidris alpina*) 267

East Africa, national parks 103–4
East coast biogeographic provinces (Australia) 56
Eastern Mediterranean, cliff plants 85
Ecological gradients 10
Ecological island 84
Ecological processes 82–8
  extinction 85–8, *323*
  immigration 82–5, *323*
Ecological succession 30, 69–91
  primary 70
  secondary 70
Ecoregions (Bailey's) 123–4

Ecosystem
   fragility concept   29–30
   long-term survival   102
   minimum disturbance by man   25
   models   90
   modification by man   26
   period of development as
      criterion   31
   re-creation of different types   91
   resilience concept   30
   stability   30
   urban   31
   vegetation structure   31
Ecotypes, rarity assessment of   169
Ecuador
   endemism areas   31
   protected areas   103
Education value (as a criterion)   9
   (table), 13 (table)
Effective population size   97
Element ranking   116–18
Emberiza cirlus (cirl bunting)   *250*
Environmental stochasticity   320
Ephemeroptera   289
Endemicity (endemism)   31, 32
Endymion non-scriptus (bluebell)   151,
   238
Environmental Impact Statement on
   Water extraction in South
   Kennemerland (1981)   169
Environments of South Australia (Laut *et*
   al., 1977)   55–6
Epilobium hirsutum   83
Epipactis atrorubens   74 (table)
Equilibrium island biogeographical
   theory   322–4
Essex woodlands   *210*
Estuarine birds   254
Eucalyptus   56, *58*
   distribution along altitudinal
      gradient   *62, 63*
   minimum area of forest   18
   species   *58*, 65
Eucryphia moorei   *58*
European Nature Conservation Year
   (1970)   164
Evaluation   297–337
   definition   5

diversity role   307
ecological vulnerability   309–10
monetary terms   153
North Yorkshire exercises   299–306
objectives   5, 44, 248–51
operational framework   310–13
representativeness   310; *see also*
   Representativeness
requisites   6–8
site area importance   307–8
threat of destruction   309–10
Extinction   85–8, *323*
Eysey Farm   231 (table), 236

Falco columbarius (merlin)   267
Falco peregrinus (peregrine falcon)   23,
   90, 248, 267
Farming and Wildlife Advisory
   Groups (Britain)   232, 245
Farmland birds (Britain)   267–8
Feathertail glider (*Acrobates*
   pygmaeus)   64
Federal Committee on Ecological
   Reserves   112
Federal Committee on Research
   Natural Areas   113
Fieldfare (*Turdus pilaris*)   267
Fire, occasional   316–18
Firecrest (*Regulus ignicapillus*)   250
Floral statistics   167
Florida (USA), insularization   317
Floristic indices
   limestone pavements   81
   marl pits   81
Food and Agriculture
   Organization   103
Forest evaluation *see* Woodland and
   forest evaluation
Forest type definition   59
Fragility   9 (table), 13 (table), 29, 142,
   189
Fraxinus excelsior (ash)   151
Functional units   *60; see also* Area

Galapagos Islands national park   104
Gallinago gallinago (snipe)   268
Game ranching   104

Game reserves (East Africa), species–area relationship 21
Gamma–beta–alpha (top–down) approach 130
Genetic diversion within species/populations 97
Gentianella germanica 74 (table)
Geological formations, vegetation communities on 65
Giant pied-billed grebe (atitlán) 101
Glaciation effects 23
Gleason equation 20
Gliders 63–5
Golden eagle (*Aquila chrysaetos*) 252, 255, 267
Golden lion tamarin (*Leontopithecus rosalia*) 318–19
Golden plover (*Pluvialis apricaria*) 267
Gorse (*Ulex spp.*) 151
Grasshoppers 275, 276
Grassland 191, 230 (table), 233 (table)
Great Basin of North America, mammalian species in 21
Great Britain *see* Britain
Green parakeet (*Psittacula eupatria wardi*) 329
Greenshank (*Tringa nebularia*) 267
Grey squirrel (*Sciurus caroliensis*) 205
Grey wagtail (*Motacilla cinerea*) 266
Grizzly bear (*Ursus arctos*) 321
Guatemala, atitlán (giant pied-billed grebe) refuge 101
Gymnocarpium robertiana 32 (table)
Gymnogyps californianus (California condor) 329

Habitat 119, 190
    patches in matrix of another habitat 21
Habitat Evaluation Procedure (USA) 121–5
Habitat Evaluation System (USA) 120, 128–9
Habitat gradient model (Short) 129–130
Habitat Quality Index curve *128*
Habitat Suitability Index 121–4

Hardwood woodchip industry (Australia) 18
Heather (*Calluna vulgaris*) 26–7, 151
Heath hen (*Tympanuchus cupido attwateri*) 319
Heathland 230 (table), 233 (table)
    high-altitude dwarf-shrub 70
    Rhacomitrium moss-lichen 70
Heavy metals concentrations 71, 73
Hebridean machair 268
Hedges 233
    loss of 226
Heiligenhafen criteria 263
Hen harrier (*Circus cyanens*) 267
Herminium monorchis 74 (table)
Herniaria glabra 74 (table)
Higher plants, species–area relationship 21
Highland Boundary Fault 194, 195
Homoptera, survey 279
Honey buzzard (*Pernis apivorus*) 250
Hopewell House Farm (Yorkshire) 6–8, 237–42
    botanical survey 238–9
    invertebrate survey 239–41
    mammal survey 239
    multi-purpose plan 242
    ornithological survey 239
    plant list for field 6–8
    single-purpose plan 238–42
Hornungia petraea 74 (table)
Hortobágy National Park (Hungary), fauna 281
Hour-squares 167, 168
Huxley Committee (1947) 137, 138, 147, 183
Hydrologic units (Seaber) 124
Hylocichla furescens (veery) 125
Hypochaeris maculata 74 (table)

Immigration 82–5, 323
Inbreeding depression, avoidance of 154
Index (indices)
    diversity 15–16, 264
    evenness 264
    Wetland 264

Industrial habitats 70–1
Industrial waste tips 71
International organizations 103,
    105–7, 152
International Union for the
    Conservation of Nature and
    Natural Resources 103, 336
  Commission on National Parks and
    Protected Areas 103, 105
  Species Survival Commission 105
Interrelationship richness (pattern
    richness) 97–8, 102
Intrinsic appeal 32
Invertebrates 271–93
  area 278
  collecting methods 284–6
  collecting success/efficiency 282–4
  criteria 274–9
  diversity 278–9
  equivalence, collecting-efficiency
      and habitat type 286–8
  Mann-Whitney U-test 288–9
  means obtained with sets of
      traps 288–9
  ranking of sites 289–92
  rarity 276–8
  sampling problems 282
  surveys
    comprehensive 279–81
    sample 281–92
Ireland, breeding seabirds 257, 258–9
Isatis tinctoria (woad) 27

Joshua tree (Yucca brevifolia) 329
Juncus spp. 239
Juncus alpinus (alpine rush) 23
Juniperus communis (juniper) 21

Kenya
  'Ahmed' (elephant) 104
  Amboseli Reserve 104
  Lake Nakuru National Park 104
Kingfisher (Alcedo atthis) 266
Kingley Vale Reserve (Sussex) 204
Kirtland's warbler (Dendroica
    kirtlandii) 329

Komodo National Park
    (Indonesia) 100–1
Krakatau, colonization by orchids 84

Lagopus lagopus scoticus (grouse) 27
Lake Malawi National Park
    (Malawi) 101
Lake Nakuru National Park
    (Kenya) 104
Land classification 49–53
  climate 53 (table)
  complexity 51–2
  floristics 53 (table)
  information 51–2
  land form 53 (table)
  lithology 53 (table)
  numerical 49–51
  scale 51–2
  soils 53 (table)
  vegetation 53 (table)
Lapwing (Vanellus vanellus) 267, 268
Lead mines 71
Leblanc waste heaps 71, 90–1
Leontopithecus rosalia (golden lion
    tamarin) 318–19
Lime (Tilia cordata) 204, 205
Lime beds (Cheshire) 71
Limestone pavements
    (Yorkshire) 32–42
  floristic indices 81
  plant species 32 (table), 35–42
  species–area relationship 40–2
  which to conserve? 35–42
Limosa limosa (blacktailed godwit) 251
Lincolnshire woods 185, 203–4
Ling (Calluna vulgaris) 26–7, 151
Little ringed plover (Charadrius
    dubius) 228
Loblolly pine (Pinus taeda) 316
Loch Leven (Scotland) 195
Longleaf pine (Pinus palustris) 316
Lord Howe wood rail (Tricholimnas
    sylvestris) 256
Lords-and-ladies (Arum
    maculatum) 238
Lotus angustissimus 74 (table)
Lowest common denominators of
    nature preserves 116

Lowland waders (Britain)   268
Loxia scotica (Scottish crossbill)   265
Lullula arborea (woodlark)   *250*
Lycaena dispar (large copper
    butterfly)   23
Lychnis viscaria   74 (table)

Machair, Hebridean   268
Magnesian (Permian) limestone
    grassland   78
Magpie robin (*Copsychus
    sechellarum*)   255, 329
Maianthemum bifolium (may lily)   207
Mammals, conservation
    evaluation   12
Man-environment interaction   25–6
Management considerations (as a
    criterion)   9 (table), 13 (table)
Mann–Whitney U-test   288–9
Man orchid (*Aceras
    anthropophorum*)   77
Marine parks   102–3
Marl pits, floristic indices   81
Marsh warbler (*Acrocephalus
    palustris*)   250
Marsupials, arboreal   63–5
Masai Mara   104
May lily (*Maianthemum bifolium*)   207
Melanargia galathea   87
Merlin (*Falco columbarius*)   267
Mesostigmata   88, *89*
Migratory species protection   101
Migratory wildfowl   13 (table)
Milvus milvus (red kite)   *250*
Minimum population size   17, 19,
    318–22
Minimum viable size   17, 18–19,
    318–22
Mining subsidence   71
Minsmere Marsh (Britain)   145
Minuartia verna   71
Mixed landscape evaluation   209
Models   90, 125–6
Molluscs   274–5, 279
Monarch butterfly   101
Monitor lizard (*Varanus
    komodoensis*)   100–1
Monetary terms, wildlife evaluation
    in   153

Moorland   26–7, 28
    agricultural environment   233
        (table)
    rotational burning   27
Moorland birds (Britain)   266–7
Moss heath   151
Motacilla cinerea (grey wagtail)   266
Mountain birds (Britain)   266–7
Mount Lofty Ranges (Australia),
    remnant woodlots   307
Murray-Darling Drainage
    Basin   56–7, 66
Muscari neglectum   74 (table)
National Nature Reserve (GB)   136
National Nature Reserve Series   138
National Parks and Access to the
    Countryside Act 1949 (England
    and Wales)   182
National Science Foundation   112
National Vegetation
    Classification   154
Natural catastrophe   320
Naturalness   9 (table), 13 (table),
    25–7, 141, 171–2, 187–8, 205, 274–6
Nature Conservancy (USA)   114–18
Nature Conservancy Council
    (GB)   73, 135–56
    habitat categories   229–32
    international importance of
        sites   150–2
    local planning authority
        application   182
    strategies   228–9
Nature Conservation Review   137
    changing requirement for site
        safeguard   148–9
    key sites   143–8
    number of sites   155
    scientific assessor   145, 147
Nature preserve *see* nature reserve
Nature refuge *see* nature reserve
Nature reserve
    animal   100–1
    design   315–37
    equilibrium island biogeographical
        theory   322–4, 335
    large   102, 154
    minimum viable population
        sizes   318–22

peninsula effect   330–3, 335
selection   136–9
shape   330–3
SLOSS (single large or several
        small)   324–30, 335, 336
vegetation classification
        (Tansley's)   139
see also site(s)
Nature resource   4, 136
Nepal, agriculture   225 (table)
Netherlands   161–80
birds, breeding/overwintering
        172–3
ecotypes, rarity assessment of   169
evaluation procedures   173–80
general ecological model   164
heathland bird communities   257
historical review   164–6
naturalness   171–2
plant alliances (associations)   167,
        172
rarity   166–9
replaceability   170–1
representativeness   169–70
species richness   172
wetlands   172–3, 263–4
Netherlands Society for Landscape
        Ecological Research   166
New Zealand
agriculture   225 (table)
'Blossom' (elephant seal)   104
Norfolk (England)
reedswamps   147
valley fens   145–7
Norfolk broads   141, 145, 147
North York Moors National Park   28,
        299
North Yorkshire (England)   28,
        297–306
Norway, land birds   259–60
Nyctea scandiaca (snowy owl)   260

Ophrys apifera (bee orchid)   87, 88
Orchids
Krakatau colonization by   84
public interest   208
Orchis majalis (broad-leaved marsh
        orchid)   174
Orchis militaris   74 (table)

Ornithological evaluation   247–69
attributes   251–7
breeding bird(s) (West German)
        260–1
breeding bird communities in
        lakes/mines (southern
        Sweden)   257, 259
breeding seabirds (Ireland)   257,
        258–9
classification of sites (Britain)   263
farmland birds (Britain)   267–8
land birds (Norway)   259–60
lowland waders (Britain)   268
moorland/mountain birds
        (Britain)   266–7
non-breeding wildfowl
        (England)   257, 258
objectives   248–51
relative importance of different
        species   261–2
river systems (Britain)   266
wetlands, non-breeding
        birds   262–4
woodland birds   264–5
Orobanche elatior (knapweed
        broomrape)   77
Orobanche purpurea   75 (table)
Orthoptera, survey   279
Osprey (*Pandion haliaetus*)   250, 260

Pandion haliaetus (osprey)   250, 260
Papua New Guinea
commercial production of crocodiles
        and insects   104
geographical regions   47
volcanoes   84
Parus cristatus (crested tit)   265
PATREC   126
Pattern richness (interrelationship
        richness)   98, 102
Peat bogs   185
Peatlands   192
Peppermints (Eucalypts)   65
Peregrine falcon (*Falco peregrinus*)   23,
        90, 248, 267
Permian (Magnesian) limestone
        grassland   78
Pernis apivorus (honey buzzard)   250
Petauroides volaris (greater glider)   64

Petaurus australis (yellow bellied glider) 64
Petaurus breviceps (sugar glider) 64
Petaurus norfolcensis (squirrel glider) 64
Peucedanum palustre 75 (table)
Phoridae 279
Phillipines, national park proposal 106
Phyllitis scolopendrium 35, 37
Physical Quality of Life Index (Morris), tropical countries 98
Phytogeographic regions, Prance's classification 103
Phytophagous insects, in *Juniperus communis* 21
Phytosociological system 139
Pinus palustris (longleaf pine) 316
Pinus sylvestris (Caledonian Scots pine) 264–5
Pinus taeda (loblolly pine) 316
Pits, disused *see* Disused quarries and chalk pits
Plant alliances (associations) 172
Platyarthrus hoffmanneseggi 87, 88
Pleistocene refugia (Brazil) 102
Pluvialis apricaria (golden plover) 267
Poa bulbosa 75 (table)
Pollen 4, 26, 274–5
Polygonatum odoratum 32 (table)
Popularity poll 9, 13–14
Population density 16
Population size (as a criterion) 13 (table), 262–4
Position in ecological/geographical unit (criterion) 9 (table), 13 (table)
Possums 63–5
Potamogeton sp. 191
Potamogeton epihydrus (American pondweed) 27
Potential value (as a criterion) 9 (table), 13 (table), 71–91
Predators, range requirement 17, 18
prediction, from case histories 90
primrose (*Primula vulgaris*) 238
Primula vulgaris (primrose) 238
'Procedures', Cushwa's 127

Prunus spinosa (blackthorn) 207
Pseudocheirus peregrinus (ringtail possum) 64
Psittacula eupatria wardi (green parakeet) 329
Puerto Rican parrot (*Amazona vittata*) 320
Pulsatilla vulgaris (pasque flower) 75 (table), 76, 77
Pyrola rotundifolia 75 (table)
Pyrrhocorax pyrrhocorax (chough) 90

Quantification 249
Quarries, disused *see* Disused quarries and chalk pits
Quarrying, 'holes' resulting from 71
Quercus petraea (sessile oak) 4
Quercus robur (pedunculate oak) 4

Ramsar Convention 152, 262, 263
Rarefaction 16
Rare species, site with 29
Rarity 7, 8, 9 (table), 13 (table), 22–5, 141–2, 166–9, 185–6
  concept 308–9
  geographical distribution 23
  grid squares 24, 25, 153
  Netherlands studies 166–9
  vascular plants 153
Rarity line 23
  higher plants in Britain 22
Recessive alleles, diseases caused by 318–19
Recorded history 9 (table), 13 (table), 142
Red-cockaded woodpecker (*Dendrocopus borealis*) 316
Red kite (*Milvus milvus*) 250
Redshank (*Tringa totanus*) 268
Redwing (*Turdus iliacus*) 250, 267
Regulus ignicapillus (firecrest) 250
Relation theory (van Leeuwen) 163
Replaceability (as a criterion) 9 (table), 13 (table), 170–1
Representativeness 9 (table), 10, 13 (table), 27–9, 310
  Netherlands studies 169–70

Representativeness, assessment
  of   45–67, 142–3
  analysis methods   54–65
  Australian methods   46–9
  continental (between region)
    scales   54–7
  within region scales   58–65
  see also land classification
Reseda luteola   87
Reserve *see* Nature reserve
Rhacomitrium lanuginosum   151
Ribes spicatum   32 (table)
Ringtail possum (*Pseudocheirus
  peregrinus*)   64
River systems (Britain)   266
Rotterdam harbour expansion   164
Royal Society for the Protection of
  Birds   245

Saguaro cactus (*Cargenia
  gigantea*)   329
Sample size   16
Sand dune systems   70
Sarawak forest trees   208
Scientific value (as a criterion)   9
  (table), 13 (table)
Sciurus caroliensis (grey squirrel)   205
Scotland   192–7
  Highland Boundary Fault   194, 195
  Regions   192–7
  Shelducks nesting on Firth of
    Forth   254
Scottish crossbill (*Loxia scotica*)   265
Sea turtles, national park for (Costa
  Rica)   101
Seed dispersal   84–5
Sepsidae   279
Serengeti plains (Tanzania)   104
Seseli libanotis   75 (table)
Sesleria albicans (blue moor grass)   23,
  25, 39
Seychelles Islands   329–40
Shannon Index   15, 17
Shannon-Weaver index   15
Shelduck (*Tadorna tadorna*)   254
Shingle beaches   70
Shrub crown cover   125
Silsoe exercise (1960)   234

Silvicultural gene bank   13 (table)
Simpson's index   15, 17
Simulium larvae/pupae   283–4
Site(s)
  area importance   307–8
  background considerations of
    safeguard strategy   136
  basis for selection   140
  comparative evaluation   140
  conservation principle   138
  definition   136
  evaluation developments   152–5
  field recording   139–40
  individual, assessment of   141–3
  minimum standards   137
  number to be protected   219–20
  scoring systems   152–3
  survey of candidates   139
  see also Nature reserve
Sites of Special Scientific Interest
    (SSSIs)   73, 136, 140, 182–3
  changing requirement   148–50
  management by owners   228
  Moore's revision   153–4
  minimum acceptable
    standards   153–4
  scale   189–90
Smithsonian Institute   112
Snipe (*Gallinago gallinago*)   268
Snowy owl (*Nyctea scandiaca*)   260
Social behaviour dysfunction   320
Soil mites   11, 88, *89*
Somerset levels   243
Specially Protected Areas   10
South America, endemic birds   31
South coast project (Australia)   59–61
Southern Uplands (GB)   147
Species
  endangered   255
  multiplication factors   167
  native/non-native   4
  relative importance (Maine)   261–2
  threatened with extinction   101,
    255
  vulnerable   255
Species–area relationship   17, 20–2,
  79, 90, 325–7
Species richness   7, 8, 17

conservation for 101–3
Netherlands 172
vascular plants 206
woodland/forest evaluation 205–7
Sphagnum sp. 192
Spring-tails (Collembola) 88, *89*, 241
Standing Conference on Countryside Sports 243
Starling (*Sturnus vulgaris*) 267
Stethophyma grossum 277
Structure of communities 31
Sturnus vulgaris (starling) 267
Subsahelian Africa, per capita food supplies 98
Succession
definition 70
evaluation 69–91
stage as criterion 13 (table)
Survey methodology, in relation to evaluation 216–18
Sweden
agriculture 225 (table)
breeding bird communities in lakes/mines 257, 259
Sylvia undata (Dartford warbler) *250*

Tadorna tadorna (shelduck) 254
Tanzania, Serengeti plains 104
Taxus baccata (yew) 204
Tennessee (USA), Habitat Evaluation System in 129
Testate rhizopods 276
Tetrix ceporoi 277
Thalictrum minus 35, 36
Thlaspi alpestre 71
Thlaspi perfoliatum 75 (table)
Threat of human interference (as a criterion) 9 (table), 13 (table)
Tilia cordata (lime) 204, 205
Top–down (gamma–beta–alpha) approach 130
Tortuguero National Park (Costa Rica) 101
Trap-efficiency 282
Tricholimnas sylvestris (Lord Howe wood rail) 256
Trichosurus vulpecula (bushtail possum) *64*

Tringa nebularia (greenshank) 267
Tringa totanus (redshank) 268
Trinidad and Tobago
agriculture 255 (table)
national park 104
Tropical areas 95–110
agriculture 99, 225 (table)
changes in values 107
definitions of 'tropics' 96
evaluation procedures 108
land evaluation 100–5
land use 98–9
megadevelopment 99–100
priorities 108
selection criteria 108–9
Tropical ecosystems 97
Tropical forests 208–9
Tropical rain forests 96, 97
lowland, neotropical birds 102
species richness 97
Tropical species 96, 97
Turdus iliacus (redwing) *250*, 267
Turdus pilaris (fieldfare) 267
Tuross-Deua National Park (proposed) 61 (table)
Tympanuchus cupido attwateri (heath hen) 319
Type localities 31–2
Typicalness 9 (table), 13 (table), 27, 29, 142, 143, 188–9; *see also* Representativeness

Ulex europaeus 151
Ulex gallii 151
Ulex minor 151
UNESCO Man and Biosphere Programme 103, 152
Uniqueness (as a criterion) 9 (table), 13 (table)
United Nations Environment Programme 103, 105
United States of America 111–33
agriculture 255 (table)
database for management 127
Department of Agriculture Forest Service 113
Department of Interior National Park Service (USDI) 113, 119–22, 129

diversity, evaluating and managing for 129–30
evaluation methods 111–33
Habitat Evaluation Procedure 121–5
Habitat Evaluation System 128–9
mixed forest ecosystem model 90
National Environmental Policy Act (1969) 118
National Forest Management Act (1976) 129
National Natural Landmarks 113–14
Natural Heritage Programmes 115–18
Nature Conservancy 114–18
research natural areas 112–13
relative importance of different species (Maine) 261–2
species-habitat model validation 130
Wildlife and Fish Habitat Relationships System 126–7
Uplands 192
Ursus arctos (grizzly bear) 321

Vanellus vanellus (lapwing) 267, 268
Varanus komodoensis (monitor lizard) 100–1
Vascular plants, species in broadleaf woodland *216, 217*
Veery (*Hylocichla furescens*) 125
Vegetation associations, North York Moors National Park 28
Vegetation communities, frequency on different geological formations *65*
Vegetation differentiation (Australia) 47
Vegetation classification (Tansley's) 139
Vegetation layers 172
Vegetation structure 31
Venezuela, protected areas 103
Verbascum lychnitis 75 (table)
Viola persicifolia (V. stagnina) 85
Virgin forest, definition 25
Vulnerability 189
Vulpin unilateralis 75 (table)

Waders, Danish Wadden Sea *253*

Walberswick Marsh (England) 145
Waterfowl, ornithological evaluation 262–4
Watershed protection 107
Welchman's Hall Gully (Barbados) 106
West Dean Estate (Sussex) 231 (table)
West Germany
  agriculture 225 (table)
  breeding birds 260–1
Wetland 191–2, 230 (table)
  agricultural environment 233
  MAR list 262
  Netherlands 172–3, 263–4
  ornithological evaluation of non-breeding birds 262–3
Wetland Index 264
Wharram Quarry Nature Reserve 85–8, 90
Wicken Fen (England) 85, 145
Wildfowl
  non-breeding (Southern England) 257, 258
  populations 137
Wildlife
  conservation aim 272–3
  definition 4
  categories of protected area *148*
  husbandry relationship 242–3
Wildlife and Countryside Act 1981 (England and Wales) 148–9, 183
Wildlife and Fish Habitat Relationships System (USA) 126–7, 130
Wildlife conservation sites, criteria for evaluation 71–3
Wildlife reservoir potential (as a criterion) 9 (table), 13 (table)
Wiregrass (*Aristidia stricta*) 316
Woad (*Isatis tinctoria*) 27
Woodchip concession area 63
Woodland 190, 191, 230 (table)
  agricultural environment 233
  flora/fauna in 212
  future-natural features 205
  past-natural features 205
  Scotland 191
Woodland and forest evaluation 201–21

between wood comparisons
    209–10
British semi–natural woodlands
    212–20
definitions 202–4
judgement 220
long-term change assessment 212
mixed landscapes 209
naturalness 205
past records 208
rare species in 207–8
special localities 203
species richness 205–7
tropical forests 208–9
within wood comparisons 210–12
Woodland birds (Britain) 264–5
Woodland indicator species 185
Woodlark (*Lullola arborea*) *250*
Woodpecker life-form model 125–6

Woodwalten Fen (England) 145
World Conservation Strategy (IUCN,
    1980) 244
World Heritage Convention sites 152

Yellow warbler (*Dendroica
    petechia*) 122
Yew (*Taxus baccata*) 204
Yorkshire Dales, limestone
    pavements 32–42
Yorkshire Wolds, disused quarries
    and chalk pits 78–82
Yucca brevifolia (Joshua tree) 329

Zambia, agriculture 255 (table)
Zinc mines 71
Zosterops mayottensis semiflava
    (chestnut-flanked white-
    eye) 329